About IFPRI

The International Food Policy Research Institute (IFPRI), established in 1975, provides research-based policy solutions to sustainably reduce poverty and end hunger and malnutrition. The Institute conducts research, communicates results, optimizes partnerships, and builds capacity to ensure sustainable food production, promote healthy food systems, improve markets and trade, transform agriculture, build resilience, and strengthen institutions and governance. Gender is considered in all of the Institute's work. IFPRI collaborates with partners around the world, including development implementers, public institutions, the private sector, and farmers' organizations. IFPRI is a member of the CGIAR Consortium.

About IFPRI's Peer Review Process

IFPRI books are policy-relevant publications based on original and innovative research conducted at IFPRI. All manuscripts submitted for publication as IFPRI books undergo an extensive review procedure that is managed by IFPRI's Publications Review Committee (PRC). Upon submission to the PRC, the manuscript is reviewed by a PRC member. Once the manuscript is considered ready for external review, the PRC submits it to at least two external reviewers who are chosen for their familiarity with the subject matter and the country setting. Upon receipt of these blind external peer reviews, the PRC provides the author with an editorial decision and, when necessary, instructions for revision based on the external reviews. The PRC reassesses the revised manuscript and makes a recommendation regarding publication to the director general of IFPRI. With the director general's approval, the manuscript enters the editorial and production phase to become an IFPRI book.

East African Agriculture and Climate Change

A Comprehensive Analysis

Edited by Michael Waithaka, Gerald C. Nelson,
Timothy S. Thomas, and Miriam Kyotalimye

A peer-reviewed publication

International Food Policy Research Institute
Washington, DC

International Food Policy Research Institute
2033 K Street, NW
Washington, DC 20006-1002, USA
Telephone: +1-202-862-5600
www.ifpri.org

DOI: http://dx.doi.org/10.2499/9780896292055

Library of Congress Cataloging-in-Publication Data

East African agriculture and climate change : a comprehensive analysis /
    Michael Waithaka ... [et al.].
        p.   cm. — (Climate change in Africa)
    Includes bibliographical references and index.
    ISBN 978-0-89629-205-5 (alk. paper)
    1. Agriculture—Africa, Eastern.   2. Climatic changes—Africa, Eastern.
    3. Land degradation—Africa, Eastern.   4. Forests and forestry—Africa,
    Eastern.   5. Africa, Eastern—Population.   I. Waithaka, Michael.
    II. International Food Policy Research Institute.   III. Series: Climate
    change in Africa.
    S472.A354E27   2013
    334'68309687—dc23   2013002209

Cover design: Carolyn Hallowell
Book layout: Princeton Editorial Associates Inc., Scottsdale, Arizona

Contents

Figures

Tables

Foreword

As the world's population grows from around 7 billion in 2012 to around 9 billion by 2050, the population of Africa south of the Sahara is likely to surge from around 850 million today to around 1.7 billion in 2050. East Africa alone will make up more than 44 percent of the population of Africa south of the Sahara and almost 9 percent of the world's population in 2050. Most of the people accounting for this population increase are expected to live in urban areas and to have higher incomes than currently is the case, which will result in increased demand for food. In the best of circumstances, the challenge of meeting this demand in a sustainable manner will be enormous. When one takes into account the effects of climate change (higher temperatures, shifting seasons, more frequent and extreme weather events, flooding, and drought) on food production, that challenge grows even more daunting. The global food price spikes of 2008, 2010, and 2012 are harbingers of a troubled future for global food security.

At the end of 2010, IFPRI published *Food Security, Farming, and Climate Change to 2050: Scenarios, Results, Policy Options,* a research monograph by Gerald C. Nelson and a team of IFPRI researchers that quantitatively assessed the additional challenges to sustainable food security that climate change would bring, focusing on global outcomes but also including national and subnational results. Three years later, Nelson and a group of leading agriculturists and climate change researchers have written this monograph, which draws out those national results based on a detailed global model and enhances them with country-specific analysis and insights for the countries of East Africa.

The second of three publications (covering West, East, and southern Africa) that make up IFPRI's *Climate Change in Africa* series provides the most comprehensive analysis to date on the scope of climate change as it relates to food security in East Africa, including who will be most affected and what policymakers can do to facilitate adaptation. Augmenting the text are dozens of detailed maps that provide graphical representations of the range of food security challenges and the special threats from climate change.

Using comprehensive empirical analysis, the authors have placed climate change in the forefront of national development issues and have suggested that policymakers take into account (1) the need for more research and development programs across diverse disciplines; (2) the urgency of rehabilitating degraded agricultural land to enhance agricultural productivity; (3) the need for improved coordination and implementation of climate change policies and strategies; and (4) the need to build capacities in terms of human resources, institutions, and infrastructure. It is becoming increasingly clear to policymakers in the developing world that neither food security nor climate change can be viewed in isolation. *East African Agriculture and Climate Change* will be indispensable to a wide range of readers, including policymakers, development workers, and researchers who tackle these inextricably linked issues.

Shenggen Fan
Director General, International Food Policy Research Institute

Acknowledgments

The editors of this monograph and the authors of the individual chapters thank the following organizations for their financial support: the GIZ (Gesellschaft für Internationale Zusammenarbeit); the EU through its support for the Climate Change, Agriculture, and Food Security Research Program of the CGIAR; and the Bill and Melinda Gates Foundation and their respective home institutions—ASARECA (Association for Strengthening Agricultural Research in Eastern and Central Africa), and IFPRI—for encouraging them to undertake this work. We also give special recognition to Michael Waithaka of ASARECA. He identified counterpart national scientists to undertake the national reports and provided invaluable intellectual leadership in managing the challenging process of coordinating and supporting many different authors while leading the development of the regional overview chapter. Any errors or omissions remain the responsibility of the authors.

Abbreviations and Acronyms

A1B	greenhouse gas emissions scenario that assumes fast economic growth, a population that peaks midcentury, and the development of new and efficient technologies, along with a balanced use of energy sources
AR4	Fourth Assessment Report of the Intergovernmental Panel on Climate Change
ASAL	arid and semiarid land
B1	greenhouse gas emissions scenario that assumes a population that peaks midcentury (like the A1B), but with rapid changes toward a service and information economy, and the introduction of clean and resource-efficient technologies
CNR	climate model developed by National Meteorological Research Center, France
CNRM-CM3	National Meteorological Research Center–Climate Model 3
CSIRO	Commonwealth Scientific and Industrial Research Organisation, Australia
CSIRO MARK 3	climate model developed at the Australia Commonwealth Scientific and Industrial Research Organisation
DRC	Democratic Republic of Congo
DSSAT	Decision Support Software for Agrotechnology Transfer

ECHAM 5	fifth-generation climate model developed at the Max Planck Institute for Meteorology (Hamburg)
FPU	food production unit
GCM	general circulation model
GDP	gross domestic product
GIS	geographic information system
HCENR	Higher Council for Environment and Natural Resources
HPI	human poverty index
HWSD	*Harmonized World Soil Data Base*
IDP	internally displaced person
IFPRI	International Food and Policy Research Institute
IMPACT	International Model for Policy Analysis of Agricultural Commodities and Trade
INECN	National Institute for Nature Conservation (Burundi)
IPCC	Intergovernmental Panel on Climate Change
IUCN	International Union for the Conservation of Nature
MDG	Millennium Development Goal
MIROC 3.2	Model for Interdisciplinary Research on Climate, developed at the University of Tokyo Center for Climate System Research
NAPA	National Adaptation Programme of Action
NCCRS	*National Climate Change Response Strategy*
R&D	research and development
SPAM	Spatial Production Allocation Model
SPNN	Southern Peoples Nations and Nationalities
UN	United Nations
UNEP	United Nations Environment Programme
UNFCCC	United Nations Framework Convention on Climate Change
UNHS	Uganda National Household Survey
UNPOP	United Nations Department of Economic and Social Affairs–Population Division

Chapter 1

OVERVIEW

Michael Waithaka, Timothy S. Thomas, Miriam Kyotalimye,
and Gerald C. Nelson

This monograph considers potential impacts of climate change adaptation on agriculture in 10 countries of eastern and central Africa: Burundi, Democratic Republic of Congo (DRC), Eritrea, Ethiopia, Kenya, Madagascar, Rwanda, Sudan, Tanzania, and Uganda.[1] Each of these countries is treated separately in one of the chapters that follow this overview chapter and a chapter on methodology and precede the chapter in which we summarize our findings and draw some conclusions. In the course of this monograph we present tables, graphs, and maps that provide an overview for each country, showing historical trends in demographics, poverty, and income, along with a brief summary of current land cover and land use, with a focus on agricultural uses.

In this monograph we make two main analytical contributions. One is from using Decision Support Software for Agrotechnology Transfer (DSSAT) crop modeling software together with climate model results to look at the effects of climate change on yield without any adaptation. The other is from inputting projections regarding demography and gross domestic product (GDP) into a global partial equilibrium food and agriculture model, the International Model for Policy Analysis of Agricultural Commodities and Trade (IMPACT), along with climate change and technological change, to get a more complete picture of what the future may hold when adaptation and progress are accounted for. Our focus is on the cropping sector, and although we acknowledge that the livestock sector is important and we make references to it throughout the monograph, it is not one of our objectives to treat it here.

Agriculture drives the economies of these countries and contributes to 43 percent of their GDP annually (Omamo et al. 2006). Agriculture in Burundi, DRC, Ethiopia, Sudan, and Tanzania accounts for more than 50 percent of GDP, while in Kenya, Eritrea, and Madagascar it accounts for less than

---

1 The new republic of South Sudan separated from Sudan on July 9, 2011. This was after the analysis had been completed for this monograph, which was based on the old Sudan.

30 percent. Countries with relatively large (small) national economies also have relatively large (small) agricultural economies (Omamo et al. 2006). The largest economies are those of Kenya, Tanzania, Uganda, DRC, Ethiopia, and Madagascar; the smallest are those of Eritrea and Burundi. Eritrea has little agricultural land; Kenya's structural transformation toward a less agriculture-based economy is more advanced than in other countries in the region; Madagascar's large agricultural potential remains mostly untapped. Out of the division of the former country of Sudan, prospects for the two countries may favor the new Sudan due to its smaller population, its oil reserves, and its high potential for irrigated agriculture. Southern Sudan may stay in the league of small economies for a while as it lays foundations and develops institutions for the economy. Discovery of oil in Uganda may change the economic setting for that country drastically in a few years.

Most agricultural production relies on rainfall and is spread across diverse agroecological zones that can be differentiated by topography and soil types, most of which are highly degraded or eroded and face poor soil fertility management and continuous cropping (Sanchez et al. 1997). The main mode of production is dominated by smallholders. Population growth in the 10 countries is among the highest in the world and poses challenges of food insecurity, which is already severe.

Recent trends and the current performance of agriculture in the 10 countries expose a region that is progressively less able to meet the needs of its burgeoning population. With agriculture looming so large in most of the economies of eastern and central Africa, sluggish growth in agricultural productivity translates into sluggish overall growth and generally low per capita income levels. High levels of agricultural importation—particularly of staples—appear to be only partially filling the consumption needs of a population lacking purchasing power, resulting in extensive adult and child malnutrition and towering child mortality rates. In the face of climate change, adaptation is essential for sustained economic growth in these countries. Arable areas in the region are under severe pressure to increase their productivity to feed a rapidly increasing human population. Sharp increases in the prices of staple foods such as cereals since 2008 as well as high price volatility are already hitting the poorest consumers, who spend a large proportion—about 50 to 70 percent—of their income on food and have limited capacity to adjust quickly to rapid price increases (Karugia et al. 2011). Because it is not possible to increase the area under production in the higher-potential areas, effective technologies and interventions are required to increase farm productivity and enhance sustainability and thereby improve human well-being.

Labor productivity in eastern and central Africa has declined to the extent that the region produced less per worker in 2000 than it did four decades earlier (Omamo et al. 2006). This regionwide contraction has been based on contractions in DRC, Kenya, Madagascar, and Tanzania. However, labor productivity in Ethiopia, Rwanda, Sudan, and Uganda has recovered substantially in recent years. Given these trends in agricultural productivity in eastern and central Africa, it is not surprising that the average yields of the region's major crops currently fall well below those elsewhere in Africa and even further below global levels. Only for cassava, beans, coffee, and tea do the yields of eastern and central Africa compare favorably with average African and global yields.

These challenges are redefining many of the problems facing agricultural policymakers in eastern and central Africa and thus the kinds of policy solutions required. Most of these forces have roots and expressions that extend beyond national boundaries, implying the need for broad perspectives and regional responses.

The Intergovernmental Panel on Climate Change (IPCC), Climate Change and Agriculture, and Food Security

In the Fourth Assessment Report of the IPCC, Working Group 1 reports that "climate is often defined as 'average weather.' Climate is usually described in terms of the mean and variability of temperature, precipitation, and wind over a period of time, ranging from months to millions of years (the classical period is 30 years)" (Le Treut et al. 2007, 96).

The increase in greenhouse gas emissions is raising average temperatures. The consequences include changes in precipitation patterns, more and more extreme weather events, and shifting seasons. The accelerating pace of climate change, combined with global population and income growth, threatens food security everywhere.

Agriculture is vulnerable to climate change in a number of dimensions. Higher temperatures eventually reduce yields of desirable crops and tend to encourage weed and pest proliferation. Greater variations in precipitation patterns increase the likelihood of short-run crop failures and long-run production declines. Although there might be gains in some crops in some regions of the world, the overall effects of climate change on agriculture are expected to be negative, threatening global food security. The expected effects are

- direct, on crops and livestock productivity domestically;

- indirect, on the availability or prices of food domestically and in international markets; and

- indirect, on income from agricultural production at both the farm and the country levels.

Review of Current Regional Trends

This section provides an overview of the starting point for an assessment of the potential vulnerability of eastern and central African agriculture to climate change. It looks at recent population and income developments to provide a backdrop to potential futures. Two key indicators of well-being are reviewed—under-five mortality and life expectancy at birth. The current climate situation is discussed, along with the role of regional programs in supporting food security.

Economic and Demographic Indicators

Population

The current population of the 10 countries of eastern and central Africa is 359 million people. Between 1988 and 2008 the populations in 9 of the 10 countries (excluding DRC) increased by a staggering 74 percent, to 277 million people, signifying an annual growth rate of 3.7 percent (Table 1.1). Growth rates between 1990 and 2008 were generally high, with Uganda (which was recovering from many years of internal conflicts), Ethiopia, and Tanzania leading. The low rates in Rwanda and Burundi may be explained by the genocide that occurred in those two countries in that period. The big countries in terms of area—Ethiopia, Tanzania, and Sudan—are also the most populous, whereas Rwanda, Burundi, and Eritrea—which are relatively small—have the lowest populations. In general, there is growing urbanization, resulting in higher population densities in cities and major towns (Figure 1.1). The highest rates of growth are in Sudan, Madagascar, and Tanzania. Rapid urbanization in the region is posing a great challenge to governments in providing basic amenities for the inhabitants.

In all the countries, there is a general pattern of high population densities in and around urban areas. There is also a generally higher population density in the highlands and along lakes and the coastal belt (see Figure 1.1).

TABLE 1.1 Population of East Africa, annualized growth rate, and percent urban, 1988 and 2008

Country	Total population (millions)		Annualized growth rate, 1990–2008 (%)	Percent urban	
	1988	2008		1988	2008
Burundi	5.39	8.07	2.49	6	10
DR Congo	35.49	64.21	4.05	28	34
Eritrea	3.05	5.00	3.20	16	21
Ethiopia	45.21	80.71	3.93	12	17
Kenya	21.91	38.53	3.79	18	22
Madagascar	10.63	19.11	3.99	23	30
Rwanda	7.01	9.72	1.93	5	18
Sudan	25.86	41.35	2.99	25	43
Tanzania	23.89	42.48	3.89	18	26
Uganda	16.49	31.66	4.60	10	13

Source: *World Development Indicators* (World Bank 2009).
Note: DR Congo = Democratic Republic of Congo.

FIGURE 1.1 Population distribution in East Africa, 2000 (persons per square kilometer)

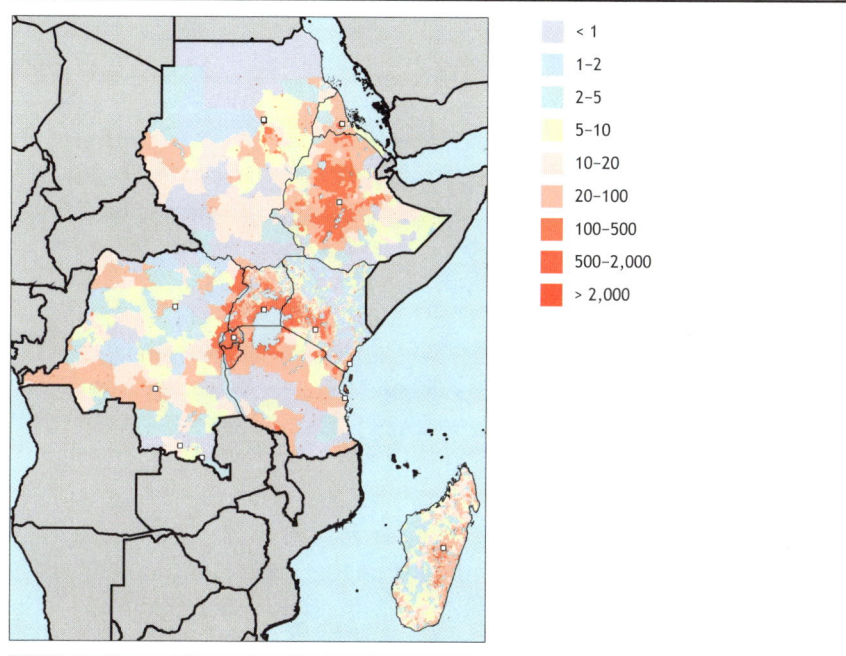

Source: CIESIN et al. (2004).

Income

Per capita GDP in the 10 countries of eastern and central Africa is low by world standards; the highest is US$532 in Sudan and the lowest US$111 in Burundi (Table 1.2).[2] The highest growth rates between 1998 and 2008 were in Uganda, Sudan, Ethiopia, and Tanzania; growth was stagnant in Kenya and declined in Burundi and Madagascar. The growth rate in Uganda can be attributed to a good economic environment after years of conflicts. For Sudan the growth coincided with the extraction of oil. The Kenya situation can be explained by slow economic growth between 1998 and 2002, which reversed thereafter. The cases of Burundi and Madagascar can be explained by genocide and protracted political standoff, respectively. In general, adverse climatic conditions may also have contributed to slow growth in most countries.

Across the region, there has been an increasing decline in the share of agriculture in overall GDP (see Table 1.2). This can be attributed to the rapid expansion of other sectors, such as oil in Sudan, construction and service sectors such as mobile communications in Uganda and Kenya, and tourism in Kenya and Madagascar. In addition, the relative stagnation in crop productivity and production as a result of inadequate use of desired inputs and inadequate mechanization coupled with adverse weather conditions has also contributed to the decline of the agricultural sector.

Well-being Indicators (Regional)

Under-five mortality declined by 22 percent between 1988 and 2008 in eight of the nine countries of eastern and central Africa where data were available (Table 1.3). The only exception was Kenya, where a rise was attributed to the effects of structural adjustment programs of the early 1990s, which saw a decline in public health services. Although significant declines have been noted in Eritrea, Ethiopia, and Madagascar, the average rate of infant mortality, 126 per 1,000 in 2008, was still high. Life expectancy at birth generally improved across the nine countries with the exception of Kenya, where it declined. The greatest improvement was in Rwanda, where life expectancy jumped from 32 years to 50 between 1988 and 2008. The majority of the countries have a life expectancy of between 50 and 60 years. The general decrease in under-five mortality and increase in life expectancy are due to increasing numbers of health campaigns, improvement in the provision of health services, and probably a decline in the prevalence of major diseases such as malaria and HIV/AIDS.

---

2 All dollar figures in the book are constant 2000 U.S. dollars.

TABLE 1.2 Income of East Africans (GDP per capita and share of GDP from agriculture), 1988 and 2008

	GDP per capita (constant 2000 US$)		Share of GDP from agriculture (%)	
Country	1998	2008	1998	2008
Burundi	153.13	111.31	54.25	n.a.
DR Congo	99.62	98.51	48.71	38.12
Eritrea	212.15	147.46	23.37	23.47
Ethiopia	134.72	189.80	53.86	42.70
Kenya	441.24	463.72	29.89	21.27
Madagascar	286.16	270.78	33.49	25.20
Rwanda	244.52	313.20	39.22	34.64
Sudan	265.25	532.12	41.52	25.80
Tanzania	256.26	362.36	44.76	n.a.
Uganda	172.15	348.09	56.71	22.67

Source: *World Development Indicators* (World Bank 2009).
Note: DR Congo = Democratic Republic of Congo; GDP = gross domestic product; n.a. = not available; US$ = US dollars.

TABLE 1.3 Under-five mortality and life expectancy at birth in East Africa, 1988 and 2008

	Under-five mortality (deaths per 1,000)		Life expectancy at birth (years)	
Country	1988	2008	1988	2008
Burundi	188.7	179.5	45.98	50.23
DR Congo	n.a.	n.a.	n.a.	n.a.
Eritrea	147.0	70.4	48.57	57.87
Ethiopia	204.4	118.6	47.16	54.98
Kenya	97.0	121.2	59.38	54.06
Madagascar	168.0	111.8	51.39	60.11
Rwanda	195.3	180.6	31.90	49.81
Sudan	125.2	108.6	52.75	57.95
Tanzania	157.2	115.6	50.66	55.36
Uganda	174.6	130.4	47.55	52.37

Source: *World Development Indicators* (World Bank 2009).
Note: DR Congo = Democratic Republic of Congo; n.a. = not available.

FIGURE 1.2   Poverty in East Africa, circa 2005 (percentage of population below US$2 per day)

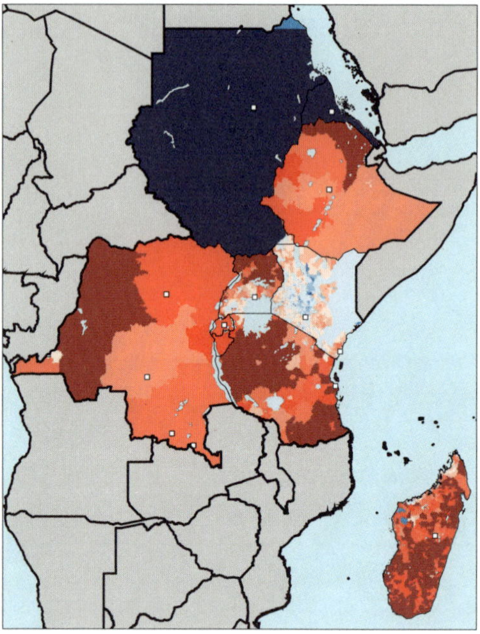

Legend:
- 0 (or no data)
- < 10
- 10–20
- 20–30
- 30–40
- 40–50
- 50–60
- 60–70
- 70–80
- 80–90
- 90–95
- > 95

Source: Wood et al. (2010).
Note: Based on 2005 US$ (US dollars) and on purchasing power parity value.

Poverty is widespread in eastern and central Africa; in a majority of the countries, more than 50 percent of the population live on less than US$2 per day (Figure 1.2). The situation in Sudan is inconclusive because of a lack of data. However, pockets in Kenya, Uganda, and Tanzania have 20–40 percent of the population living on less than US$2 per day.

Climate, Land Use, and Agriculture

At a glance, most parts of the 10 countries of eastern and central Africa appear to be arid and receive less than 1,000 millimeters of rainfall in a year (Figure 1.3). However, large swaths of DRC, Madagascar, and Ethiopia receive large amounts of rainfall (ranging up to 3,300 millimeters per year). At the same time, large parts of Sudan, Ethiopia, and Kenya receive less than 700 millimeters of rainfall in a year.

The average daily maximum temperature in the warmest month is relatively moderate at the equator, which cuts the northern quarter of DRC and

FIGURE 1.3   Annual average precipitation in East Africa, 1950–2000 (millimeters per year)

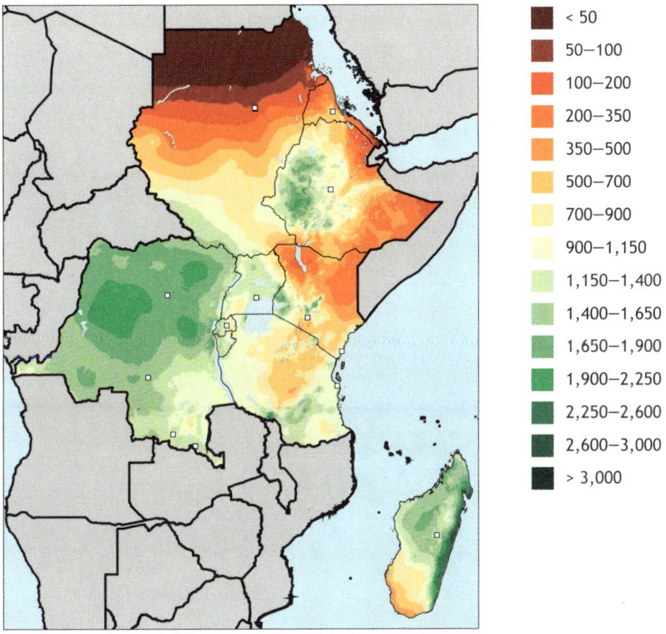

■	< 50
■	50–100
■	100–200
■	200–350
■	350–500
■	500–700
■	700–900
■	900–1,150
■	1,150–1,400
■	1,400–1,650
■	1,650–1,900
■	1,900–2,250
■	2,250–2,600
■	2,600–3,000
■	> 3,000

Source: WorldClim version 1.4 (Hijmans et al. 2005).

half of Uganda and Kenya (Figure 1.4), except in the semiarid northeast of Kenya. However, moving away from the equator, particularly northward, we see the temperatures rise, except in the highlands, up to 44°C. The other parts, which include the highlands of Ethiopia and Eritrea, have moderate temperature ranges from 24° to 33°C to near-tropical temperatures of less than 21°C in the highlands.

The agroecology of eastern and central Africa ranges from bare lands in the north to dense tree cover in DRC, Uganda, southern Tanzania, pockets of southwestern Ethiopia, and the costal belts of Tanzania and Kenya (Figure 1.5). In between are large swaths of shrubs and cultivated areas. As a result of high population pressure in areas with high agricultural potential, natural forests and sparse herbaceous cover areas—the savannahs that are home to wildlife and are suitable for livestock rearing—are under threat of being converted into croplands. In general, plantation tree crops as well as root crops dominate the humid coastal areas, whereas cereals become predominant northward. The arid parts of Madagascar, northern Tanzania, northern and eastern Ethiopia, and central Sudan are home to people with pastoral

FIGURE 1.4   Annual daily maximum temperature in East Africa, 1950–2000 (°C)

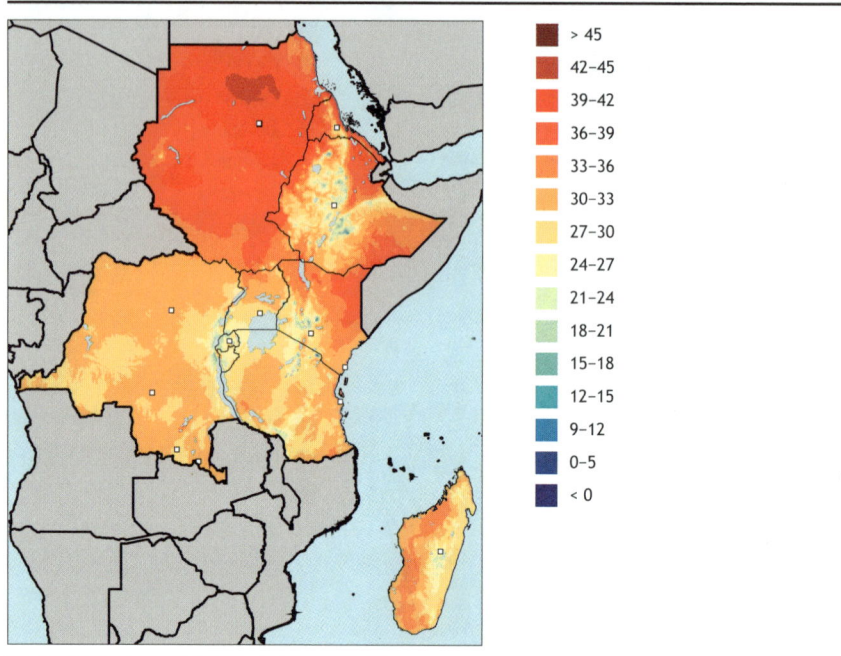

■	> 45
■	42–45
■	39–42
■	36–39
■	33–36
■	30–33
■	27–30
■	24–27
■	21–24
■	18–21
■	15–18
■	12–15
■	9–12
■	0–5
■	< 0

Source: WorldClim version 1.4 (Hijmans et al. 2005).

livelihoods and have pockets of endemic biodiversity. Many governments are increasingly supporting initiatives aimed at conserving natural habitats and protecting key natural resources including forests (Figure 1.6). The protected areas are home to many species of plants and animals and are important for tourism. Several of these sites are among the world heritage sites (http://whc. unesco.org/en/list). The naming of Maasai Mara game reserve in Kenya as one of the seven new wonders of the world in an ABC Television poll in 2006 has boosted tourism.

In general, major cities are linked within each country of eastern and central Africa (Figure 1.7). The longest artery is the highway that runs from the coast of Kenya, then cuts westward through the country to enter Uganda, where a branch runs to the north, linking Uganda to Sudan, and another runs to the west, linking DRC, Rwanda, and Burundi. A similar artery runs from the coast of Tanzania; it cuts through the mainland and has a branch into Kenya and another into Burundi, with links to the south of Tanzania. Although the major cities and towns, especially those on the key arteries,

FIGURE 1.5 Land cover and land use in East Africa, 2000

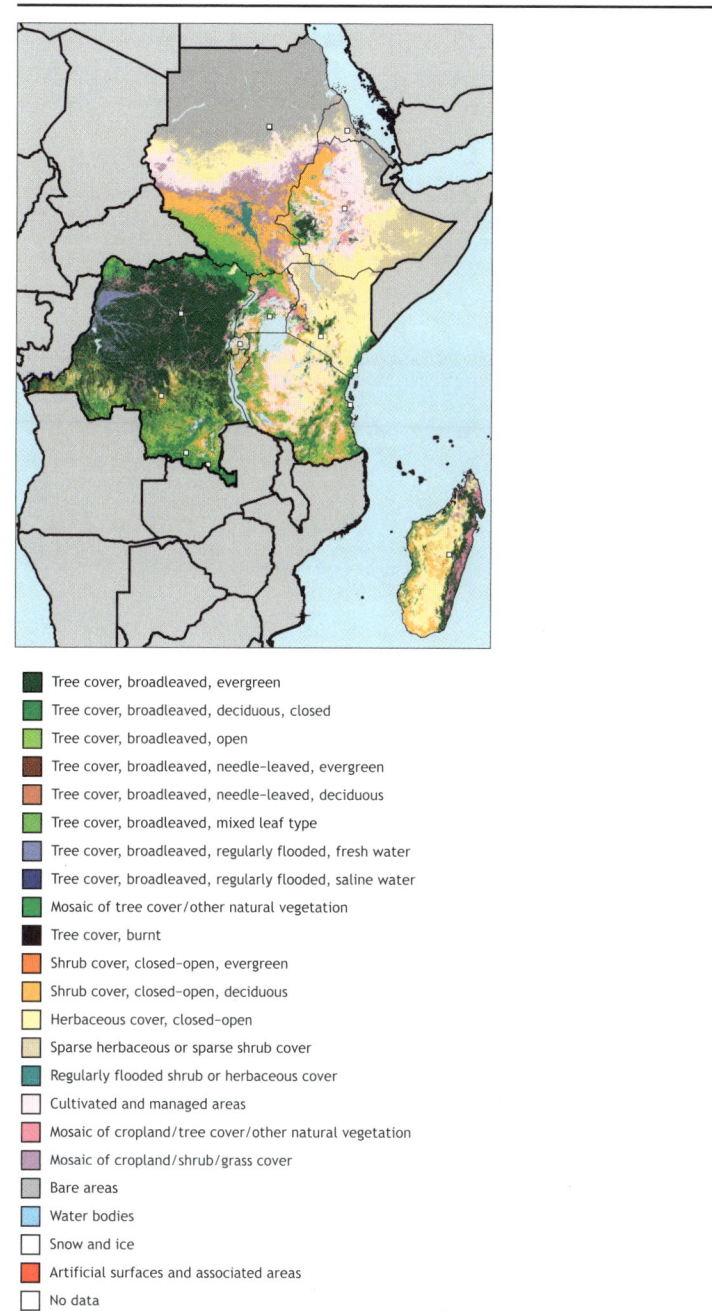

■ Tree cover, broadleaved, evergreen
■ Tree cover, broadleaved, deciduous, closed
■ Tree cover, broadleaved, open
■ Tree cover, broadleaved, needle-leaved, evergreen
■ Tree cover, broadleaved, needle-leaved, deciduous
■ Tree cover, broadleaved, mixed leaf type
■ Tree cover, broadleaved, regularly flooded, fresh water
■ Tree cover, broadleaved, regularly flooded, saline water
■ Mosaic of tree cover/other natural vegetation
■ Tree cover, burnt
■ Shrub cover, closed-open, evergreen
■ Shrub cover, closed-open, deciduous
■ Herbaceous cover, closed-open
■ Sparse herbaceous or sparse shrub cover
■ Regularly flooded shrub or herbaceous cover
□ Cultivated and managed areas
■ Mosaic of cropland/tree cover/other natural vegetation
■ Mosaic of cropland/shrub/grass cover
■ Bare areas
■ Water bodies
□ Snow and ice
■ Artificial surfaces and associated areas
□ No data

Source: GLC2000 (Bartholome and Belward 2005).

FIGURE 1.6   Protected areas in East Africa, 2009

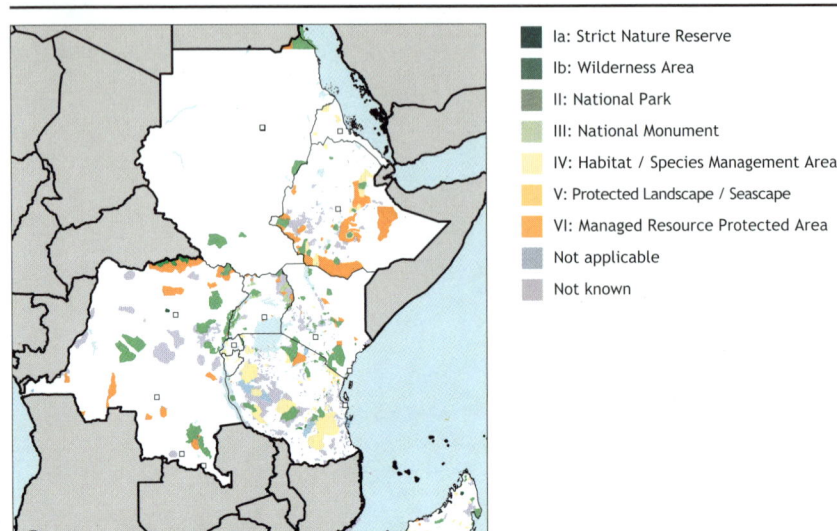

■	Ia: Strict Nature Reserve
■	Ib: Wilderness Area
■	II: National Park
■	III: National Monument
■	IV: Habitat / Species Management Area
■	V: Protected Landscape / Seascape
■	VI: Managed Resource Protected Area
■	Not applicable
■	Not known

Sources: Protected areas are from the World Database on Protected Areas (UNEP and IUCN 2009). Water bodies are from the World Wildlife Fund's Global Lakes and Wetlands Database (Lehner and Döll 2004).

are well served and have relatively short travel times, the same is not true for most rural and arid areas. Most rural areas, where agriculture is the key activity, and arid areas, which are home to pastoral activities, have travel times of 3 to 26 hours. Landlocked countries such as Burundi, Ethiopia, Rwanda, and Uganda face added travel costs for their goods (imports or exports) that have to go by sea. Current efforts by the East African Community and the Common Market for Eastern and Southern Africa to build and refurbish regional road networks along arteries will go a long way to open and facilitate regional trade.

Tables 1.4–1.7 show the major crops grown in eastern and central Africa. The major cereals are sorghum, maize, rice, wheat, and millet (Table 1.4). Sorghum occupies the largest area among the cereal crops, followed very closely by maize. Both these crops and wheat are grown in all countries. The dominant producers are Sudan for sorghum, Tanzania for maize, and Ethiopia for wheat. Beans, groundnuts, and cowpeas are the major legumes cultivated in the region (Table 1.5). Beans and groundnuts are grown in all countries in the

FIGURE 1.7 Travel time in East Africa, circa 2000

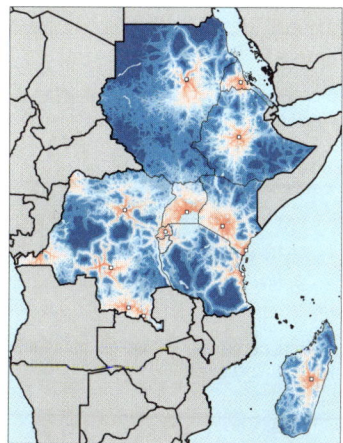

To cities of 500,000 or more people

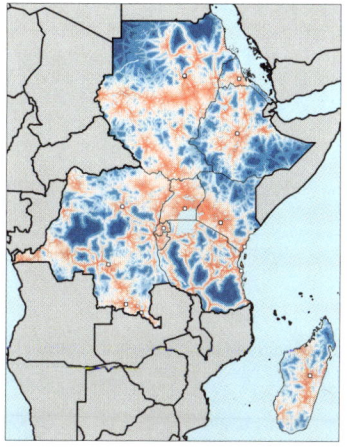

To cities of 100,000 or more people

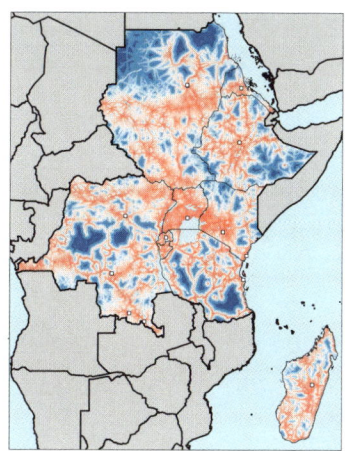

To towns and cities of 25,000 or more people

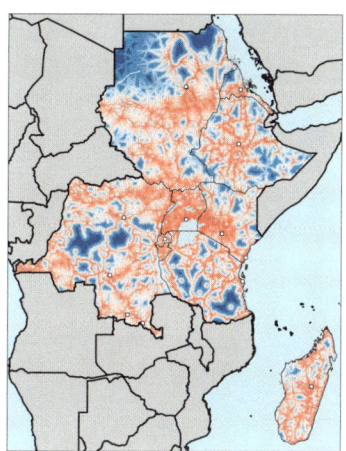

To towns and cities of 10,000 or more people

- Urban location
- < 1 hour
- 1–3 hours
- 3–5 hours
- 5–8 hours
- 8–11 hours
- 11–16 hours
- 16–26 hours
- > 26 hours

Source: Authors' calculations.

TABLE 1.4   Average harvest area of leading agricultural commodities in East Africa, grains, 2006–2008 (hectares)

Country	Barley	Maize	Millet	Rice (paddy)	Sorghum	Wheat	Total
Burundi	0	115,000	10,333	20,833	65,667	9,667	221,500
DR Congo	780	1,483,594	56,778	418,781	9,425	6,763	1,976,121
Eritrea	47,667	18,333	57,000	0	260,667	15,833	399,500
Ethiopia	1,000,708	1,662,679	388,106	8,507	1,488,642	1,452,725	6,001,366
Kenya	17,892	1,734,496	106,327	18,766	141,152	127,243	2,145,876
Madagascar	0	276,667	0	1,243,667	2,000	4,400	1,526,733
Rwanda	0	111,612	5,000	14,011	172,099	22,991	325,713
Sudan	0	57,169	2,296,111	6,685	6,542,557	253,477	9,155,998
Tanzania	2,000	3,066,667	265,000	723,333	896,667	54,000	5,007,667
Uganda	0	841,667	438,000	120,000	314,333	10,667	1,724,667
Total	1,069,047	9,367,884	3,622,654	2,574,584	9,893,209	1,957,766	28,485,141

Source: FAOSTAT (FAO 2010).
Notes: DR Congo = Democratic Republic of Congo. All values are based on the three-year average for 2006-2008.

TABLE 1.5   Average harvest area of leading agricultural commodities in East Africa, pulses and nuts, 2006–2008 (hectares)

Country	Dry beans	Cashew nuts	Cowpeas	Groundnuts	Pigeon peas	Soybeans	Total
Burundi	230,000	0	0	12,833	2,000	3,700	248,533
DR Congo	207,492	0	116,015	474,769	9,880	33,492	841,648
Eritrea	3,000	0	0	2,533	0	0	5,533
Ethiopia	206,159	0	0	35,146	0	6,826	248,131
Kenya	827,885	2,000	146,764	17,000	182,381	2,504	1,178,534
Madagascar	83,767	16,400	4,600	54,933	0	50	159,750
Rwanda	358,794	0	0	16,732	0	42,788	418,314
Sudan	7,258	0	109,510	715,427	0	0	832,195
Tanzania	716,667	92,333	150,000	413,333	67,500	5,000	1,444,833
Uganda	871,667	0	72,333	236,333	87,000	146,667	1,414,000
Total	3,512,688	110,733	599,222	1,979,040	348,761	241,027	6,791,472

Source: FAOSTAT (FAO 2010).
Notes: DR Congo = Democratic Republic of Congo. All values are based on the three-year average for 2006-2008.

TABLE 1.6   Average harvest area of leading agricultural commodities in East Africa, root crops, bananas, and plantains, 2006–2008 (hectares)

Country	Bananas	Cassava	Plantains	Potatoes	Sweet potatoes	Yams	Total
Burundi	336,667	64,333	0	10,000	29,000	1,700	541,700
DR Congo	84,284	1,859,203	268,178	20,194	47,202	19,789	2,298,850
Eritrea	0	0	0	2,467	0	0	2,467
Ethiopia	34,809	0	0	61,798	56,327	32,753	185,687
Kenya	39,538	58,928	39,538	118,783	66,278	858	323,923
Madagascar	58,333	316,790	0	37,947	125,467	0	538,537
Rwanda	10	112,953	375,431	139,917	138,575	1,500	768,386
Sudan	2,300	6,000	0	15,698	650	57,000	81,648
Tanzania	480,000	673,333	308,000	125,333	505,000	1,483	2,093,150
Uganda	135,000	382,667	1,678,333	93,333	587,000	0	2,876,333
Total	1,170,942	3,474,208	2,669,481	625,470	1,655,498	115,083	9,710,682

Source: FAOSTAT (FAO 2010).
Notes: DR Congo = Democratic Republic of Congo. All values are based on the three-year average for 2006-2008.

TABLE 1.7   Average harvest area of leading agricultural commodities in East Africa, other crops, 2006–2008 (hectares)

Country	Cocoa beans	Coffee	Seed cotton	Sesame seeds	Sugar cane	Sunflowers	Total
Burundi	0	22,667	5,100	0	2,500	0	30,267
DR Congo	18,500	81,872	60,000	11,022	39,307	0	210,702
Eritrea	0	0	0	51,833	0	0	51,833
Ethiopia	0	370,794	84,333	205,635	21,494	0	682,256
Kenya	0	162,667	84,133	26,000	56,096	13,000	341,896
Madagascar	6,170	121,700	12,667	0	82,000	0	222,537
Rwanda	0	34,000	0	0	4,433	0	38,433
Sudan	0	0	150,965	1,470,110	68,667	71,873	1,761,615
Tanzania	6,833	110,000	415,000	120,000	22,333	82,333	756,500
Uganda	30,433	250,000	91,667	280,667	35,000	170,333	858,100
Total	61,937	1,153,699	903,865	2,165,267	331,830	337,540	4,954,139

Source: FAOSTAT (FAO 2010).
Notes: DR Congo = Democratic Republic of Congo. All values are based on the three-year average for 2006-2008.

FIGURE 1.8   Yield (metric tons per hectare) and harvest area density (hectares) for rainfed maize in East Africa, 2000

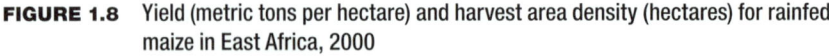

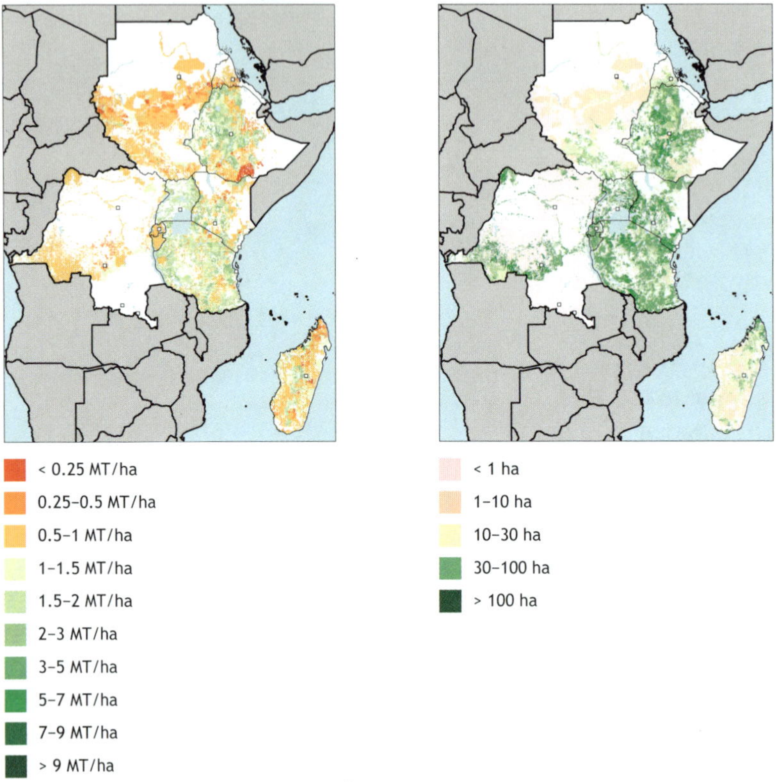

■ < 0.25 MT/ha	■ < 1 ha
■ 0.25–0.5 MT/ha	■ 1–10 ha
■ 0.5–1 MT/ha	■ 10–30 ha
■ 1–1.5 MT/ha	■ 30–100 ha
■ 1.5–2 MT/ha	■ > 100 ha
■ 2–3 MT/ha	
■ 3–5 MT/ha	
■ 5–7 MT/ha	
■ 7–9 MT/ha	
■ > 9 MT/ha	

Source: SPAM (Spatial Production Allocation Model) (You and Wood 2006; You, Wood, and Wood-Sichra 2006, 2009).
Note: ha = hectare; MT/ha = metric tons per hectare.

region. Uganda has the largest area under beans, while Sudan has the largest area under groundnuts. Cassava and sweet potatoes are the major root crops grown and consumed in eastern and central Africa (Table 1.6). The largest producer of cassava is DRC, while Uganda leads in the production of plantains and sweet potatoes. The major cash crops in the region are sesame, coffee, and cotton (Table 1.7). The largest producer of sesame is Sudan, while Ethiopia leads in the production of coffee and Tanzania in the production of cotton.

Figures 1.8–1.12 show the distribution and yield of major cereal crops grown in eastern and central Africa. Rainfed maize is produced across the region (Figure 1.8). Millet and sorghum are produced mainly in Sudan, Ethiopia,

FIGURE 1.9   Yield (metric tons per hectare) and harvest area density (hectares) for rainfed millet in East Africa, 2000

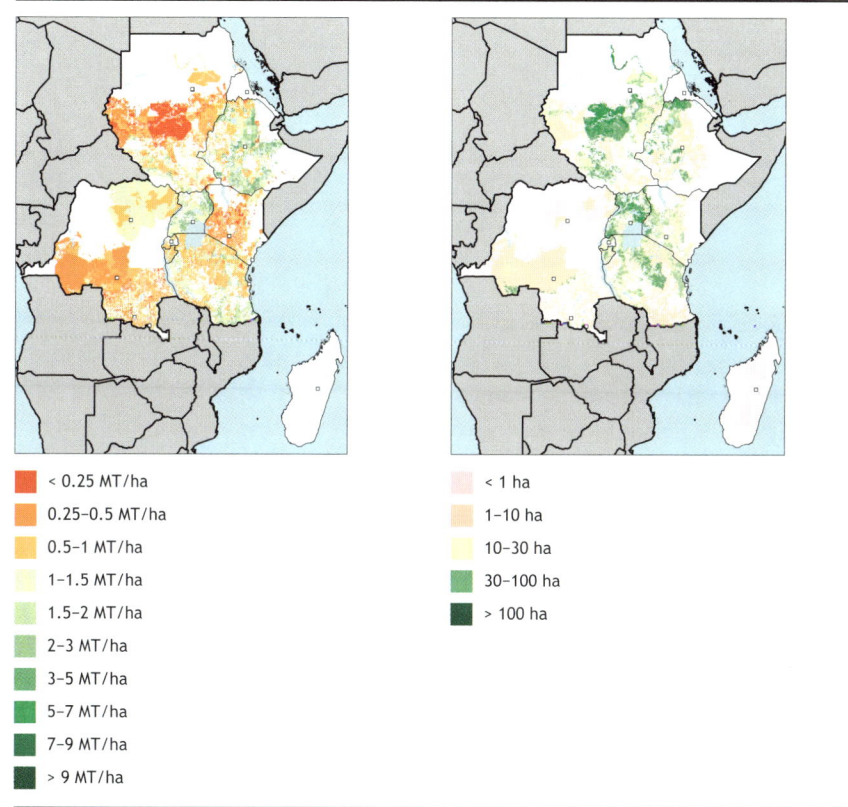

▮ < 0.25 MT/ha	▮ < 1 ha
▮ 0.25–0.5 MT/ha	▮ 1–10 ha
▮ 0.5–1 MT/ha	▮ 10–30 ha
▮ 1–1.5 MT/ha	▮ 30–100 ha
▮ 1.5–2 MT/ha	▮ > 100 ha
▮ 2–3 MT/ha	
▮ 3–5 MT/ha	
▮ 5–7 MT/ha	
▮ 7–9 MT/ha	
▮ > 9 MT/ha	

Source: SPAM (Spatial Production Allocation Model) (You and Wood 2006; You, Wood, and Wood-Sichra 2006, 2009).
Note: ha = hectare; MT/ha = metric tons per hectare.

Uganda, Kenya, and Tanzania (Figures 1.9 and 1.10). Similar to maize production, rice production is concentrated in Tanzania, Madagascar, Uganda, and Sudan (Figure 1.11). Rainfed wheat is grown in large portions of Ethiopia and Madagascar and in more limited areas of Kenya, Tanzania, Rwanda, Burundi, and DRC (Figure 1.12). Rainfed cereal yields are still very low in the region. Rice and maize yield an average of between 1.2 and 1.4 tons per hectare, while sorghum yields around 1 ton per hectare and millet yields about 0.6 ton per hectare.[3] Figure 1.13 shows the data for a number of irrigated crops.

---

3 All tons in this book are metric tons.

FIGURE 1.10   Rainfed sorghum projections for East Africa for yield, area, and production, 2010 and 2050

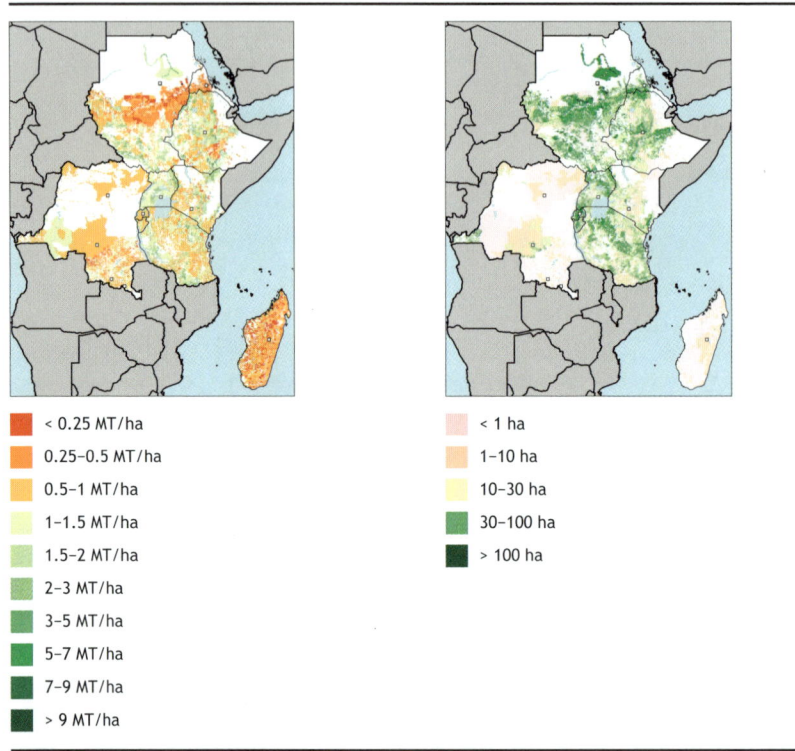

■ < 0.25 MT/ha	■ < 1 ha
■ 0.25–0.5 MT/ha	■ 1–10 ha
■ 0.5–1 MT/ha	■ 10–30 ha
■ 1–1.5 MT/ha	■ 30–100 ha
■ 1.5–2 MT/ha	■ > 100 ha
■ 2–3 MT/ha	
■ 3–5 MT/ha	
■ 5–7 MT/ha	
■ 7–9 MT/ha	
■ > 9 MT/ha	

Source: SPAM (Spatial Production Allocation Model) (You and Wood 2006; You, Wood, and Wood-Sichra 2006, 2009).

Notes: DR Congo = Democratic Republic of Congo; Min (minimum) represents the smallest projected number from the simulations based on the CSIRO A1B, CSIRO B1, MIROC A1B, and MIROC B1 climate model/scenarios combined with the pessimistic, baseline, and optimistic scenarios. Max (maximum) represents the largest of the twelve simulated values. A1B = greenhouse gas emissions scenario that assumes fast economic growth, a population that peaks midcentury, and the development of new and efficient technologies, along with a balanced use of energy sources; B1 = greenhouse gas emissions scenario that assumes a population that peaks midcentury (like the A1B), but with rapid changes toward a service and information economy and the introduction of clean and resource efficient technologies; CSIRO = climate model developed at the Australia Commonwealth Scientific and Industrial Research Organisation; MIROC = Model for Interdisciplinary Research on Climate, developed at the University of Tokyo Center for Climate System Research; ha = hectares; MT = metric tons.

Scenarios for the Future

Population and Income Scenarios

All scenarios for the future, described further in Chapter 2, include a significant increase in the populations of the 10 countries of eastern and central Africa by 2050 (Table 1.8). In the pessimistic scenario, populations of all countries in the region will more than double. A similar outcome occurs in the baseline scenario for all countries except Burundi and Sudan. In the optimistic scenario, the population doubles only in DRC, Ethiopia, Tanzania, and Uganda. Income per capita improves significantly in the optimistic scenario, with increases of up to

FIGURE 1.11   Rainfed rice projections for East Africa for yield, area, and production, 2010 and 2050

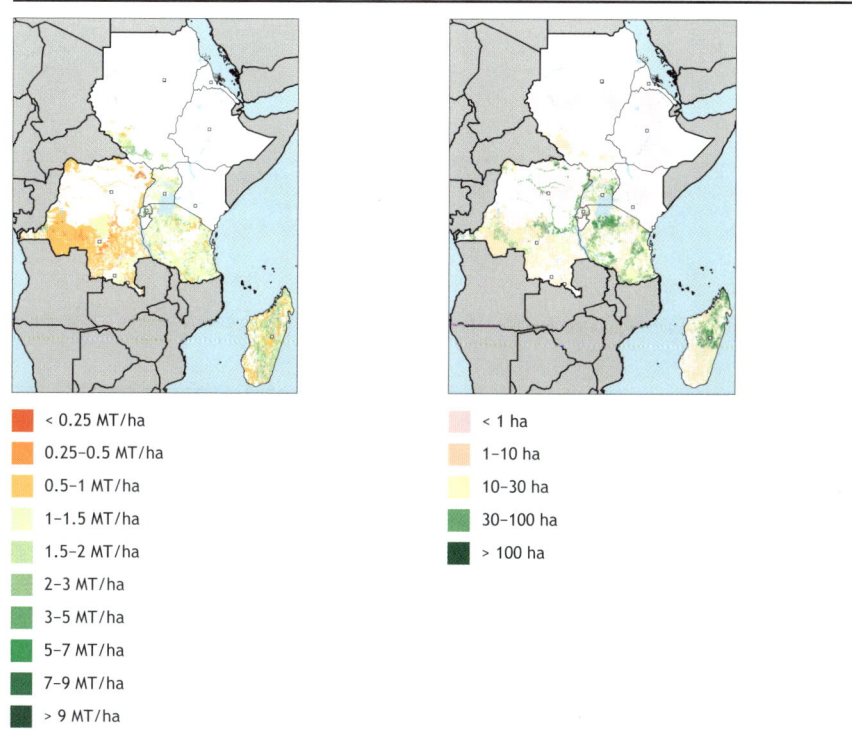

▮ < 0.25 MT/ha	▮ < 1 ha
▮ 0.25–0.5 MT/ha	▮ 1–10 ha
▮ 0.5–1 MT/ha	▮ 10–30 ha
▮ 1–1.5 MT/ha	▮ 30–100 ha
▮ 1.5–2 MT/ha	▮ > 100 ha
▮ 2–3 MT/ha	
▮ 3–5 MT/ha	
▮ 5–7 MT/ha	
▮ 7–9 MT/ha	
▮ > 9 MT/ha	

Source: SPAM (Spatial Production Allocation Model) (You and Wood 2006; You, Wood, and Wood-Sichra 2006, 2009).

Notes: DR Congo = Democratic Republic of Congo; Min (minimum) represents the smallest projected number from the simulations based on the CSIRO A1B, CSIRO B1, MIROC A1B, and MIROC B1 climate model/scenarios combined with the pessimistic, baseline, and optimistic scenarios. Max (maximum) represents the largest of the twelve simulated values. A1B = greenhouse gas emissions scenario that assumes fast economic growth, a population that peaks midcentury, and the development of new and efficient technologies, along with a balanced use of energy sources; B1 = greenhouse gas emissions scenario that assumes a population that peaks midcentury (like the A1B), but with rapid changes toward a service and information economy and the introduction of clean and resource efficient technologies; CSIRO = climate model developed at the Australia Commonwealth Scientific and Industrial Research Organisation; MIROC = Model for Interdisciplinary Research on Climate, developed at the University of Tokyo Center for Climate System Research; ha = hectares; MT = metric tons.

nine times in Burundi and the least increase, of four times, in Sudan. These numbers are further reflected in the baseline scenario. In the pessimistic scenario, the highest increase is a little less than fourfold in Burundi and Tanzania.

Biophysical Analysis

Predictions for rainfall vary between the general circulation models (GCMs) (Figure 1.14).[4] Generally, for the A1B scenario, the CSIRO model predicts the driest future and MIROC predicts the wettest future, but there is significant

---

4   See Chapter 2 for details on how these scenarios were produced.

FIGURE 1.12   Rainfed wheat projections for East Africa for yield, area, and production, 2010 and 2050

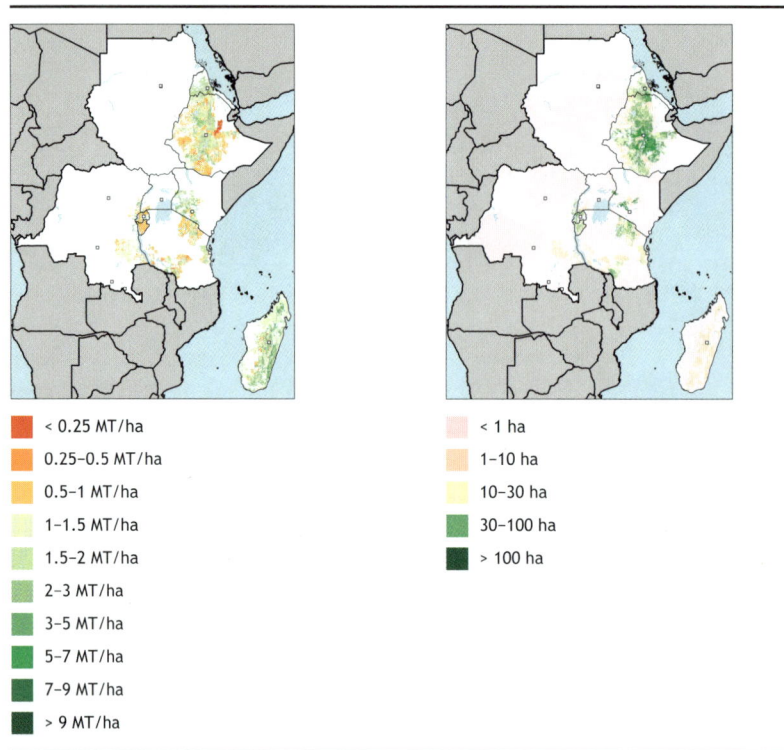

■ < 0.25 MT/ha	■ < 1 ha
■ 0.25–0.5 MT/ha	■ 1–10 ha
■ 0.5–1 MT/ha	□ 10–30 ha
■ 1–1.5 MT/ha	■ 30–100 ha
■ 1.5–2 MT/ha	■ > 100 ha
■ 2–3 MT/ha	
■ 3–5 MT/ha	
■ 5–7 MT/ha	
■ 7–9 MT/ha	
■ > 9 MT/ha	

Source: SPAM (Spatial Production Allocation Model) (You and Wood 2006; You, Wood, and Wood-Sichra 2006, 2009).

Notes: DR Congo = Democratic Republic of Congo; Min (minimum) represents the smallest projected number from the simulations based on the CSIRO A1B, CSIRO B1, MIROC A1B, and MIROC B1 climate model/scenarios combined with the pessimistic, baseline, and optimistic scenarios. Max (maximum) represents the largest of the twelve simulated values. A1B = greenhouse gas emissions scenario that assumes fast economic growth, a population that peaks midcentury, and the development of new and efficient technologies, along with a balanced use of energy sources; B1 = greenhouse gas emissions scenario that assumes a population that peaks midcentury (like the A1B), but with rapid changes toward a service and information economy and the introduction of clean and resource efficient technologies; CSIRO = climate model developed at the Australia Commonwealth Scientific and Industrial Research Organisation; MIROC = Model for Interdisciplinary Research on Climate, developed at the University of Tokyo Center for Climate System Research; ha = hectares; MT = metric tons.

spatial variation in each GCM, and in each location there are variations between GCMs.[5] Simple averages tell us that all but the CSIRO model predict an average increase in rainfall for the region, and even the CSIRO model

---

5   CSIRO and MIROC are acronyms for two of the general circulation models (GCMs) discussed in this book. CSIRO is a climate model developed at the Australia Commonwealth Scientific and Industrial Research Organisation. MIROC is the Model for Interdisciplinary Research on Climate, developed at the University of Tokyo Center for Climate System Research. The A1B scenario is a greenhouse gas emissions scenario that assumes fast economic growth, a population that peaks midcentury, and the development of new and efficient technologies, along with a balanced use of energy sources.

FIGURE 1.13 Harvest area density (hectares) and yields (metric tons per hectare) of crops under irrigation in East Africa, 2000

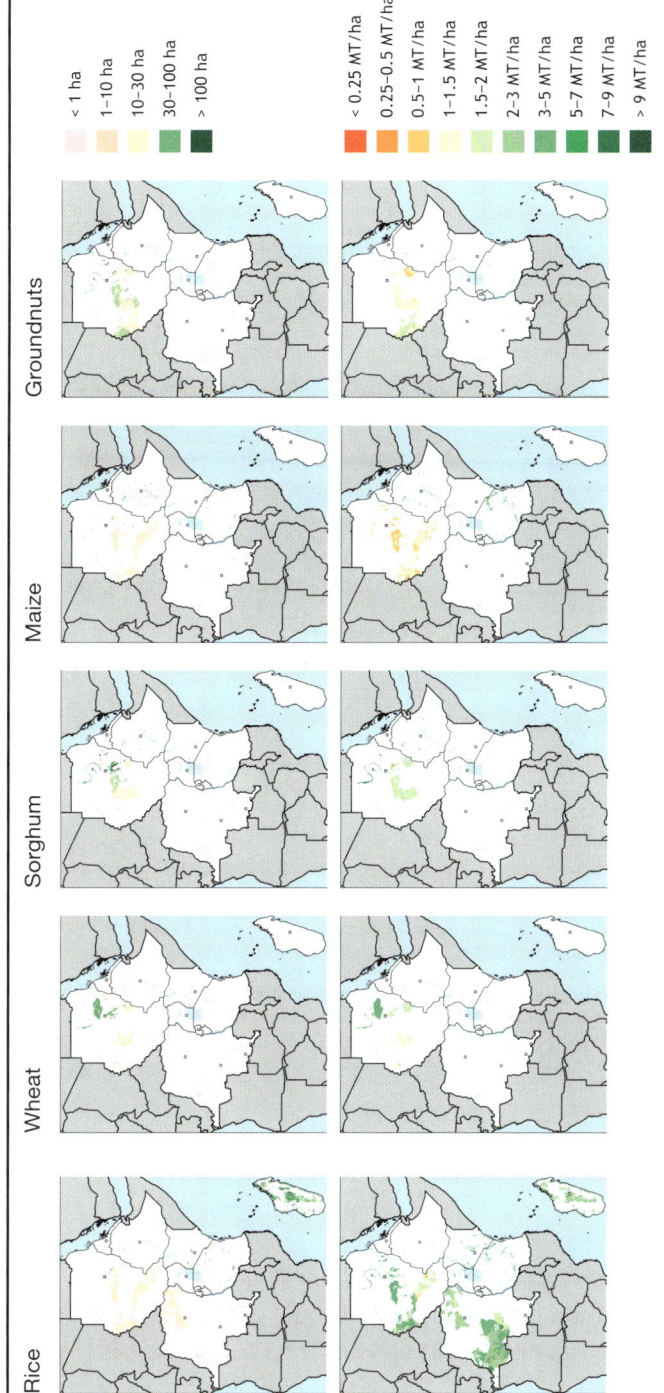

Legend (harvest area density):
< 1 ha
1–10 ha
10–30 ha
30–100 ha
> 100 ha

Legend (yields):
< 0.25 MT/ha
0.25–0.5 MT/ha
0.5–1 MT/ha
1–1.5 MT/ha
1.5–2 MT/ha
2–3 MT/ha
3–5 MT/ha
5–7 MT/ha
7–9 MT/ha
> 9 MT/ha

Rice   Wheat   Sorghum   Maize   Groundnuts

Source: SPAM (Spatial Production Allocation Model) (You and Wood 2006; You, Wood, and Wood-Sichra 2006, 2009).

Note: ha = hectare; MT/ha = metric tons per hectare.

TABLE 1.8   Summary statistics for assumptions on East Africa's population and per capita GDP used in the IMPACT model, 2010 and 2050

Category	2010	2050 Pessimistic	2050 Baseline	2050 Optimistic
Population (thousands)				
Burundi	8,519	16,814	14,846	13,006
Democratic Republic of Congo	67,827	166,249	147,512	130,013
Eritrea	5,224	12,198	10,787	9,458
Ethiopia	84,976	196,245	173,811	152,720
Kenya	40,863	97,541	85,410	74,187
Madagascar	20,146	48,694	42,693	37,155
Rwanda	10,277	24,829	22,082	19,498
Sudan	43,192	86,371	75,884	66,140
Tanzania	45,040	124,020	109,450	95,884
Uganda	33,796	102,678	91,271	80,573
Income per capita (2000 US$)				
Burundi	153	569	973	1,450
Democratic Republic of Congo	103	277	440	715
Eritrea	201	505	955	1,379
Ethiopia	138	323	720	1,037
Kenya	407	543	2,255	3,286
Madagascar	252	654	1,195	1,741
Rwanda	300	468	1,583	2,268
Sudan	151	320	372	680
Tanzania	275	1,013	1,310	2,416
Uganda	433	1,156	2,563	3,667

Sources: Computed from GDP data from the World Bank Economic Adaptation to Climate Change project (World Bank 2010), from the Millennium Ecosystem Assessment (2005) reports, and from population data from the United Nations (UNPOP 2009). Notes: 2010 income per capita is for the baseline scenario. GDP = gross domestic product; IMPACT = International Model for Policy Analysis of Agricultural Commodities and Trade (International Food Policy Research Institute); US$ = US dollars.

predicts that rainfall, on average, will be virtually unchanged. But locally, the models can tell a different story. For example, in southern Madagascar, the models agree that rainfall will not increase and will more than likely decrease.

Recent analysis by Funk et al. (2008) and Williams and Funk (2011) suggest that these global models are too optimistic about precipitation in the future, suggesting that when the impact of Indian Ocean sea-surface temperature change on precipitation is taken into consideration, East Africa will be

FIGURE 1.14   Change in mean annual precipitation in East Africa, 2000–2050, A1B scenario (millimeters)

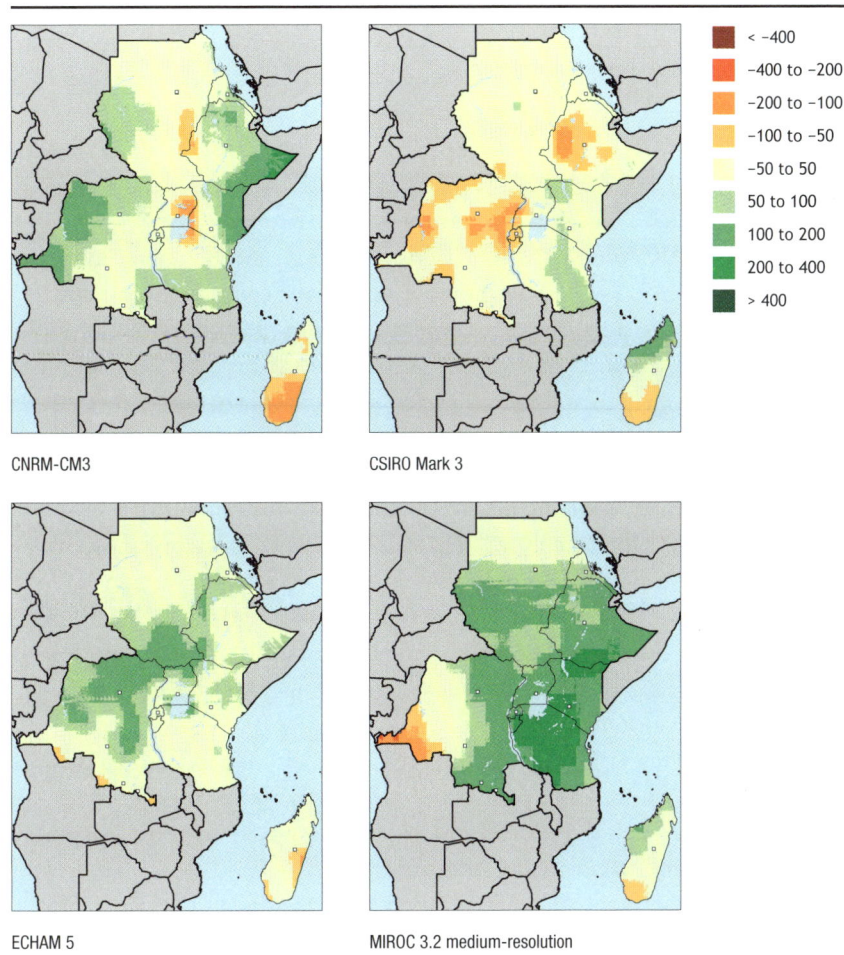

CNRM-CM3

CSIRO Mark 3

ECHAM 5

MIROC 3.2 medium-resolution

Legend:
- < −400
- −400 to −200
- −200 to −100
- −100 to −50
- −50 to 50
- 50 to 100
- 100 to 200
- 200 to 400
- > 400

Source: Authors' calculations based on Jones, Thornton, and Heinke (2009).

Notes: A1B = greenhouse gas emissions scenario that assumes fast economic growth, a population that peaks midcentury, and the development of new and efficient technologies, along with a balanced use of energy sources; CNRM-CM3 = National Meteorological Research Center–Climate Model 3; CSIRO = climate model developed at the Australia Commonwealth Scientific and Industrial Research Organization; ECHAM 5 = fifth-generation climate model developed at the Max Planck Institute for Meteorology (Hamburg); GCM = general circulation model; MIROC = the Model for Interdisciplinary Research on Climate, developed by the University of Tokyo Center for Climate System Research.

much drier, particularly in the "long rains" period of March to June. If these authors' projections prove to be more realistic than the results based on the four GCMs that we use, the reader should expect the worst-case or minimum-yield results presented throughout this monograph.

Predictions of changes in the average daily maximum temperature of the warmest month are shown in Figure 1.15. Generally speaking, the CSIRO and MIROC models show the coolest future climates, averaging changes over the entire region of 1.3°C and 1.5°C, respectively. Of the two, the CSIRO model has smaller spatial variation, reflecting a narrower range of temperature changes over the entire region. Both the CNRM-CM3 and the ECHAM 5 GCMs show an increase of 2.1°C on average, with the ECHAM 5 model exhibiting a smaller range of temperatures over the region.[6]

Both the CSIRO and MIROC GCMs result in a general increase in maize yields of 5–25 percent of baseline in most parts of the countries of east and central Africa and yield losses of 5–25 percent in large parts of DRC, Ethiopia, Tanzania, and northern Uganda (Figure 1.16). We also note increases in some areas, mostly in the highlands of Kenya and Ethiopia. These likely reflect an increase in temperature that would permit maize to be grown in areas in which it is currently too cold to grow maize well.

Based on both the CSIRO and MIROC climate outcomes with the A1B SRES scenario, sorghum yields will decline by 5–25 percent across most of eastern and central Africa (Figure 1.17).[7] The models also show a gain in baseline area of 5–25 percent in parts of western DRC and the highlands of Ethiopia, Kenya, Tanzania, and Sudan. All but the CNRM model show losses in a band across central Sudan. This may reflect temperatures rising too much for sorghum to continue being grown there.

Figure 1.18 shows the results of the work done with DSSAT crop modeling for rainfed wheat. On average across all models, yields are reduced by around 8 percent, with the CSIRO model predicting relatively little loss and the ECHAM 5 model predicting the greatest loss.

Figure 1.19 shows the results for rainfed rice. On average, the rice yield under climate change is projected to increase by just over 3 percent, with all of the GCMs reporting very similar mean yields with climate change.

Figure 1.20 shows the results for rainfed groundnuts. The results vary across GCMs. For example, yields fall by around 7 percent on average for the CNRM model, in large part due to losses in Sudan. On the other hand, yields rise by around 8 percent on average in the MIROC model.

---

6 CNRM-CM3 is National Meteorological Research Center–Climate Model 3; ECHAM 5 is a fifth-generation climate model developed at the Max Planck Institute for Meteorology (Hamburg).
7 SRES stands for the Special Report on Emissions Scenarios of the IPCC.

FIGURE 1.15 Change in monthly mean maximum daily temperature in East Africa for the warmest month, 2000–2050, A1B scenario (°C)

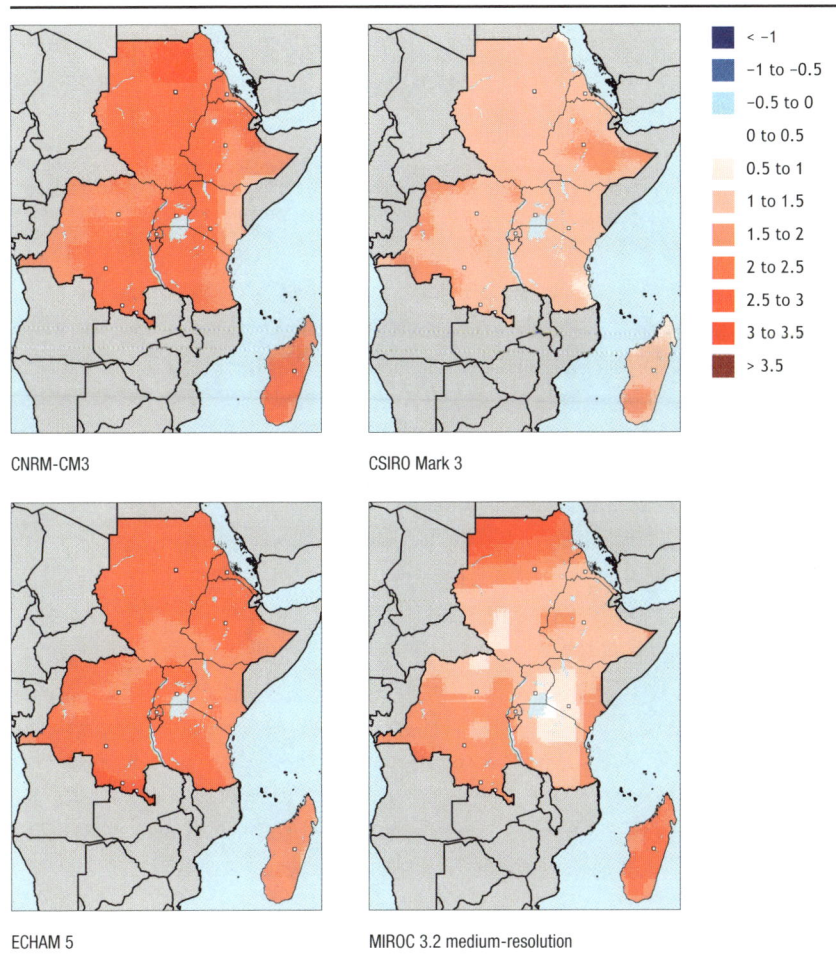

■	< −1
■	−1 to −0.5
■	−0.5 to 0
	0 to 0.5
	0.5 to 1
■	1 to 1.5
■	1.5 to 2
■	2 to 2.5
■	2.5 to 3
■	3 to 3.5
■	> 3.5

CNRM-CM3

CSIRO Mark 3

ECHAM 5

MIROC 3.2 medium-resolution

Source: Authors' calculations based on Jones, Thornton, and Heinke (2009).

Notes: A1B = greenhouse gas emissions scenario that assumes fast economic growth, a population that peaks midcentury, and the development of new and efficient technologies, along with a balanced use of energy sources; CNRM-CM3 = National Meteorological Research Center–Climate Model 3; CSIRO = climate model developed at the Australia Commonwealth Scientific and Industrial Research Organization; ECHAM 5 = fifth-generation climate model developed at the Max Planck Institute for Meteorology (Hamburg); GCM = general circulation model; MIROC = the Model for Interdisciplinary Research on Climate, developed at the University of Tokyo Center for Climate System Research.

FIGURE 1.16   Yield change under climate change: Rainfed maize in East Africa, 2000–2050, A1B scenario

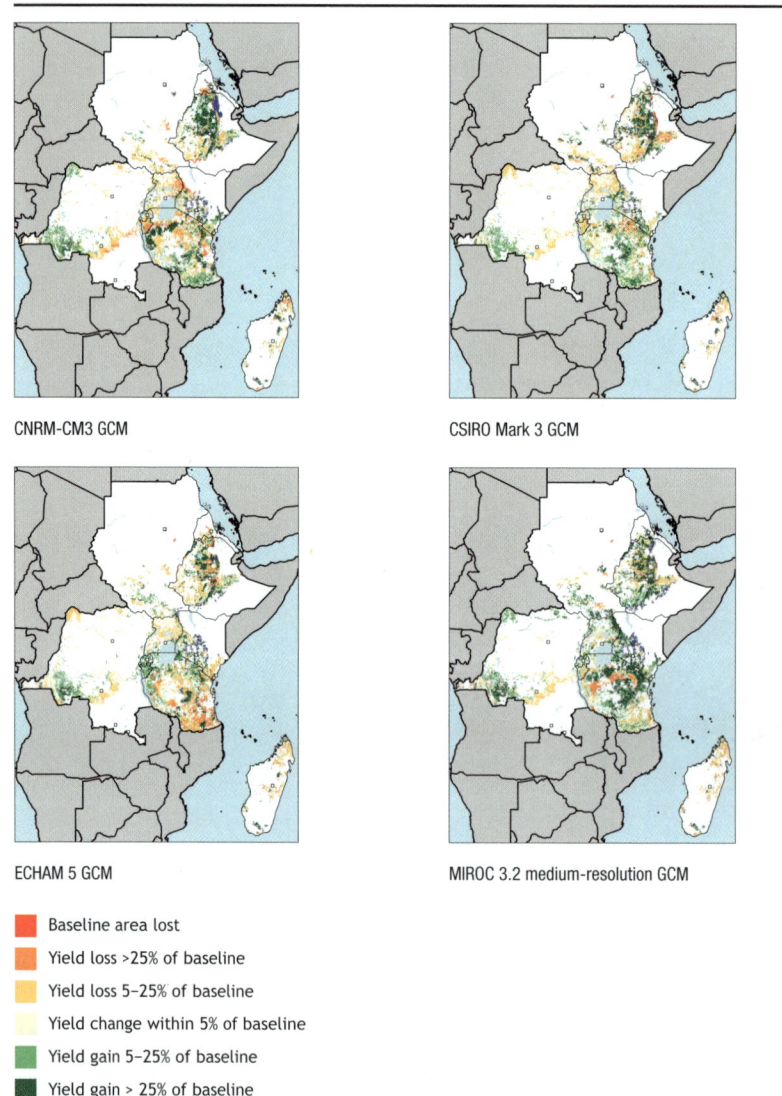

CNRM-CM3 GCM

CSIRO Mark 3 GCM

ECHAM 5 GCM

MIROC 3.2 medium-resolution GCM

■ Baseline area lost
■ Yield loss >25% of baseline
■ Yield loss 5–25% of baseline
□ Yield change within 5% of baseline
■ Yield gain 5–25% of baseline
■ Yield gain > 25% of baseline
■ New area gained

Source: Authors' calculations.

Notes: A1B = greenhouse gas emissions scenario that assumes fast economic growth, a population that peaks midcentury, and the development of new and efficient technologies, along with a balanced use of energy sources; CNRM-CM3 = National Meteorological Research Center–Climate Model 3; CSIRO = climate model developed at the Australia Commonwealth Scientific and Industrial Research Organisation; ECHAM 5 = fifth-generation climate model developed at the Max Planck Institute for Meteorology (Hamburg); GCM = general circulation model; MIROC 3.2 = the Model for Interdisciplinary Research on Climate, developed at the University of Tokyo Center for Climate System Research.

FIGURE 1.17 Yield change under climate change: Rainfed sorghum in East Africa, 2000–2050, A1B scenario

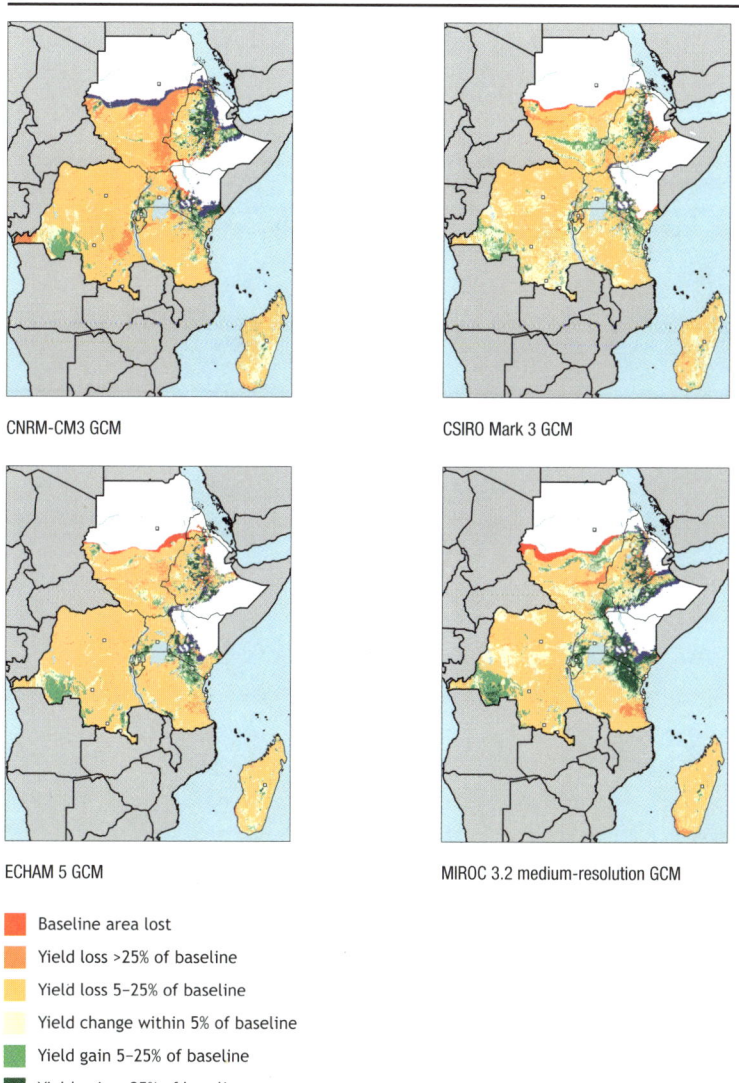

CNRM-CM3 GCM

CSIRO Mark 3 GCM

ECHAM 5 GCM

MIROC 3.2 medium-resolution GCM

- Baseline area lost
- Yield loss >25% of baseline
- Yield loss 5–25% of baseline
- Yield change within 5% of baseline
- Yield gain 5–25% of baseline
- Yield gain > 25% of baseline
- New area gained

Source: Authors' calculations.

Notes: A1B = greenhouse gas emissions scenario that assumes fast economic growth, a population that peaks midcentury, and the development of new and efficient technologies, along with a balanced use of energy sources; CNRM-CM3 = National Meteorological Research Center–Climate Model 3; CSIRO = climate model developed at the Australia Commonwealth Scientific and Industrial Research Organisation; ECHAM 5 = fifth-generation climate model developed at the Max Planck Institute for Meteorology (Hamburg); GCM = general circulation model; MIROC 3.2 = the Model for Interdisciplinary Research on Climate, developed at the University of Tokyo Center for Climate System Research.

FIGURE 1.18 Yield change under climate change: Rainfed wheat in East Africa, 2000–2050, A1B scenario

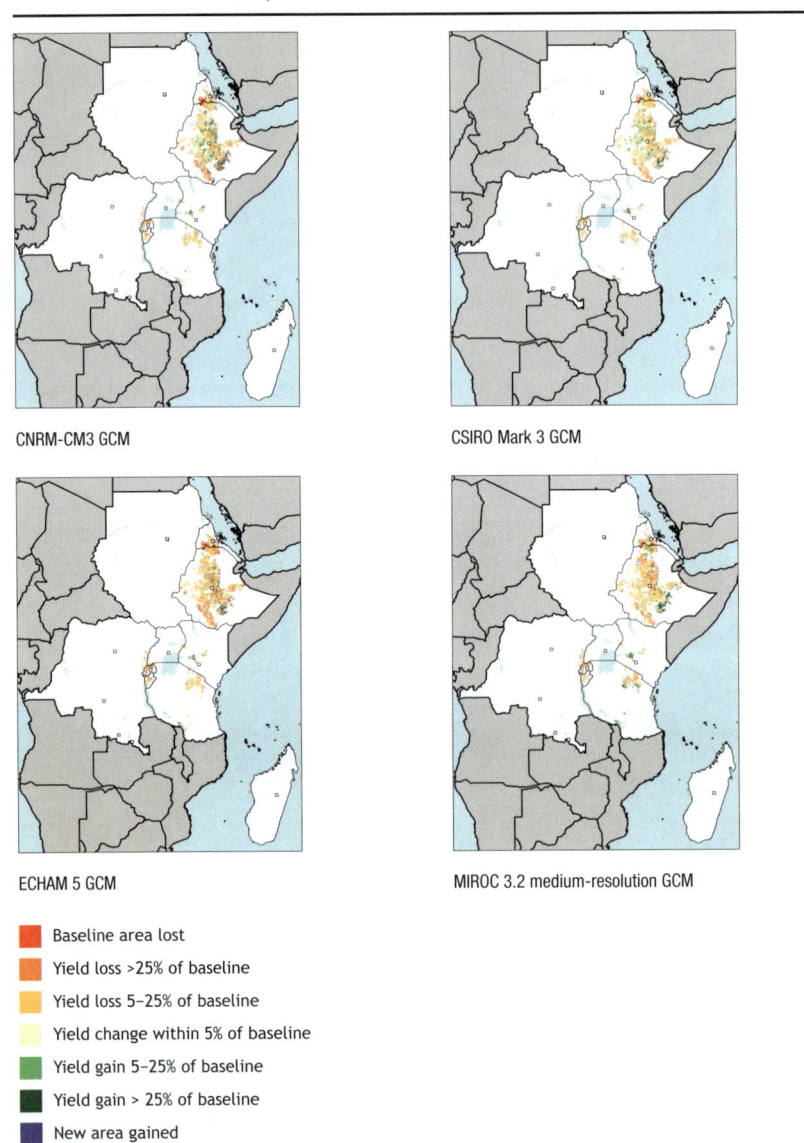

CNRM-CM3 GCM

CSIRO Mark 3 GCM

ECHAM 5 GCM

MIROC 3.2 medium-resolution GCM

■ Baseline area lost
■ Yield loss >25% of baseline
■ Yield loss 5–25% of baseline
□ Yield change within 5% of baseline
■ Yield gain 5–25% of baseline
■ Yield gain > 25% of baseline
■ New area gained

Source: Authors' calculations.

Notes: A1B = greenhouse gas emissions scenario that assumes fast economic growth, a population that peaks midcentury, and the development of new and efficient technologies, along with a balanced use of energy sources; CNRM-CM3 = National Meteorological Research Center–Climate Model 3; CSIRO = climate model developed at the Australia Commonwealth Scientific and Industrial Research Organisation; ECHAM 5 = fifth-generation climate model developed at the Max Planck Institute for Meteorology (Hamburg); GCM = general circulation model; MIROC 3.2 = the Model for Interdisciplinary Research on Climate, developed at the University of Tokyo Center for Climate System Research.

FIGURE 1.19   Yield change under climate change: Rainfed rice in East Africa, 2000–2050, A1B scenario

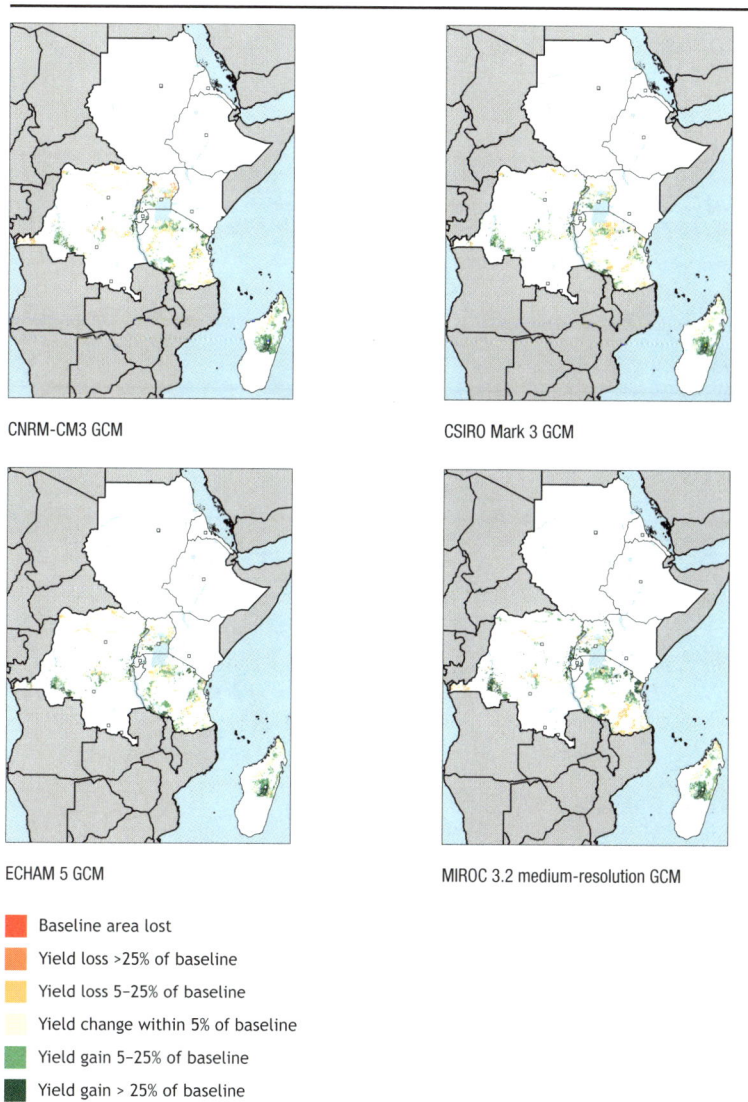

CNRM-CM3 GCM

CSIRO Mark 3 GCM

ECHAM 5 GCM

MIROC 3.2 medium-resolution GCM

■ Baseline area lost
■ Yield loss >25% of baseline
■ Yield loss 5–25% of baseline
□ Yield change within 5% of baseline
■ Yield gain 5–25% of baseline
■ Yield gain > 25% of baseline
■ New area gained

Source: Authors' calculations.

Notes: A1B = greenhouse gas emissions scenario that assumes fast economic growth, a population that peaks midcentury, and the development of new and efficient technologies, along with a balanced use of energy sources; CNRM-CM3 = National Meteorological Research Center–Climate Model 3; CSIRO = climate model developed at the Australia Commonwealth Scientific and Industrial Research Organisation; ECHAM 5 = fifth-generation climate model developed at the Max Planck Institute for Meteorology (Hamburg); GCM = general circulation model; MIROC 3.2 = the Model for Interdisciplinary Research on Climate, developed at the University of Tokyo Center for Climate System Research.

FIGURE 1.20   Yield change under climate change: Rainfed groundnuts in East Africa, 2000–2050, A1B scenario

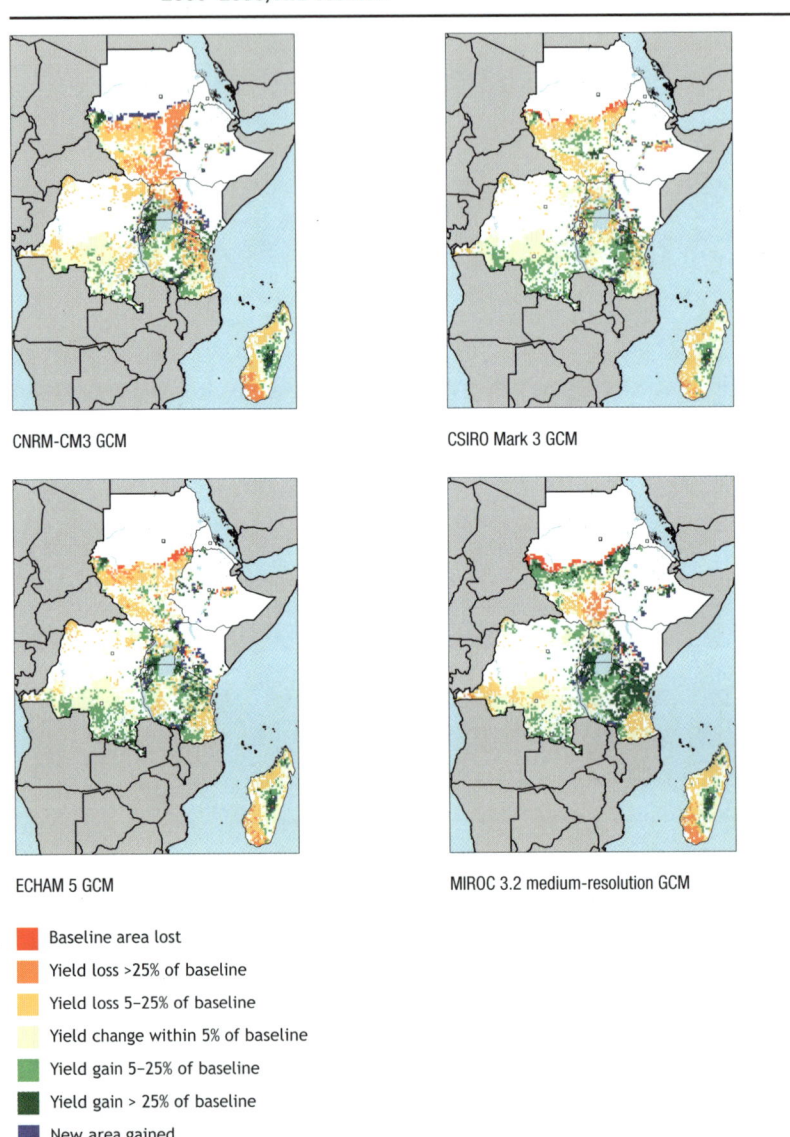

CNRM-CM3 GCM

CSIRO Mark 3 GCM

ECHAM 5 GCM

MIROC 3.2 medium-resolution GCM

- ■ Baseline area lost
- ■ Yield loss >25% of baseline
- ■ Yield loss 5–25% of baseline
- ■ Yield change within 5% of baseline
- ■ Yield gain 5–25% of baseline
- ■ Yield gain > 25% of baseline
- ■ New area gained

Source: Authors' calculations.

Notes: A1B = greenhouse gas emissions scenario that assumes fast economic growth, a population that peaks midcentury, and the development of new and efficient technologies, along with a balanced use of energy sources; CNRM-CM3 = National Meteorological Research Center–Climate Model 3; CSIRO = climate model developed at the Australia Commonwealth Scientific and Industrial Research Organisation; ECHAM 5 = fifth-generation climate model developed at the Max Planck Institute for Meteorology (Hamburg); GCM = general circulation model; MIROC 3.2 = the Model for Interdisciplinary Research on Climate, developed at the University of Tokyo Center for Climate System Research.

Regional Agricultural Outcomes

Prices of major foodcrops are presented in Table 1.9. World market prices for maize, rice, sorghum, and wheat increase in all scenarios, whereas the price of millet is less in 2050. In 2050, prices for millet, rice, sorghum and wheat are higher under the pessimistic scenario than under the optimistic scenario.

Production of maize (Table 1.10), millet (Table 1.11), and sorghum (Table 1.12) increases by 2050 in all countries of eastern and central Africa except Ethiopia, Madagascar, and Uganda. The area under cultivation of both millet and sorghum increases except in Burundi for millet. The productivity of all three crops increases, mainly due to improved management practices, because the projected yields in 2050 are already attainable with existing local and improved varieties of these crops.

Vulnerability Outcomes

In the optimistic scenario, the number of malnourished children decreases for all the countries in eastern and central Africa (Table 1.13). Under the pessimistic scenario, the number increases in all countries. Keep in mind that an increase in numbers still reflects a decrease in the proportion of children under five who are malnourished, because by 2050 the population has increased by a large margin (Table 1.14).

TABLE 1.9   Global commodity prices, 2010 and 2050 (2000 US$ per metric ton)

| | Model price, 2010 | 2050 | | | | | |
| | | Pessimistic | | Baseline | | Optimistic | |
Crop		Min	Max	Min	Max	Min	Max
Maize	111	209	265	216	272	200	253
Millet	341	305	327	291	307	267	283
Rice	239	433	441	378	388	323	328
Sorghum	145	184	193	184	193	175	184
Wheat	146	218	252	222	254	206	236

Source: Based on analysis conducted for Nelson et al. (2010).

Notes: Min (minimum) represents the smallest projected number from the simulations based on the CSIRO A1B, CSIRO B1, MIROC A1B, and MIROC B1 climate model/scenario combinations. Max (maximum) represents the largest of the four simulated values. A1B = greenhouse gas emissions scenario that assumes fast economic growth, a population that peaks midcentury, and the development of new and efficient technologies, along with a balanced use of energy sources; B1 = greenhouse gas emissions scenario that assumes a population that peaks midcentury (like the A1B), but with rapid changes toward a service and information economy and the introduction of clean and resource efficient technologies; CSIRO = climate model developed at the Australia Commonwealth Scientific and Industrial Research Organisation; MIROC = Model for Interdisciplinary Research on Climate, developed at the University of Tokyo Center for Climate System Research.

TABLE 1.10   Maize changes in East Africa under the baseline scenario, 2010 and 2050

	2010			2050					
		Area		Yield (MT/ha)		Area (thousands of ha)		Production (MT)	
Country	Yield (MT/ha)	(thousands of ha)	Production (MT)	Min	Max	Min	Max	Min	Max
Burundi	0.94	117	10	1.68	2.11	83	88	140	186
DR Congo	0.74	1,590	1,171	1.18	1.28	1,639	1,718	1,943	2,201
Eritrea	0.34	13	4	0.63	0.68	12	15	8	10
Ethiopia	1.87	1,809	3,385	2.13	2.49	1,437	1,565	3,060	3,891
Kenya	1.69	1,747	2,958	2.72	3.64	1,726	1,860	4,824	6,775
Madagascar	1.46	227	332	2.39	2.50	109	119	261	297
Rwanda	0.78	118	92	1.21	1.63	123	130	148	212
Sudan	0.72	79	57	0.96	1.01	87	90	84	90
Tanzania	2.51	1,066	2,671	2.56	3.02	1,068	1,137	2,740	3,428
Uganda	1.55	819	1,266	4.87	5.77	733	771	3,578	4,453

Source: Based on analysis conducted for Nelson et al. (2010).

Notes: DR Congo = Democratic Republic of Congo; Min (minimum) represents the smallest projected number from the simulations based on the CSIRO A1B, CSIRO B1, MIROC A1B, and MIROC B1 climate model/scenario combinations. Max (maximum) represents the largest of the four simulated values. A1B = greenhouse gas emissions scenario that assumes fast economic growth, a population that peaks midcentury, and the development of new and efficient technologies, along with a balanced use of energy sources; B1 = greenhouse gas emissions scenario that assumes a population that peaks midcentury (like the A1B), but with rapid changes toward a service and information economy and the introduction of clean and resource efficient technologies; CSIRO = climate model developed at the Australia Commonwealth Scientific and Industrial Research Organisation; MIROC = Model for Interdisciplinary Research on Climate, developed at the University of Tokyo Center for Climate System Research.

A measure of vulnerability somewhat related to child malnutrition is mean calorie consumption, which we report in each of the country chapters and explain in Chapter 2. These values are forecast for the future within IMPACT, which is a worldwide partial equilibrium model dealing with food and agriculture. In some of the low-income, high-population scenarios for the future, we find mean calorie consumption falling despite rises in per capita incomes. This result is in part due to the fact that the price of food also rises, and the negative price effects on consumption counteract the positive income effects. However, in examining the model more carefully, we also noted that the own-price elasticities in the model were set too high in some countries, and this also influenced the falling calorie consumption result for those countries.

Conclusions and Policy Recommendations

Countries in eastern and central Africa are highly vulnerable to climate, primarily due to their reliance on rainfed agriculture, high population growth rates that average 3.7 percent, and endemic poverty that affects more than 50 percent of

TABLE 1.11   Millet changes in East Africa under the baseline scenario, 2010 and 2050

Country	2010 Yield (MT/ha)	2010 Area (thousands of ha)	2010 Production (MT)	2050 Yield (MT/ha) Min	2050 Yield (MT/ha) Max	2050 Area (thousands of ha) Min	2050 Area (thousands of ha) Max	2050 Production (MT) Min	2050 Production (MT) Max
Burundi	1.01	10	10	2.41	2.93	7	7	17	21
DR Congo	0.67	66	44	2.16	2.22	100	107	217	235
Eritrea	0.49	40	20	1.28	1.44	55	69	75	96
Ethiopia	1.19	372	444	3.37	4.14	467	485	1,573	2,007
Kenya	0.46	125	57	1.50	1.82	164	177	255	315
Madagascar	0.00	0	0	0.00	0.00	0	0	0	0
Rwanda	0.69	6	4	1.34	1.57	8	9	12	14
Sudan	0.36	2,132	761	0.58	0.65	2,507	2,695	1,507	1,709
Tanzania	0.63	265	165	2.07	2.35	350	372	737	859
Uganda	1.38	477	660	3.06	3.32	602	644	1,889	2,091

Source: Based on analysis conducted for Nelson et al. (2010).

Notes: DR Congo = Democratic Republic of Congo; Min (minimum) represents the smallest projected number from the simulations based on the CSIRO A1B, CSIRO B1, MIROC A1B, and MIROC B1 climate model/scenario combinations. Max (maximum) represents the largest of the four simulated values. A1B = greenhouse gas emissions scenario that assumes fast economic growth, a population that peaks midcentury, and the development of new and efficient technologies, along with a balanced use of energy sources; B1 = greenhouse gas emissions scenario that assumes a population that peaks midcentury (like the A1B), but with rapid changes toward a service and information economy and the introduction of clean and resource efficient technologies; CSIRO = climate model developed at the Australia Commonwealth Scientific and Industrial Research Organisation; MIROC = Model for Interdisciplinary Research on Climate, developed at the University of Tokyo Center for Climate System Research.

the population. Although the countries are gradually shifting from agriculture to industrial, construction, and service sectors, agriculture still looms large as the economic driver and is hence vulnerable to climate change.

Climate change predictions point to increased rainfall and a rise in temperatures. Rainfall is also predicted to be more erratic and violent, further disrupting predominantly rainfed agricultural production systems. The future climate will also affect the productive infrastructure and exacerbate the constraints on other livelihood systems. Climate change response options that are noted in the National Adaptation Programmes of Action are costly and are yet to be implemented.

Climate scenarios show a general increase in maize yields in most parts of the countries of east and central Africa and yield losses in large parts of DRC, Ethiopia, Tanzania, and northern Uganda. On the other hand, they show that sorghum yields will decline across eastern and central Africa, while there will be gains in western DRC and the highlands of Ethiopia, Kenya, Tanzania, and Sudan.

TABLE 1.12   Sorghum changes in East Africa under the baseline scenario, 2010 and 2050

	2010			2050					
		Area		Yield (MT/ha)		Area (thousands of ha)		Production (MT)	
Country	Yield (MT/ha)	(thousands of ha)	Production (MT)	Min	Max	Min	Max	Min	Max
Burundi	1.31	66	86	2.10	2.41	70	72	147	173
DR Congo	0.69	112	77	1.49	1.54	187	192	280	295
Eritrea	0.31	243	76	0.89	1.04	352	449	320	454
Ethiopia	1.60	1,605	2,568	2.98	3.79	2,229	2,318	6,650	8,776
Kenya	0.80	169	135	1.96	2.38	272	283	542	672
Madagascar	0.50	2	1	1.21	1.23	5	5	6	6
Rwanda	1.01	222	225	1.95	2.41	330	339	644	816
Sudan	0.75	6,450	4,814	0.86	1.00	7,611	7,846	6,6827	7,653
Tanzania	0.96	1,012	976	3.03	3.43	1,462	1,508	4,448	5,170
Uganda	1.36	365	497	2.95	3.22	561	575	1,658	1,852

Source: Based on analysis conducted for Nelson et al. (2010).

Notes: DR Congo = Democratic Republic of Congo; Min (minimum) represents the smallest projected number from the simulations based on the CSIRO A1B, CSIRO B1, MIROC A1B, and MIROC B1 climate model/scenario combinations. Max (maximum) represents the largest of the four simulated values. A1B = greenhouse gas emissions scenario that assumes fast economic growth, a population that peaks midcentury, and the development of new and efficient technologies, along with a balanced use of energy sources; B1 = greenhouse gas emissions scenario that assumes a population that peaks midcentury (like the A1B), but with rapid changes toward a service and information economy and the introduction of clean and resource efficient technologies; CSIRO = climate model developed at the Australia Commonwealth Scientific and Industrial Research Organisation; MIROC = Model for Interdisciplinary Research on Climate, developed at the University of Tokyo Center for Climate System Research.

These changes point to possible expansion in the crop production zones for staple crops and livestock. The merits of this change include enhancement of the food security of communities in the new production areas, although an adverse impact is possible if food prices drop as a result of production increases, which would possibly harm farmers who are net producers if their gains in production are not sufficient to compensate for drops in unit prices of their output.

IMPACT, which includes climate effects as well as demographic and income effects, predicts that output of the majority of the foodcrops will rise on account of acreage expansion in some cases, but more generally because of technological advancement, including increased fertilizer use. Although not captured in this analysis, disease pressure will increase, especially for coffee, cassava, and plantains. In the face of climate change, policies and investments to promote agricultural growth, with a focus on smallholder productivity, are required.

The occurrence of the global food crisis has renewed attention to agriculture and spurred increased investment in the sector (Fan, Torero, and Headey 2011). Public policy should ensure that small farmers have opportunities to

TABLE 1.13 Number of malnourished children in East Africa, 2010 and 2050 (thousands)

Country	2010	2050					
		Pessimistic		Baseline		Optimistic	
		Min	Max	Min	Max	Min	Max
Burundi	757	956	978	794	813	645	661
DR Congo	4,320	6,175	6,647	5,034	5,435	3,841	4,188
Eritrea	190	191	206	126	139	78	90
Ethiopia	7,110	9,058	9,330	7,277	7,514	5,925	6,139
Kenya	1,466	1,654	1,805	783	905	436	542
Madagascar	731	809	858	407	443	139	167
Rwanda	474	682	708	473	495	359	380
Sudan	2,240	2,332	2,383	2,012	2,055	1,524	1,565
Tanzania	2,156	2,700	2,887	2,236	2,391	1,553	1,682
Uganda	1,548	2,141	2,272	1,353	1,467	849	953

Source: Based on analysis conducted for Nelson et al. (2010).

Notes: DR Congo = Democratic Republic of Congo; Min (minimum) represents the smallest projected number from the simulations based on the CSIRO A1B, CSIRO B1, MIROC A1B, and MIROC B1 climate model/scenario combinations. Max (maximum) represents the largest of the four simulated values. A1B = greenhouse gas emissions scenario that assumes fast economic growth, a population that peaks midcentury, and the development of new and efficient technologies, along with a balanced use of energy sources; B1 = greenhouse gas emissions scenario that assumes a population that peaks midcentury (like the A1B), but with rapid changes toward a service and information economy and the introduction of clean and resource-efficient technologies; CSIRO = climate model developed at the Australia Commonwealth Scientific and Industrial Research Organisation; MIROC = Model for Interdisciplinary Research on Climate, developed at the University of Tokyo Center for Climate System Research.

increase their productivity and income. Investments by national governments, as well as global and regional institutions, should focus on improved smallholder access to inputs such as seeds and fertilizer—through lower transport and marketing costs, improved market infrastructure, and greater competition, as well as financial and extension services and weather-based crop insurance.

Governments and institutions should strongly promote new agricultural technologies suitable for smallholders through increased investment in crop breeding and livestock research. Rural infrastructure should also be strengthened to increase access to markets.

Capacity building, in terms of human, institutional, and infrastructure capacity as well as financial support, is a prerequisite for addressing the issue of climate change. There should be improved capacity and training on modeling for climate change, crop modeling, and remote sensing. The capacity for early warning systems has to be developed and strengthened in order to mitigate climate change.

Inadequate access to and use of land, fragmentation of plots, and scattered parcels of land has contributed to land degradation. It is important to provide

TABLE 1.14 Share of malnourished children in East Africa, 2010 and 2050 (percent)

| | | 2050 | | | | | |
| | | Pessimistic | | Baseline | | Optimistic | |
Country	2010	Min	Max	Min	Max	Min	Max
Burundi	46.9	42.6	43.6	40.1	41.0	37.1	38.1
DR Congo	33.1	31.7	34.1	29.1	31.4	25.2	27.5
Eritrea	22.9	16.9	18.2	12.6	13.9	8.9	10.2
Ethiopia	50.0	45.5	46.9	41.3	42.6	38.3	39.6
Kenya	22.2	18.6	20.3	10.1	11.6	6.5	8.0
Madagascar	22.1	16.8	17.9	9.7	10.5	3.8	4.6
Rwanda	28.8	29.0	30.1	22.6	23.7	19.4	20.5
Sudan	42.7	36.2	37.0	35.6	36.3	30.9	31.7
Tanzania	33.3	28.0	29.9	26.3	28.1	20.8	22.5
Uganda	22.3	16.0	17.0	11.4	12.3	8.1	9.1
Total	35.0	30.2	31.8	26.3	27.7	22.4	23.9

Source: Based on analysis conducted for Nelson et al. (2010).

Notes: DR Congo = Democratic Republic of Congo. Min (minimum) represents the smallest projected share from the simulations based on the CSIRO A1B, CSIRO B1, MIROC A1B, and MIROC B1 climate model/scenario combinations. Max (maximum) represents the largest of the four simulated values. A1B = greenhouse gas emissions scenario that assumes fast economic growth, a population that peaks midcentury, and the development of new and efficient technologies, along with a balanced use of energy sources; B1 = greenhouse gas emissions scenario that assumes a population that peaks midcentury (like the A1B), but with rapid changes toward a service and information economy and the introduction of clean and resource-efficient technologies; CSIRO = climate model developed at the Australia Commonwealth Scientific and Industrial Research Organisation; MIROC = Model for Interdisciplinary Research on Climate, developed at the University of Tokyo Center for Climate System Research.

equitable land tenure by ensuring long periods of tenancy for proper management of land. A number of environment-related policies have been drafted, yet they have to be supported with relevant and coherent laws and regulations.

References

Bartholome, E., and A. S. Belward. 2005. "GLC2000: A New Approach to Global Land Cover Mapping from Earth Observation Data." *International Journal of Remote Sensing* 26 (9): 1959–1977.

CIESIN (Center for International Earth Science Information Network, Columbia University), Columbia University, IFPRI (International Food Policy Research Institute), World Bank, and CIAT (Centro Internacional de Agricultura Tropical). 2004. *Global Rural–Urban Mapping Project, Version 1 (GRUMPv1)*. Palisades, NY, US: Socioeconomic Data and Applications Center (SEDAC), Columbia University. http://sedac.ciesin.columbia.edu/gpw.

Fan, S., M. Torero, and D. Headey. 2011. *Urgent Actions Needed to Prevent Recurring Food Crises.* IFPRI Policy Brief. March 16. Washington, DC: International Food Policy Research Institute.

FAO (Food and Agriculture Organization). 2010. FAOSTAT. Rome. http://faostat.fao.org.

Funk, C., M. Dettinger, J. Michaelsen, J. Verdin, M. Brown, et al., 2008. "Warming of the Indian Ocean Threatens Eastern and Southern African Food Security but Could Be Mitigated by Agricultural Development." *Proceedings of the National Academy of Sciences U.S.A.* 105 (32): 11081–11086.

Jones, P. G., P. K. Thornton, and J. Heinke. 2009. "Generating Characteristic Daily Weather Data Using Downscaled Climate Model Data from the IPCC's Fourth Assessment." Project report for the International Institute for Land Reclamation and Improvement, Wageningen, the Netherlands. Accessed May 7, 2010. www.ccafs-climate.org/pattern_scaling/.

Karugia, J., J. Wanjiku, J. Waithaka, and S. Babu. 2011. "Persistence of High Food Prices in Eastern Africa: What Role for Policy?" Unpublished paper. IFPRI (International Food Policy Research Institute), Washington, DC, and ASARECA (Association for Strengthening Agricultural Research in Eastern and Central Africa), Kampala, Uganda.

Lehner, B., and P. Döll. 2004. "Development and Validation of a Global Database of Lakes, Reservoirs and Wetlands." *Journal of Hydrology* 296 (1–4): 1–22.

Le Treut, H., R. Somerville, U. Cubasch, Y. Ding, C. Mauritzen, A. Mokssit, T. Peterson, and M. Prather. 2007. "Historical Overview of Climate Change." In *Climate Change 2007: The Physical Science Basis; Contribution of Working Group I to the Fourth Assessment Report of the Intergovernmental Panel on Climate Change,* edited by S. Solomon, D. Qin, M. Manning, Z. Chen, M. Marquis, K. B. Averyt, M. Tignor, and H. L. Miller. New York: Cambridge University Press.

Millennium Ecosystem Assessment. 2005. *Ecosystems and Human Well-being: Synthesis.* Washington, DC: Island Press. www.maweb.org/en/Global.aspx.

Nelson, G. C., M. W. Rosegrant, A. Palazzo, I. Gray, C. Ingersoll, R. Robertson, S. Tokgoz, et al. 2010. *Food Security, Farming, and Climate Change to 2050: Scenarios, Results, Policy Options.* Washington, DC.: International Food Policy Research Institute.

Omamo, S. W., X. Diao, W. Wood, J. Chamberlain, L. You, S. Benin, U. Wood-Sichara, and A. Tatwangire. 2006. *Strategic Priorities for Agricultural Development in Eastern and Central Africa.* IFPRI Report 150. Washington, DC: International Food Policy Research Institute. http://ifpri.org/pubs/ABSTRACT/rr150.asp#dl.

Sanchez, P. A., K. D. Shepherd, M. J. Soule, F. M. Place, R. J. Buresh, A.-M. N. Izac, A. U. Mokwunye, F. R. Kwesiga, C. G. Ndiritu, and P. L. Woomer. 1997. "Soil Fertility Replenishment in Africa: An Investment in Natural Resource Capital." In *Replenishing Soil Fertility in Africa,* edited by R. J. Buresh, P. A. Sanchez, and F. Calhoun. Madison, WI, US: Soil Science Society of America.

UNEP (United Nations Environment Programme) and IUCN (International Union for the
   Conservation of Nature). 2009. World Database on Protected Areas (WDPA). Annual release.
   http://wdpa.org/protectedplanet.aspx.

UNPOP (United Nations Department of Economic and Social Affairs–Population Division). 2009.
   World Population Prospects: The 2008 Revision. New York. http://esa.un.org/unpd/wpp/.

Williams, A. P., and C. Funk. 2011. "A Westward Extension of the Warm Pool Leads to a Westward
   Extension of the Walker Circulation, Drying Eastern Africa." *Climate Dynamics:* 1–19.

Wood, S., G. Hyman, U. Deichmann, E. Barona, R. Tenorio, Z. Guo, et al. 2010. "Sub-national Poverty
   Maps for the Developing World Using International Poverty Lines: Preliminary Data Release."
   Washington, DC: Harvest Choice and International Food Policy Research Institute. http://labs
   .harvestchoice.org/2010/08/poverty-maps/.

World Bank. 2009. *World Development Indicators.* Accessed May 2011. http://data.worldbank.org/
   data-catalog/world-development-indicators.

————. 2010. *Economics of Adaptation to Climate Change: Synthesis Report.* Washington, DC.
   http://climatechange.worldbank.org/content/economics-adaptation-climate-change-study
   -homepage.

You, L., and S. Wood. 2006. "An Entropy Approach to Spatial Disaggregation of Agricultural
   Production." *Agricultural Systems* 90 (1–3): 329–347.

You, L., S. Wood, and U. Wood-Sichra. 2006. "Generating Global Crop Distribution Maps: From
   Census to Grid." Paper presented at the International Association of Agricultural Economists
   Conference, Brisbane, Australia, August 11–18.

————. 2009. "Generating Plausible Crop Distribution and Performance Maps for Sub-Saharan
   Africa Using a Spatially Disaggregated Data Fusion and Optimization Approach." *Agricultural
   Systems* 99 (2–3): 126–140.

Chapter 2

METHODOLOGY

Gerald C. Nelson, Amanda Palazzo, Daniel Mason-d'Croz,
Richard Robertson, and Timothy S. Thomas

M odeling the impacts of climate change on agriculture presents a complex challenge arising from the wide-ranging processes underlying the working of markets, ecosystems, and human behavior. The analytical framework used in this monograph integrates modeling components that range from the macro to the micro to model a range of processes, from those driven by economics to those that are essentially biological in nature. This chapter brings together in one place the technical details associated with models used in this monograph along with other technical information that is common to most or all of the chapters. Figure 2.1 provides a diagram of the links among the three models used: the International Food Policy Research Institute's (IFPRI's) International Model for Policy Analysis of Agricultural Commodities and Trade (IMPACT) model (Rosegrant et al. 2008) a partial equilibrium agriculture model that emphasizes policy simulations; a hydrology model incorporated into IMPACT; and the Decision Support Software for Agrotechnology Transfer (DSSAT) crop model suite (Jones et al. 2003) that estimates yields of crops in varying management systems and climate change scenarios.

General Circulation Models (GCMs) and Climate Scenarios

GCMs model the physics and chemistry of the atmosphere and its interactions with oceans and the land surface. Several GCMs have been developed independently around the world. For the Fourth Assessment Report (AR4) of the Intergovernmental Panel on Climate Change (IPCC), 23 GCMs made some model results publicly available. Results from four are used in this monograph.

This chapter draws heavily on Nelson et al. (2010).

FIGURE 2.1   The International Model for Policy Analysis of Agricultural Commodities and Trade (IMPACT) modeling framework

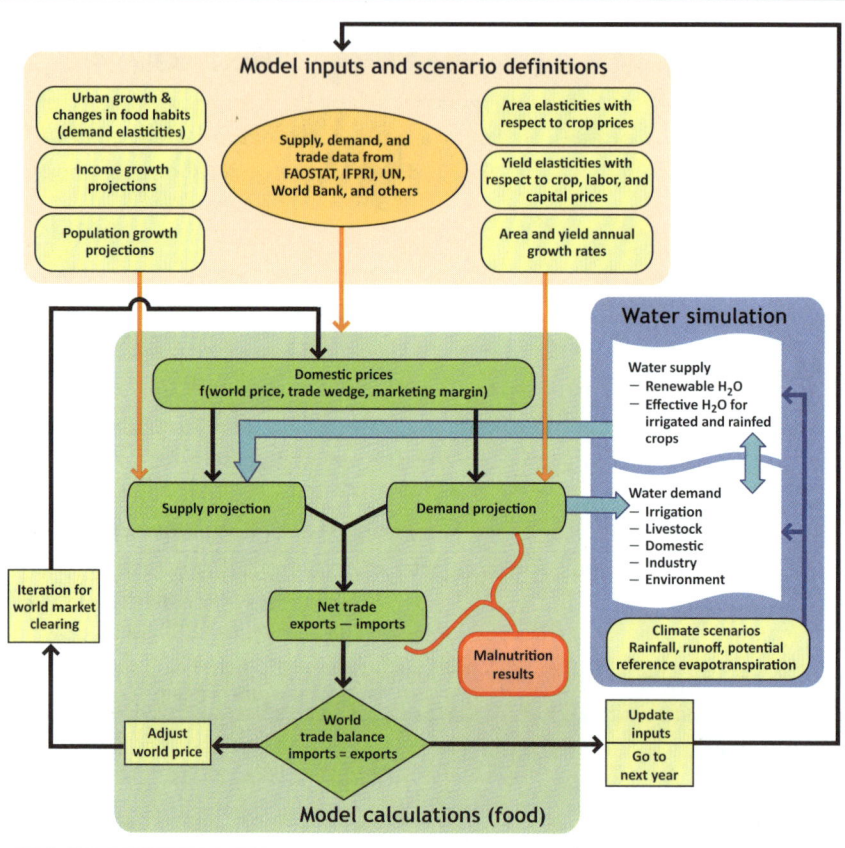

Source: Nelson et al. (2010)

Note: FAOSTAT = *FAOSTAT Database on Agriculture* (FAO 2010); IFPRI = International Food Policy Research Institute; UN = United Nations.

The GCMs create estimates of precipitation and temperature values around the globe often at something close to 2-degree intervals (about 200 kilometers at the equator) for most models. This is very coarse and may hide important differences on a more local scale. In order to have finer resolution, it is common to "downscale" the data. Data downscaled by Jones, Thornton, and Heinke (2009) provide precipitation and temperature data at a 5 arc-minute resolution (around 9 kilometers at the equator, smaller away from it, which is called "10-kilometer resolution" in this monograph).

Greenhouse gas emissions alter atmospheric chemistry, ultimately increasing temperatures and altering precipitation patterns. AR4 had three scenarios of greenhouse gas emissions pathways: B1, A1B, and A2.[1] Scenario B1 was a low-emissions scenario, which by 2011 does not appear to be very realistic. Scenarios A2 and A1B are higher emission scenarios, with similar trajectories through 2050 but different ones after 2050. Because this monograph is primarily concerned with changes between now and 2050, we elected to focus on scenario A1B when presenting the biophysical effects on crop yields, but we used both B1 and A1B in the IMPACT model in order to give a wider range of scenario outcomes.

To illustrate the range of potential effects on crops, we used results from four GCMs, CNRM-CM3, CSIRO Mark 3, ECHAM 5, and MIROC 3.2 medium resolution.[2] For inputs into the IMPACT model, results from only two GCMs were used, the CSIRO Mark 3 and the MIROC 3.2 medium-resolution models. The rationale for doing that can be seen more clearly in Table 2.1, in which we see that the lowest levels of precipitation change and lowest levels of temperature change are given by the CSIRO GCM and the highest levels of precipitation and temperature change are given by the MIROC GCM.

In the country analyses in the other chapters of this monograph, we display two kinds of maps that show spatially differentiated predictions of the GCMs. One shows changes in annual rainfall, and the other shows changes in the mean daily maximum temperature for the warmest month. The changes in the latter are determined by finding the month in 2000 with the highest mean daily maximum temperature and subtracting that mean value from the mean value for the month with the highest mean daily maximum temperature for 2050.

The Spatial Production Allocation Model (SPAM)

SPAM is a set of raster datasets showing harvest area, production, and yield for 20 crops or aggregates of crops and for three management systems

---

1   B1 is a greenhouse gas emissions scenario that assumes a population that peaks midcentury (like A1B), but with rapid changes toward a service and information economy, and the introduction of clean and resource-efficient technologies. A1B is a greenhouse gas emissions scenario that assumes fast economic growth, a population that peaks midcentury, and the development of new and efficient technologies, along with a balanced use of energy sources. A2 is a greenhouse gas emissions scenario that assumes a very heterogeneous world with continuously increasing global population and regionally oriented economic growth that is more fragmented and slower than in other storylines (Nakicenovic et al. 2000).

2   CNRM-CM3 is National Meteorological Research Center–Climate Model 3. CSIRO Mark 3 is a climate model developed at the Australia Commonwealth Scientific and Industrial Research Organisation. ECHAM 5 is a fifth-generation climate model developed at the Max Planck Institute for Meteorology in Hamburg. MIROC 3.2 is the Model for Interdisciplinary Research on Climate, developed at the University of Tokyo Center for Climate System Research.

TABLE 2.1   GCM and SRES scenario global average changes, 2000–2050

GCM	SRES scenario	Change between 2000 and 2050 in the annual averages			
		Precipitation (percent)	Precipitation (millimeters)	Minimum temperature (°C)	Maximum temperature (°C)
CSIRO	**B1**	**0.0**	**0.1**	**1.2**	**1.0**
CSIRO	**A1B**	**0.7**	**4.8**	**1.6**	**1.4**
CSIRO	A2	0.9	6.5	1.9	1.8
ECHAM 5	B1	1.6	11.6	2.1	1.9
CNRM-CM3	B1	1.9	14.0	1.9	1.7
ECHAM 5	A2	2.1	15.0	2.4	2.2
CNRM-CM3	A2	2.7	19.5	2.5	2.2
ECHAM 5	A1B	3.2	23.4	2.7	2.5
MIROC	A2	3.2	23.4	2.8	2.6
CNRM-CM3	A1B	3.3	23.8	2.6	2.3
MIROC	**B1**	**3.6**	**25.7**	**2.4**	**2.3**
MIROC	**A1B**	**4.7**	**33.8**	**3.0**	**2.8**

Source: Nelson et al. (2010).

Notes: In this table and elsewhere in the text, a reference to a particular year for a climate realization, such as 2000 or 2050, in fact refers to mean values around that year. For example, the data described as 2000 in this table are representative of the period 1950–2000. The data described as 2050 are representative of the period 2041–2060. GCM scenario combinations in bold are the ones used in the climate scenario analysis. A1B = greenhouse gas emissions scenario that assumes fast economic growth, a population that peaks midcentury, and the development of new and efficient technologies, along with a balanced use of energy sources; B1 = greenhouse gas emissions scenario that assumes a population that peaks midcentury (like A1B), but with rapid changes toward a service and information economy, and the introduction of clean and resource-efficient technologies; CNRM-CM3 = climate model developed by the National Meteorological Research Center; CSIRO = climate model developed at the Australia Commonwealth Scientific and Industrial Research Organisation; ECHAM 5 = fifth-generation climate model developed at the Max Planck Institute for Meteorology (Hamburg); GCM = general circulation model; MIROC = Model for Interdisciplinary Research on Climate, developed by the University of Tokyo Center for Climate System Research; SRES = Special Report on Emissions Scenarios, a report by the Intergovernmental Panel on Climate Change that was published in 2000.

(irrigated, high-input rainfed, and low-input rainfed, with the latter two combined in this monograph to make a rainfed total). The model employs a cross-entropy approach to manage inputs with different levels of likelihood in indicating the specific locations of agricultural production (You and Wood 2006; You, Wood, and Wood-Sichra 2006, 2009).

SPAM spatially allocates crop production from large reporting units (administrative units such as province or district level) to a raster grid at a spatial resolution of 5 arc-minutes. The allocation model works by inferring likely production locations from multiple indicators that, in addition to subnational crop production statistics, also include satellite data on land cover, maps of irrigated areas, biophysical crop suitability assessments, population density, and secondary data on irrigation and rainfed production.

In some of the maps presented in this monograph, SPAM areas are reported in units of hectares per raster cell. A 5-arc-minute grid cell is just over 8,500 hectares at the equator, which is a reasonable value to use when gauging how great a proportion of the cell might be used by the crop shown in the map.

SPAM areas are used to provide weights for calculating provincial and national yield changes due to climate change in the regional overview chapter. Cells with greater current levels of that crop are weighted higher when aggregating the crop model results. This was also the approach used for aggregating crop model results to the national level for use in the IMPACT model.

DSSAT

DSSAT is a software package used for modeling crop production (Jones et al. 2003). The software "grows" the crop in daily time increments, and therefore daily weather data are required. With climate models, we have only monthly statistics on the weather. DSSAT, however, overcomes this limitation by including a weather simulator that can convert monthly statistics into simulated daily weather. In this analysis, the weather is simulated many different times and the outcome averaged over several growing seasons. The result is more of a long-term yield perspective that will not be unduly influenced by any individual stochastic extreme in the simulation.

The soil data used were adapted by Jawoo Koo and John Dimes (2010) from the *Harmonized World Soil Data Base* (HWSD ver. 1.1) by Batjes et al. (2009). They are simplified to 27 types, each with high, medium, or low levels of soil organic carbon; deep, medium, or shallow rooting depth; and major component of sand, loam, or clay. Some grid cells had more than one soil type represented, and when that was the case, the dominant type was used.

DSSAT has parameters to model different varieties of each crop. For our work, we picked what seemed like an appropriate variety and used it in all locations and time periods investigated.

DSSAT requires the user to input the planting date. For rainfed crops, it is assumed that a crop is planted in the first month of a four-month period in which monthly average maximum temperature does not exceed 37°C (about 99°F), the monthly average minimum temperature does not drop below 5°C (about 41°F), and monthly total precipitation is not less than 60 millimeters.

In the tropics, the planting month begins with the rainy season. The particular mechanism for determining the start of the rainy season at any location is to look for the block of four months that gets the most rainfall. The month before that block is called the beginning of the rainy season. For irrigated crops, the first choice is the rainfed planting month.

DSSAT has an option to include CO_2 fertilization effects at different levels of CO_2 atmospheric concentration. For this study, all results use a 369 parts per million setting, which is the concentration in the early 2000s. A short summary of the reasons for and against including CO_2 fertilization is contained in Nelson et al. (2010, 14, text and footnote):

> Plants produce more vegetative matter as atmospheric concentrations of CO_2 increase. The effect depends on the nature of the photosynthetic process used by the plant species. So-called C3 plants use CO_2 less efficiently than C4 plants, so C3 plants are more sensitive to higher concentrations of CO_2. It remains an open question whether these laboratory results translate to actual field conditions. A recent report on field experiments on CO_2 fertilization (Long et al. 2006) finds that the effects in the field are approximately 50 percent less than in experiments in enclosed containers. Another report (Zavala et al. 2008) finds that higher levels of atmospheric CO_2 increase the susceptibility of soybean plants to the Japanese beetle and of maize to the western corn rootworm. Finally, a recent study (Bloom et al. 2010) finds that higher CO_2 concentrations inhibit the assimilation of nitrate into organic nitrogen compounds. So the actual field benefits of CO_2 fertilization remain uncertain.

Some use of nitrogen fertilizer is assumed in all our crop models. For almost all countries in Africa, the level of use is 20 kilograms of nitrogen per hectare (regardless of crop). For Madagascar and parts of South Africa, the level is 100 kilograms of nitrogen per hectare (regardless of crop). Levels were set appropriately for the rest of the world in the global modeling.

DSSAT is used in two ways in this monograph. It is used directly for each country and for the region to compute yields under the climate of 2000 and the climate of 2050. DSSAT is also used to provide results for each country of the world so that IMPACT can control for climate effects. The global work and the regional work were very similar though not perfectly identical because they were produced by two different teams. One example of the differences is the spatial resolution, which was 15 arc-minutes (30 kilometers) for the global team and 5 arc-minutes (10 kilometers) for the regional team. Some of the crops for East Africa were modeled by the global team, which is why the resolution in the maps varies between crops in some cases.

IMPACT

The IMPACT model was initially developed to project global food supply, food demand, and food security to the year 2020 and beyond (Rosegrant et al. 2008). It is a partial equilibrium agricultural model with 32 crop and livestock commodities, including cereals, soybeans, roots and tubers, meats, milk, eggs, oilseeds, oilcakes and meals, sugar, and fruits and vegetables. IMPACT has 115 regions, which are usually countries (though in a few cases several countries are aggregated together, and in rare cases, a country may be subdivided), with specified supply, demand, and prices for agricultural commodities.

Large regions are further divided into major river basins. The result, portrayed in Figure 2.2, is 281 spatial units called food production units (FPUs). The model links the various countries and regions through international trade, using a series of linear and nonlinear equations to approximate the underlying

FIGURE 2.2 International Model for Policy Analysis of Agricultural Commodities and Trade (IMPACT) unit of analysis, the food production unit (FPU)

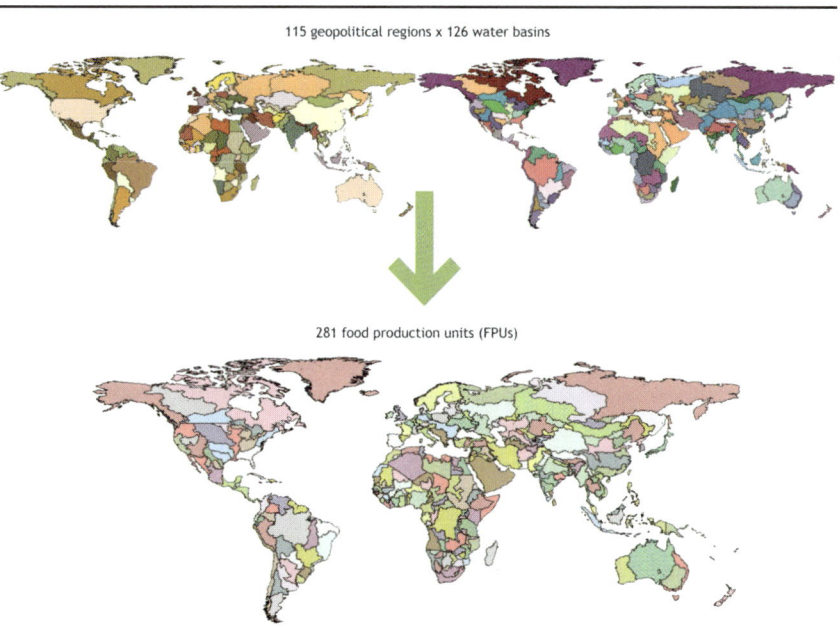

115 geopolitical regions x 126 water basins

281 food production units (FPUs)

Source: Nelson et al. (2010).

production and demand relationships. World agricultural commodity prices are determined annually at levels that clear international markets. Growth in crop production in each country is determined by crop and input prices, exogenous rates of productivity growth and area expansion, investment in irrigation, and water availability. Demand is a function of prices, income, and population growth. We distinguish four categories of commodity demand: food, feed, biofuels feedstock, and other uses.

From DSSAT to IMPACT

For input into the IMPACT model, DSSAT is run for five crops—rice, wheat, maize, soybeans, and groundnuts—at 15-arc-minute intervals for the locations where the SPAM dataset shows that each crop is currently grown. The results from this analysis are then aggregated to the IMPACT FPU level.

In extending these results to other crops in IMPACT, the primary assumption is that plants with the same photosynthetic metabolic pathway will react similarly to any given climate change effect in a particular geographic region. Millet, sorghum, sugarcane, and maize all use the C4 pathway. Millet and sugarcane are assumed to have the same productivity effects from climate change as maize in the same geographic regions. Sorghum effects for the Africa region were modeled explicitly, but for the rest of the world the maize productivity effects were assumed to apply to sorghum as well. The remainder of the crops use the C3 pathway. The climate effects for the C3 crops not directly modeled in DSSAT follow the average from wheat, rice, soy, and groundnuts from the same geographic region, with the following two exceptions. The IMPACT commodities of "other grains" and dryland legumes are directly mapped to the DSSAT results for wheat and groundnuts, respectively.

Income and Population Drivers

Differences in gross domestic product (GDP) and population growth define the overall scenarios, with all other driver values remaining the same across the three scenarios (Table 2.2). The GDP and population growth rates combine to generate the three scenarios of per capita GDP growth. The results by region are shown in Table 2.3. The baseline scenario has just over 9 billion people in 2050; the optimistic scenario results in a substantially smaller number, 7.9 billion; the pessimistic scenario results in 10.4 billion people. For developed countries, the differences among the three scenarios are relatively small, with

TABLE 2.2 Gross domestic product (GDP) and population choices for the three overall scenarios

Category	Pessimistic	Baseline	Optimistic
GDP (constant 2000 US$)	Lowest of the four GDP growth rate scenarios from the Millennium Ecosystem Assessment GDP scenarios (Millennium Ecosystem Assessment 2005) and the rate used in the baseline (next column)	Based on rates from a World Bank Economics of Adaptation to Climate Change study (World Bank 2010), updated for Africa south of the Sahara and South Asian countries	Highest of the four GDP growth rates from the Millennium Ecosystem Assessment GDP scenarios (Millennium Ecosystem Assessment 2005) and the rate used in the baseline (previous column)
Population	UN High variant, 2008 revision	UN medium variant, 2008 revision	UN low variant, 2008 revision

Source: Nelson et al. (2010).

TABLE 2.3 Global average scenario per capita dross domestic product growth rate, 1990–2000 and 2010–2050 (percent per year)

Category	1990–2000	2010–2050		
		Pessimistic	Baseline	Optimistic
Developed countries	2.7	0.74	2.17	2.56
Developing countries	3.9	2.09	3.86	5.00
Low-income developing countries	4.7	2.60	3.60	4.94
Middle-income developing countries	3.8	2.21	4.01	5.11
World	2.9	0.86	2.49	3.22

Sources: *World Development Indicators* for 1990–2000 (World Bank 2009) and Nelson et al. (2010) calculations for 2010–2050.

little overall population growth: population ranges from just over 1 billion to 1.3 billion in 2050, compared to 1 billion in 2010. For the developing countries as a group, the total 2010 population of 5.8 billion becomes 6.9–9 billion in 2050, depending on scenario.

As Table 2.4 shows, average world per capita income, beginning at $6,600 in 2010, ranges from $8,800 to $23,800 in 2050, depending on scenario.[3] The gap between average per capita income in developed and developing countries is large in 2010: developing countries' per capita income is only 5.6 percent of that in developed countries. Regardless of scenario, the relative difference is

---

3 All references to dollars are to constant 2000 US dollars.

TABLE 2.4 Summary statistics for population and per capita gross domestic project, 2010 and 2050

Category	2010	2050 Optimistic	2050 Baseline	2050 Pessimistic
Population (millions)				
World	6,870	7,913	9,096	10,399
Developed countries	1,022	1,035	1,169	1,315
Developing countries	5,848	6,877	7,927	9,083
Middle-income developing countries	4,869	5,283	6,103	7,009
Low-income developing countries	980	1,594	1,825	2,074
East Africa	361.1	879.4	777.1	681.6
Southern Africa	141.7	276.2	240.2	207.0
West Africa	300.5	697.0	618.5	545.0
Income per capita (2000 US$)				
World	6,629	23,760	17,723	8,779
Developed countries	33,700	93,975	79,427	43,531
Developing countries	1,897	13,190	8,624	3,747
Middle-income developing countries	2,194	15,821	10,577	4,531
Low-income developing countries	420	4,474	2,094	1,101
East Africa	204	565	1,161	1,778
Southern Africa	1,961	2,725	5,892	11,499
West Africa	363	816	1,695	3,185

Source: Nelson et al. (2010).
Notes: 2010 income per capita is for the baseline scenario. US$ = US dollars.

reduced over time: the developing-country income increases to between 8.6 percent and 14.0 percent of developed-country income in 2050, depending on the overall scenario. Middle- and low-income developing countries' 2010 per capita income values are 6.5 percent and 2.6 percent, respectively, of the developed-country income. By 2050, the share increases to between 10.4 percent and 16.8 percent for middle-income developing countries, depending on the overall scenario. For the low-income developing countries, however, the 2050 ratios remain low—between 2.5 percent and 4.8 percent.

The reader should be somewhat cautious in interpreting the results based on the three different scenarios. The optimistic scenario is optimistic not just for one country but for the entire world. This means that we cannot look at the impact from assuming that a single country is able to reduce its population growth rate or increase its GDP while the rest of the world continues at the same GDP and population growth rate. Rather, we have only the case in which all countries have a higher GDP and lower population growth rate. This means that changing scenarios changes supply and demand for the whole world, not just for one country.

Metrics for Human Well-being

Physical human well-being has many determinants. Calorie availability is a key element in low-income countries, where malnutrition and poverty are serious problems. Distribution, access, and supporting resources can enhance or reduce an individual's calorie availability. Similarly, child malnutrition has many determinants, including calorie intake (Rosegrant et al. 2008). The relationship used to estimate the number of malnourished children is based on a cross-country regression relationship estimated by Smith and Haddad (2000) that takes into account female access to secondary education, the quality of maternal and child care, and health and sanitation.[4] The IMPACT model provides data on per capita calorie availability by country; the other determinants are assumed to remain the same across the overall scenarios.

A number of countries included in this monograph show rising GDP per capita yet falling mean calorie consumption. This can be partially explained by noting that staple foodcrops increase in price. For example, maize prices are projected to nearly double from 2010 to 2050, rice prices are projected to be 80 percent higher, and wheat prices are expected to be more than 50 percent higher. People living on already-low GDP per capita spend a very large portion of their income on food, so income gains would be offset in part by price increases. However, the drop in consumption observed in the IMPACT model results also reflects the sensitivity to the model in response to elasticities that

---

4 Because it is a partial equilibrium model, IMPACT has no feedback mechanisms from climate change effects on productivity to income. This means that it cannot estimate directly the poverty effects of agricultural productivity declines from climate change. However, the reduced form function that relates child malnutrition to calorie availability and other determinants implicitly includes the effects of real income change on child malnutrition. Hertel, Burke, and Lobell (2010) use a general equilibrium model to estimate explicitly the effects of climate change on poverty. They find that the poverty impacts to 2030 "depend as much on where impoverished households earn their income as on the agricultural impacts themselves, with poverty rates in some non-agricultural household groups rising by 20–50 percent in parts of Africa and Asia under these price changes, and falling by equal amounts for agriculture-specialized households elsewhere in Asia and Latin America."

TABLE 2.5   Noncaloric determinants of global child malnutrition, 2010 and 2050

	Clean water access (percent)[a]		Female schooling (percent)[b]		Female relative life expectancy[c]	
Country category	2010	2050	2010	2050	2010	2050
Middle-income countries	86.8	98.4	71.6	81.7	1.066	1.060
Low-income countries	69.0	85.8	54.9	61.6	1.044	1.048

Sources: Population-weighted aggregations in Nelson et al. (2010) based on data from 2000 with expert extrapolations to 2050. Original data sources include the World Health Organization's Global Database on Child Growth and Malnutrition; the United Nations Administrative Committee on Coordination–Subcommittee on Nutrition; the World Bank's *World Development Indicators* (World Bank 2009); FAOSTAT (FAO 2010); and the United Nations Educational, Scientific, and Cultural Organization's UNESCOSTAT database. Aggregations are weighted by population shares and are based on the baseline population growth scenario.
[a]Share of population with access to safe water.
[b]Total female enrollment in secondary education (any age group) as a percentage of the female age group corresponding to national regulations for secondary education.
[c]Ratio of female to male life expectancy at birth.

were estimated with uncertainty, as well as the reality that elasticities do not necessarily function well for nonmarginal changes.

Table 2.5 shows the 2010 and 2050 values for the noncaloric determinants of child malnutrition aggregated to low- and middle-income countries. The small decline in female relative life expectancy in 2050 for the middle-income countries will be caused primarily by a decline in China, where it is expected that male life expectancy will gradually move up rather than female life expectancy moving down.

Agricultural Vulnerability to Climate Change

There are many dimensions to agricultural vulnerability to climate change: vulnerability of agricultural systems, communities, households, and individuals to climate change. Vulnerability is influenced by the degree of exposure and sensitivity to that exposure. Household-level vulnerability is most often associated with threats to livelihoods. Livelihoods can be inadequate because of resource constraints and low productivity (e.g., farmers with too little land and no access to fertilizer) or because they operate in a risky environment (e.g., droughts that cause harvest failure).

Potential impacts of climate change on vulnerability to food insecurity include both direct nutritional effects (changes in consumption quantities and composition) and livelihood effects (changes in employment opportunities and the cost of acquiring adequate nutrition). Climate change can affect each of these dimensions. This monograph focuses on the productivity effects of climate change that translate into changes in calorie availability and to the

TABLE 2.6   Mean price elasticities used for east and central African countries in IMPACT, 2010 and 2050

Food	2010		2050	
	Income	Own price	Income	Own price
Beef	0.945	−0.823	0.778	0.720
Lamb	0.536	−0.481	0.417	−0.425
Maize	0.200	−0.710	0.153	−0.689
Cassava	0.091	−0.679	−0.005	−0.632
Sorghum	0.267	−0.487	0.091	−0.401
Wheat	0.585	−0.893	0.485	−0.842

Source: Authors' calculations.

Notes: Averages are weighted averages based on national consumption of each food item in 2000. IMPACT = International Model for Policy Analysis of Agricultural Commodities and Trade.

effects on child malnutrition. At this point the methodology and data to provide quantitative estimates of livelihood vulnerability are not available.

For the calorie estimates used in this monograph, price and income elasticities from an earlier global study were used (Nelson et al. 2010). Table 2.6 presents a weighted average of these elasticities for important crops in the east and central African countries included in this monograph. The own-price elasticities are large (in absolute terms) relative to other studies. Other studies of the own-price elasticities for starchy staples report values that range from −0.05 to −0.3, whereas the elasticities in Table 2.6 are in the range of −0.40 to −0.89. The global modeling results in large price increases, which offset the effects of income increases in the pessimistic scenario and in some cases in the optimistic scenario as well.

Although it was not possible to recalibrate elasticities and generate all-new results for this monograph in a timely manner, the IMPACT model is being revised continuously, and future results will be based on improved elasticities.

Travel Time Maps

We developed databases that show simulated travel time to towns and cities of various sizes. The analysis begins with information on how long it would take someone to travel through a small region, roughly 10 kilometers on a side. This information is developed by overlaying various geographic information system (GIS) datasets, including ones for roads, rivers and other water bodies, urban areas, and international boundaries. Each feature has a particular speed associated with it, and there is a default speed for areas without detailed information.

Once the time to travel across the regions is developed, the only other data required is the location of the towns and cities of interest. We used cities and towns from two sources: CIESIN (2004) and the *World Gazetteer* (Helders 2005). ArcView 3.2 was used to calculate the shortest travel time to any point in the specified cities and towns dataset.

Box-and-Whisker Graphs

A box-and-whisker graph summarizes a variety of information for a variable in a relatively straightforward diagram. A sample box-and-whisker graph is shown in Figure 2.3. The horizontal lines at the top and bottom of the diagram are the "whiskers" and show the minimum and maximum values of the variable. The top and bottom edges of the rectangle, the "box," show the 75th and 25th percentiles, respectively, of the variable under consideration. The horizontal divider line inside the box is the median value of the data.

These graphs were generated using Stata (StataCorp 2009) with Tukey's (1977) formula for setting the upper and lower whisker values, which Stata calls "adjacent values."

Now that we have given a general overview of the models and some of the data reviewed in this monograph, we are ready to see the results of the models applied to each of the countries studied in the chapters that follow.

FIGURE 2.3   Sample box-and-whisker graph

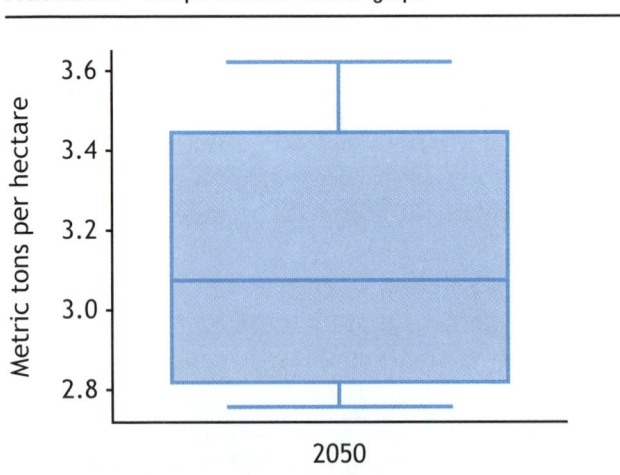

Sources: Authors, using StataCorp (2009) and Tukey (1977).

References

Batjes, N., K. Dijkshoorn, V. van Engelen, G. Fischer, A. Jones, L. Montanarella, M. Petri, S. Prieler, E. Teixeira, and X. Shi. 2009. Harmonized World Soil Database. Laxenburg, Austria: International Institute for Applied Systems Analysis (IIASA).

Bloom, A. J., M. Burger, R. Assensio, J. Salvador, and A. B. Cousins. 2010. "Carbon Dioxide Enrichment Inhibits Nitrate Assimilation in Wheat and Arabidopsis." *Science* 328: 899–902.

CIESIN (Center for International Earth Science Information Network, Columbia University), Columbia University, IFPRI (International Food Policy Research Institute), World Bank, and CIAT (Centro Internacional de Agricultura Tropical). 2004. *Global Rural–Urban Mapping Project, Version 1 (GRUMPv1)*. Palisades, NY, US: Socioeconomic Data and Applications Center (SEDAC), Columbia University. http://sedac.ciesin.columbia.edu/gpw.

Helders, S. 2005. *World Gazetteer* Database. Accessed June 7, 2007. http://world-gazetteer.com/.

Hertel, T. M., M. B. Burke, and D. B. Lobell. 2010. "The Poverty Implications of Climate-Induced Crop Yield Changes by 2030." *Global Environmental Change* 20 (4): 577–585.

Jones, J. W., G. Hoogenboom, C. H. Porter, K. J. Boote, W. D. Batchelor, L. A. Hunt, P. W. Wilkens, U. Singh, A. J. Gijsman, and J. T. Ritchie. 2003. "The DSSAT Cropping System Model." *European Journal of Agronomy* 18 (3–4): 235–265.

Jones, P. G., P. K. Thornton, and J. Heinke. 2009. "Generating Characteristic Daily Weather Data Using Downscaled Climate Model Data from the IPCC's Fourth Assessment." Project report for the International Livestock Research Institute (ILRI). Available at http://www.ccafs-climate .org/pattern_scaling.

Koo, J., and J. Dimes. 2010. HC27 Generic Soil Profile Database. Version 1, July. International Food Policy Research Institute, Washington, DC. http://hdl.handle.net/1902.1/20299.

Long, S. P., E. A. Ainsworth, A. D. B. Leakey, J. Nosberger, and D. R. Ort. 2006. "Food for Thought: Lower-than-Expected Crop Yield Stimulation with Rising CO_2 Concentrations." *Science* 312 (5782): 1918–1921. doi:10.1126/science.1114722. http://www.sciencemag.org/ cgi/content/abstract/312/5782/1918.

Millennium Ecosystem Assessment. 2005. *Ecosystems and Human Well-being: Synthesis.* Washington, DC: Island Press. http://www.maweb.org/en/Global.aspx.

Nakicenovic, N., et al. 2000. *Special Report on Emissions Scenarios: A Special Report of Working Group III of the Intergovernmental Panel on Climate Change.* Cambridge: Cambridge University Press. http://www.grida.no/climate/ipcc/emission/index.htm.

Nelson, G. C., M. W. Rosegrant, A. Palazzo, I. Gray, C. Ingersoll, R. Robertson, S. Tokgoz, et al. 2010. *Food Security, Farming, and Climate Change to 2050: Scenarios, Results, Policy Options.* Washington, DC: International Food Policy Research Institute.

Rosegrant, M. W., S. Msangi, C. Ringler, T. B. Sulser, T. Zhu, and S. A. Cline. 2008. *International Model for Policy Analysis of Agricultural Commodities and Trade (IMPACT): Model Description.* Washington, DC: International Food Policy Research Institute. http://www.ifpri.org/themes/ impact/impactwater.pdf.

Smith, L., and L. Haddad. 2000. *Explaining Child Malnutrition in Developing Countries: A Cross-Country Analysis.* Washington, DC: International Food Policy Research Institute.

StataCorp. 2009. Stata: Release 11. Statistical Software. College Station, TX, US.

Tukey, J. W. 1977. *Exploratory Data Analysis.* Reading, MA, US: Addison–Wesley.

UNPOP (United Nations Department of Economic and Social Affairs–Population Division). 2009. *World Population Prospects: The 2008 Revision.* New York. http://esa.un.org/unpd/wpp.

World Bank. 2009. *World Development Indicators.* Accessed May 2011. http://data.worldbank.org/ data-catalog/world-development-indicators.

———. 2010. *Economics of Adaptation to Climate Change: Synthesis Report.* Washington, DC. http://climatechange.worldbank.org/content/economics-adaptation-climate-change-study -homepage.

You, L., and S. Wood. 2006. "An Entropy Approach to Spatial Disaggregation of Agricultural Production." *Agricultural Systems* 90 (1–3): 329–347.

You, L., S. Wood, and U. Wood-Sichra. 2006. "Generating Global Crop Distribution Maps: From Census to Grid." Paper presented at the International Association of Agricultural Economists Conference, Brisbane, Australia, August 11–18.

———. 2009. "Generating Plausible Crop Distribution and Performance Maps for Sub-Saharan Africa Using a Spatially Disaggregated Data Fusion and Optimization Approach." *Agricultural Systems* 99 (2–3): 126–140.

Zavala, J. A., C. L. Casteel, E. H. DeLucia, and M. R. Berenbaum. 2008. "Anthropogenic Increase in Carbon Dioxide Compromises Plant Defense against Invasive Insects." *Proceedings of the National Academy of Sciences* 105 (13): 5129–5133. doi:10.1073/pnas.0800568105. http:// www.pnas.org/content/105/13/5129.abstract.

Chapter 3

BURUNDI

Juvent Baramburiye, Miriam Kyotalimye, Timothy S. Thomas, and Michael Waithaka

Burundi is a small landlocked country with a surface area of 27,834 square kilometers, of which 36 percent is arable. The overall climate is tropical humid, with rainfall and temperature heavily influenced by altitude. The country has two rainy seasons, February–May and September–November, and another short rainy period for two weeks in January. The terrain is hilly, with extensive marshlands and generally fertile lands. The rainfall varies from 2,000 millimeters at higher altitudes to 1,000 millimeters in the depressions (Table 3.1). Water and hydrological systems are abundant, although the country's forest cover is less than 5 percent. The country has five agroecological zones: the western plains of Imbo, the steep region of Mumirwa, the Congo–Nile Divide, the Central Highlands, and the depressions of Kumoso to the east and Bugesera basin to the northeast (Burundi, Ministry of Water, Environment, Land Management, and Urban Planning 2009).

Based on data from the past 60 years, Burundi has experienced alternating cycles of excess or deficit rainfall nearly every decade, as well as overall increased mean temperature, with the dry season getting longer. Past extreme

TABLE 3.1 Distribution of rainfall and temperature as a function of ecoclimatic region in Burundi, around 1950-2008

Ecoclimatic region	Percentage of total	Elevation (meters)	Average annual temperature (°C)	Average annual rainfall (millimeters)
Imbo plains	7	800–1,100	23	800–1,100
Mumirwa slopes	10	1,000–1,700	18–28	1,100–1,900
Congo–Nile Divide	15	1,700–2,500	14–15	1,300–2,000
Central Highlands	52	1,350–2,000	17–20	1,200–1,500
Kumoso and Bugesera	16	1,100–1,400	20–23	1,100–1,550

Source: Beck et al. (2010).

weather events include severe floods in 2006 and 2007 and severe droughts in 1999–2000 and in 2005 (Burundi, Ministry for Land Management, Tourism, and Environment 2007). During this time, especially hard hit were the northeastern provinces and the Bugesera depression, an area that supports a higher population density. These 4 events resulted in a high loss of annual gross domestic product (GDP), estimated between 5 and 17 percent for each event (Burundi, Ministry of Water, Environment, Land Management, and Urban Planning 2009).

Heavier and more frequent rains can damage crops and would undoubtedly increase susceptibility to erosion and landslides, especially given the extreme topographical relief in Burundi. Roads and buildings could be damaged, and siltation could negatively affect hydropower infrastructure. People resettled close to lake edges or near lowlands and marshes are likely to be flooded. Areas most vulnerable to heavy rains are the Imbo plains, the steep slopes of Mumirwa, and the Bugesera depression. Excess rain can also increase the presence of pests or diseases affecting foodcrops, livestock, and human lives. Water-borne diseases such as dysentery and cholera would likely increase. On the other hand, aquatic and avian productivity is likely to improve with the flooding of marsh areas (Burundi, Republic du, Ministère de l'Eau, de l'Environnement, de l'Aménagement du Territoire et de l'Urbanisme 2009).

Drought leads to lower levels of water in lakes and reservoirs and to decreased aquatic ecosystem productivity. According to the Institut Geographique du Burundi, the northern lakes have receded 1–2 meters within the past 5–10 years; but, given that the declines persist even in years of normal precipitation, there may be contributing factors beyond climate change, such as drainage and cultivation of marshes linked with water bodies and conversion of natural forests. Prolonged drought can lead to shortages of water for domestic and agricultural use, affecting crop and livestock production. Between 1998 and 2005, drought caused 35 percent livestock mortality and a widespread food crisis (Burundi, Republic du, Ministère de l'Eau, de l'Environnement, de l'Aménagement du Territoire et de l'Urbanisme 2009).

Agriculture is the key sector in the economy of Burundi; more than 90 percent of the population depends on agriculture as a source of food, employment, and income (Burundi, Ministry of Agriculture and Livestock 2008).

Burundi has only 140 square kilometers of irrigated land and hence remains largely dependent on rainfed agriculture (CIA 2011). The precipitation period has declined in recent years, perhaps as a sign of climate change already taking place (Burundi, Ministry of Water, Environment, Land Management, and Urban Planning 2009): the early rains currently end in April rather than May, and the later rains start in October rather than September, with farmers coping by switching to shorter-season crops such as peas. People must traverse longer distances in search of drinking water, and human health is compromised by limited access to food.

The Government of Burundi is aware of potential negative impacts of climate change on agriculture and has taken some steps toward preparing for those impacts. Burundi ratified the United Nations Framework Convention on Climate Change in 1997 and prepared a National Action Plan for Adaptation (NAPA) in 2001 (Burundi, Ministry for Land Management, Tourism, and Environment 2007). With support from the United Nations Development Programme / Global Environment Facility, Burundi also developed the second national communication on climate change (Burundi, Republic du, Ministère de l'Eau, de l'Environnement, de l'Aménagement du Territoire et de l'Urbanisme 2009), which includes a comprehensive inventory of greenhouse gas emissions, gaps, and constraints, along with the additional measures and policies needed to mitigate or adapt and an account of the financial resources and technical capabilities needed.

In terms of public health, the vulnerability of the population is high. Periods of increased rainfall and associated flooding can increase the vegetation density, generally providing suitable breeding pools for mosquito larvae. Higher temperatures also shorten the mosquito's extrinsic incubation period, increasing the transmission of malaria (Burundi, Ministry of Water, Environment, Land Management, and Urban Planning 2009). Nkurunziza, Gebhardt, and Pilz (2010) analyze the effects of climate on malaria in Burundi using monthly data on malaria epidemiology and climate for the period 1996–2007. They show a strong positive association between malaria incidence in a given month and the minimum temperature of the previous month. Increases in minimum temperatures will likely exacerbate the incidence of malaria in Burundi. Malaria is currently a major public health issue in the country, with about 2.5 million clinical cases and more than 15,000 deaths per year (Nkurunziza, Gebhardt, and Pilz 2010). Malaria is the single main cause of mortality in pregnant women and children under five years of age. The currently high rates of malaria morbidity are exacerbated by poor access to health services and by malnutrition. Control measures to address the impacts of climate change on malaria incidence will be needed.

Review of the Current Situation and Trends

Economic and Demographic Indicators

Population

The 2008 census placed Burundi's population at 8 million—69 percent under the age of 15—and its annual growth rate at 2.8 percent. Other sources give annual growth estimates of 3.4 percent in 2004 and 3.69 percent in 2010, with an average fertility rate of 6.25 children per woman (CIA 2011). Figure 3.1 shows trends for the total and rural population (left axis) as well as the share of urban population (right axis). The percentage of urban population has increased significantly since 1970.

Table 3.2 shows the total, rural, and urban growth rates from 1960 to 2008. The period 1990–2000 was characterized by slow growth in both rural and urban populations in response to the civil conflict that engulfed the country for close to sixteen years and caused massive displacement of people to neighboring countries in search of asylum.

Urban population growth rates are generally higher than rural growth rates for all years, rising sharply in the latter part of the last decade as more than

FIGURE 3.1   Population trends in Burundi: Total population, rural population, and percent urban, 1960–2008

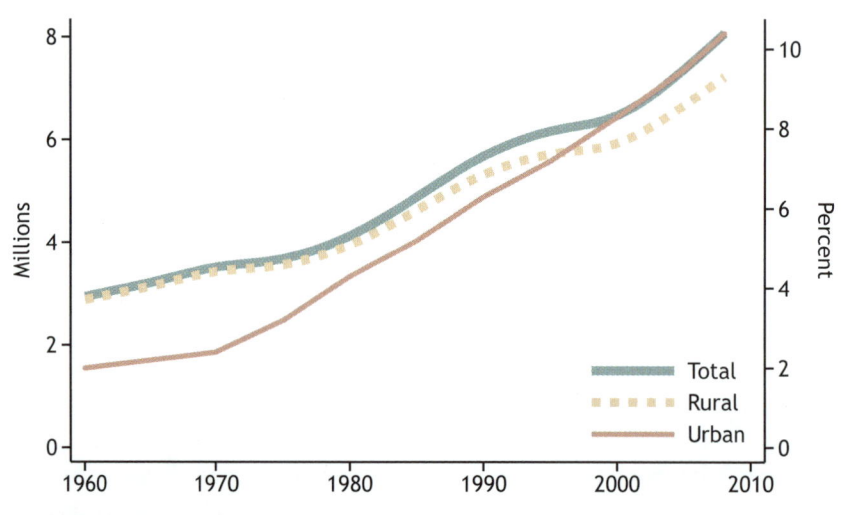

Source: *World Development Indicators* (World Bank 2009).

TABLE 3.2 Population growth rates in Burundi, 1960–2008 (percent)

Decade	Total growth rate	Rural growth rate	Urban growth rate
1960–1969	1.8	1.8	3.7
1970–1979	1.4	1.2	7.3
1980–1989	3.3	3.1	7.2
1990–1999	1.2	1.0	4.0
2000–2008	2.8	2.5	5.6

Source: Authors' calculations, based on *World Development Indicators* (World Bank 2009).

500,000 Burundians returned home after the cessation of war. Recent years have seen a surge in the rural–urban exodus in response to widespread poverty in rural areas. High population growth coupled with limited land resources and a land tenure system that favors fragmentation of plots among siblings means that rural households have less than 0.5 hectare of farmland on average. (Burundi, Ministry of Agriculture and Livestock 2008).

Households cope by encroaching on protected areas, cultivating on steep slopes without recourse to sustainable practices for highlands, and draining marshes for agricultural use. Consequently, land degradation and soil erosion in mountainous areas are key challenges. This pattern, coupled with intermittent droughts, has led to food shortages and migration to urban centers in search of alternative sources of livelihood. However, the past conflict has meant limited growth in the service and industry sectors—and consequently limited employment opportunities in the cities.

Figure 3.2 shows the population distribution in Burundi. With an average density of 289 people per square kilometer, Burundi is one of the most densely populated countries in Africa. Population densities vary across the country (Burundi, Ministry of Finance 2007). The eastern part of the country, encompassing Rutana, Ruyigi, and Cankuzo Provinces, has the lowest density, with 20–100 inhabitants per square kilometer. A second tier of 100–200 inhabitants per square kilometer occurs in the southern and northwestern provinces of Makamba, Bururi, and Cibitoke. The third tier of 200–500 inhabitants per square kilometer covers Kayanza, Ngozi, Bujumbura Rural, Kirundo, Gitega, and Bubanza Provinces. Population densities of 500–2,000 inhabitants per square kilometer occur in the capital, Bujumbura, and the main cities, such as Ngozi and Kayanza in the north, Gitega in the midlands, and Rumonge in the south.

FIGURE 3.2   Population distribution in Burundi, 2000 (persons per square kilometer)

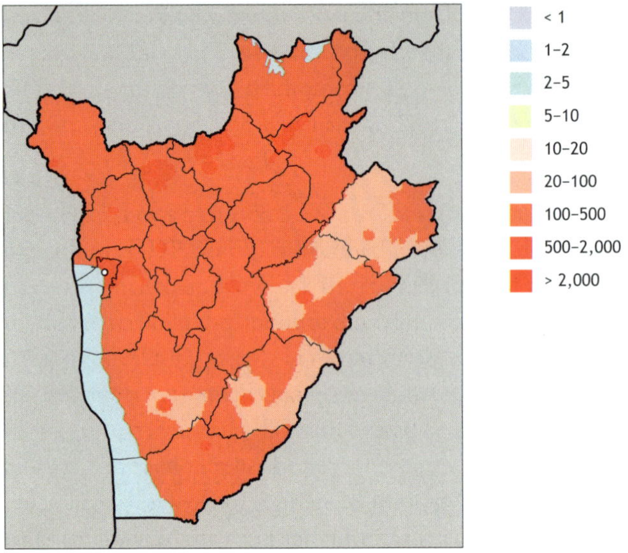

	< 1
	1–2
	2–5
	5–10
	10–20
	20–100
	100–500
	500–2,000
	> 2,000

Source: CIESIN et al. (2004).

Income

In 2009, Burundi's GDP amounted to $1.248 billion, with a real growth rate of 3.4 percent and a per capita GDP of $151. The country exported goods and services worth $63.9 million against an import bill of $402.3 million f.o.b.[1] (CIA 2011). Burundi has had a positive GDP growth rate, increasing from negative 1.3 percent in 2004 to positive 3.5 percent in 2010 (Burundi, Ministry of Agriculture and Livestock 2008), keeping pace with population growth rates; however, the rate is still below the 6 percent annual growth rate required to meet the United Nations Millennium Development Goal of halving poverty by 2015.

Figure 3.3 shows trends in per capita GDP and share of GDP from agriculture between 1960 and 2000. Annual income per capita declined suddenly, from $156 in 1991 to $110 in 1997—nearly a 33 percent reduction. It has remained stagnant since then. The decade-long reduction of the 1990s is attributed to the crisis experienced in the country around 1993, exacerbated by the trade embargo imposed on the country from 1995 to 1999. The embargo drove up the cost of inputs while reducing the availability of foreign exchange, curtailing purchasing power and consequently overall economic development.

---

1 Free on board; that is, seller pays costs of shipping.

FIGURE 3.3 Per capita GDP in Burundi (constant 2000 US$) and share of GDP from agriculture (percent), 1960–2008

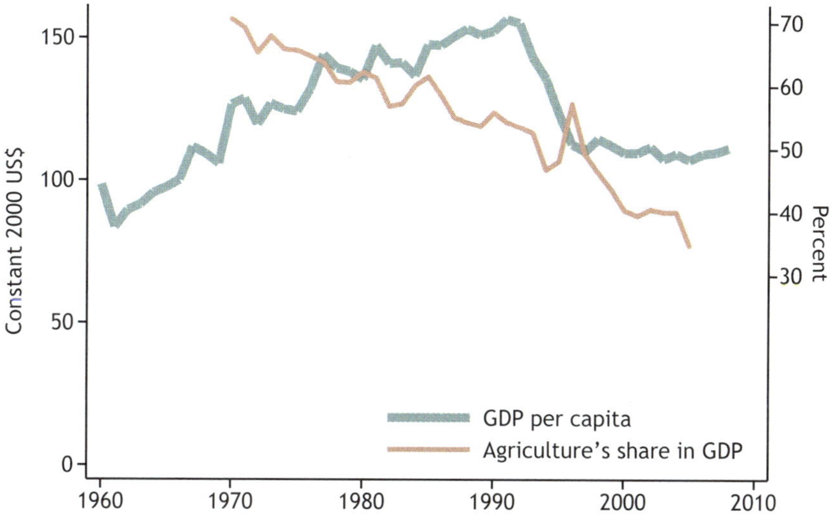

Source: *World Development Indicators* (World Bank 2009).
Note: GDP = gross domestic product; US$ = US dollars.

Agriculture is the main income earner; however, its contribution to GDP has declined over time, from over 70 percent in the 1970s to 35 percent in 2010. Total GDP increased in real terms over this period, indicating the growing importance of the service and manufacturing sectors to the economy. The performance of agriculture has been poor. Omamo et al. (2006) show that agricultural crop productivity grew by only 1.3 percent from 1983 to 1993 and declined by –0.4 percent between then and 2003. Agricultural productivity declined significantly, in part due to the prevailing conflicts but mainly due to land degradation and inefficient farming practices.

Vulnerability to Climate Change

Table 3.3 shows several indicators of a population's vulnerability and resiliency to economic shocks, including level of education, literacy, and childhood malnutrition.

A government initiative on universal primary education gives most children in Burundi the opportunity to go to primary school. However, enrollment in secondary school dropped significantly, to only 15.2 percent, as a consequence of widespread poverty. The index of literacy (59.3 percent)

TABLE 3.3   Education and nutrition statistics for Burundi, 2000s

Indicator	Year	Percent
Primary school enrollment (percent gross, three-year average)	2007	114.5
Secondary school enrollment (percent gross, three-year average)	2007	15.2
Adult literacy rate	2000	59.3
Under-five malnutrition (weight for age)	2000	38.9

Source: *World Development Indicators* (World Bank 2009).

remains lower than the regional average. Education increases an individual's resiliency to stress by enhancing the range of opportunities for income generation within and outside agriculture. Previous studies have also shown a high positive correlation between education and income levels, asset ownership, use of agricultural inputs, and credit access, all important factors enabling households to stem the effects of climate-related stress. Burundi's resilience to climate shock, as measured by the education level of its citizens, is hence considerably low.

In Burundi, 39 out of every 100 children under the age of five are malnourished in terms of weight for age. Malnutrition is especially prevalent in the drought-prone areas of the northern and eastern provinces and in camps of internally displaced persons (IDPs), and would be aggravated by further climate stress. Burundi has grappled with issues of internal displacement of its people since the coup of 1993 and the ensuing civil strife. By 2010, Burundi still had more than 100,000 IDPs, down from the peak number of 800,000 recorded in 1999. IDPs lack security of tenure in the settlements, and many are far removed from the land on which they depend for survival, leading to rapid degradation of the available soil, water, and forest resources as they attempt to meet basic needs for food, fuel, and shelter.

The lack of access to land means that many IDPs are unable to produce their own food and are dependent on humanitarian assistance. Burundi was traditionally self-sufficient in food production, but since the latest conflict and recurrent droughts the country has had to rely on food imports and international food aid in some regions, with the food import bill amounting to 12.5 percent of total imports in 2009. According to a study conducted by the World Food Program in 2004 (Burundi, Ministry of Finance 2007) the level of food vulnerability is extremely high: 61 percent of Burundi's households

FIGURE 3.4 Well-being indicators in Burundi, 1960–2008

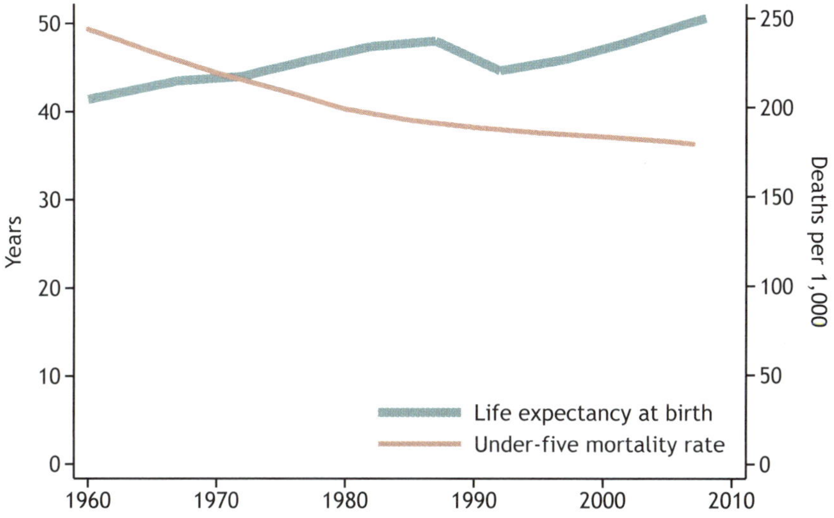

Source: *World Development Indicators* (World Bank 2009).

risk food insecurity at some point during the year as a result of weather-related events, declining soil fertility, and rising food prices.

Figure 3.4 shows trends in life expectancy at birth and under-five mortality rate. Generally life expectancy at birth has improved, rising from 40 years in the 1980s to 50–52 years in the 2000s. The infant mortality rate has declined from more than 250 cases per 1,000 births in the 1960s to fewer than 200 cases per 1,000 in the 2000s; it nevertheless remains very high. The main causes of death are malaria (40 percent) (Burundi, Ministry of Water, Environment, Land Management, and Urban Planning 2009), which mainly affects pregnant women and children under five years of age; diarrhea (3 percent); acute respiratory infections (19 percent); malnutrition; and HIV/AIDS. A healthy labor force is critical to sustaining income from agriculture and hence resiliency to climate shocks.

Wood et al. (2010) tell us that the proportion of the population living on less than $2 per day is 93 percent. The "Poverty Reduction Strategy Paper" (Burundi, Ministry of Finance 2007) shows a rural–urban divide in the incidence of poverty: the rural poor are about 49 percent below the poverty line, while the number is only 17.9 percent on average for the urban poor. The poverty level is also higher in provinces most affected by the past conflict, such as rural Bujumbura, Bubanza, Cibitoke, and Karusi, as well as the northern and eastern provinces that were hit by three years of drought. Poverty in rural areas is due to several general factors: high population pressures on overcultivated, eroded land, supporting farms of an average size of 0.5 hectares; insecurity and displacement; persistent drought; scarcity or poor quality of agricultural technologies; and limited market incentives (IFAD 2011).

Review of Land Use, Potential, and Limitations

Land Use Overview

Table 3.4 and Figure 3.5 show land use in Burundi as of 2000. Natural forest area has been largely converted to agricultural and other land uses, leaving a largely degraded landscape with a few pockets of scattered natural forest, nonnative tree plantations, and subsistence agriculture. Current estimates of forest cover range between 4.6 percent (128,375 hectares), 6 percent (180,000 hectares), and 7.4 percent (206,000 hectares) of total land area (Beck et al. 2010). These are only estimates, for there has not been a national forestry inventory since 1980. National plantation forests are estimated to cover nearly 60,000 hectares, on plots averaging 10 hectares each. There are two main causes of the current state of affairs: demographic pressure coupled with limited economic opportunities for nonagriculture or landbased livelihoods and

TABLE 3.4 Land use in Burundi, 2000 (percent of land area)

Land use	Percent of land area
Natural vegetation (including swamps and forests)	8.6
Forests	4.6
Pastures	27.8
Foodcrops (outside of swamps)	43.3
Cash crops	3.7
Cultivated swamps	2.8
Lakes	9.9
Towns	0.9

Source: Beck et al. (2010).

FIGURE 3.5 Land cover and land use in Burundi, 2000

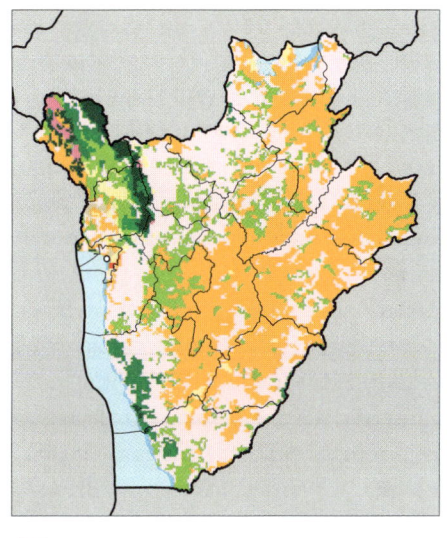

Tree cover, broadleaved, evergreen
Tree cover, broadleaved, deciduous, closed
Tree cover, broadleaved, open
Tree cover, broadleaved, needle-leaved, evergreen
Tree cover, broadleaved, needle-leaved, deciduous
Tree cover, broadleaved, mixed leaf type
Tree cover, broadleaved, regularly flooded, fresh water
Tree cover, broadleaved, regularly flooded, saline water
Mosaic of tree cover/other natural vegetation
Tree cover, burnt
Shrub cover, closed-open, evergreen
Shrub cover, closed-open, deciduous
Herbaceous cover, closed-open
Sparse herbaceous or sparse shrub cover
Regularly flooded shrub or herbaceous cover
Cultivated and managed areas
Mosaic of cropland/tree cover/other natural vegetation
Mosaic of cropland/shrub/grass cover
Bare areas
Water bodies
Snow and ice
Artificial surfaces and associated areas
No data

Source: GLC2000 (Bartholome and Belward 2005).

the country's massive wood-based energy demands. About 97 percent of fuel is derived from wood (Beck et al. 2010).

With an estimated 90 percent of Burundians directly involved in agriculture for their livelihoods (Burundi, Ministry of Agriculture and Livestock 2008), agriculture is easily the most important economic activity, mostly practiced at a subsistence level with a focus on staple crops such as cassava, bananas, corn, beans, rice, sorghum, and peanuts. Data from 2000 show a 15 percent decrease in per capita production over time, meaning that production has not kept pace with population growth (Banderembako 2006).

Protected areas provide important protection for fragile land. They also may generate income through the tourism industry and provide ecosystem services critical to the resilience of communities. Burundi established the National Institute for Nature Conservation (INECN) in 1980 to implement the decision on protected areas arising out of the Stockholm Environment Conference of 1972. INECN has since created 14 protected areas—albeit without prior consent in some cases or with inadequate compensation, leading to high levels of conflict with the affected populace. The return of people displaced by conflict also exerted great pressure on protected areas, with returnees settling and turning the areas to agricultural use in violation of the law, especially in the nature reserves of Rumonge, Vyanda, and Makamba.

Protected areas occupy a total land area of 157,700 hectares, or 5.6 percent of the country (Figure 3.6).

Figure 3.7 shows maps of travel times to urban areas of various sizes as potential markets for agricultural products and as sources for agricultural inputs and consumer goods for rural households. Most Burundians need over five hours to access urban centers of 500,000 or more, because this requires going to a neighboring country, given that this map was made with population data from 2004 and Bujumbura was thought to have fewer than 500,000 people at the time.

The country's transportation system is generally poor, with a limited feeder road network; the hilly terrain and lack of access to the sea further compound the problem. Poor infrastructure increases the number and range of middlemen, reducing farmgate margins. It also means limited access to market information (for example, on prices and high-demand areas) that would allow meaningful decisionmaking. Given poor capacities in storage and processing facilities and techniques, many farmers are shut out of lucrative markets, leading to a deepening of poverty.

FIGURE 3.6   Protected areas in Burundi, 2009

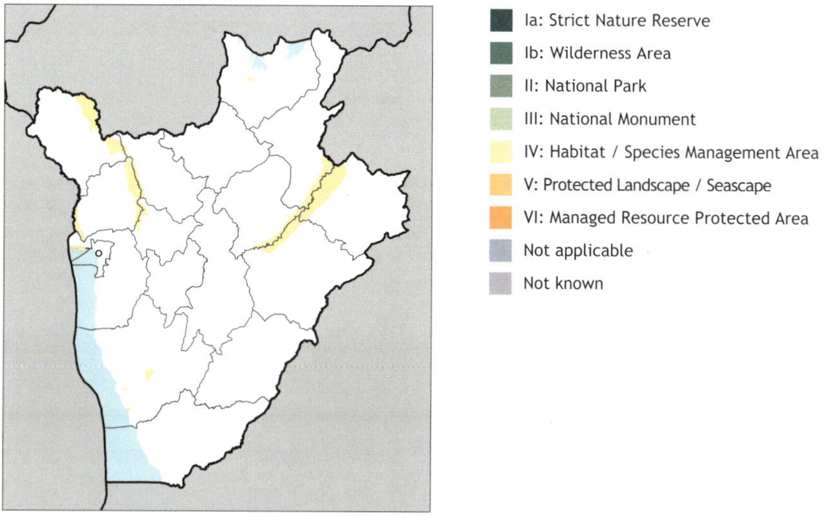

■	Ia: Strict Nature Reserve
■	Ib: Wilderness Area
■	II: National Park
■	III: National Monument
■	IV: Habitat / Species Management Area
■	V: Protected Landscape / Seascape
■	VI: Managed Resource Protected Area
■	Not applicable
■	Not known

Sources: Protected areas are from the World Database on Protected Areas (UNEP and IUCN 2009). Water bodies are from the World Wildlife Fund's Global Lakes and Wetlands Database (Lehner and Döll 2004).

Agriculture

Agriculture accounted for 45 percent of GDP in 2009 and supports more than 90 percent of the labor force, mainly as subsistence farmers. Burundi is potentially self-sufficient in food production given its climate and soil fertility. However, the civil conflicts, population growth, and soil erosion have contracted production by 30 percent in the recent past. Tables 3.5–3.7 show key agricultural commodities in terms of area harvested, value of the harvest, and food consumption.

Figures 3.8–3.12 show the estimated yield and growing areas for key crops in Burundi. Bananas and plantains, cassava, sweet potatoes and yams, and beans are the dominant crops in terms of value. Bananas are the most important staple, accounting for 29 percent of total cultivated area and 44 percent of the total value of Burundi's crop production between 2006 and 2008 (FAO 2010). Bananas are typically a smallholder crop, usually intercropped with coffee under the rainfed traditional cropping system. Yields average 7–10 tons per hectare—well below research station yields of 60 tons per hectare. Other

FIGURE 3.7   Travel time to urban areas of various sizes in Burundi, circa 2000

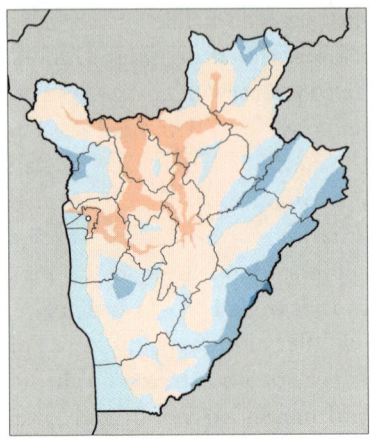

To cities of 500,000 or more people

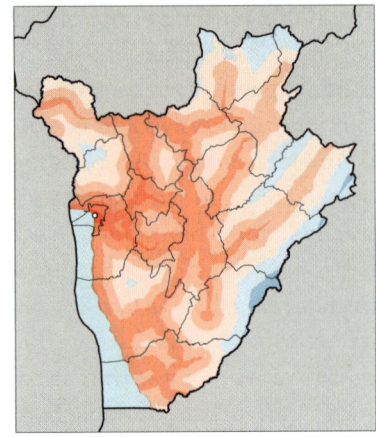

To cities of 100,000 or more people

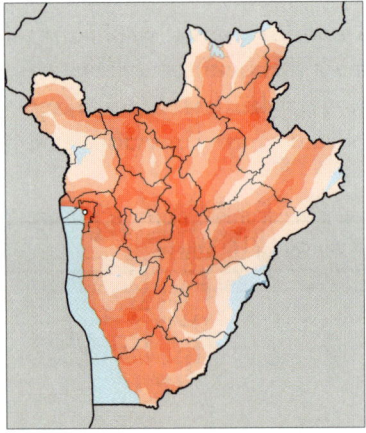

To towns and cities of 25,000 or more people

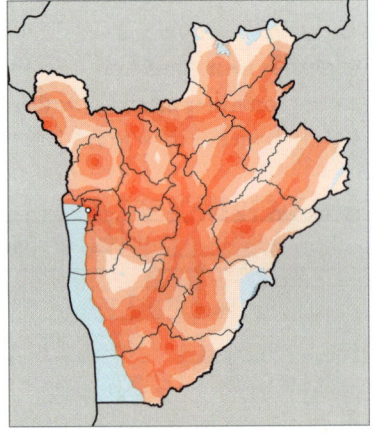

To towns and cities of 10,000 or more people

- Urban location
- < 1 hour
- 1–3 hours
- 3–5 hours
- 5–8 hours
- 8–11 hours
- 11–16 hours
- 16–26 hours
- > 26 hours

Source: Authors' calculations.

TABLE 3.5 Harvest area of leading agricultural commodities in Burundi, 2006–2008 (thousands of hectares)

Rank	Crop	Percent of total	Harvest area
	Total	100.0	1,156
1	Bananas	29.1	337
2	Beans	19.9	230
3	Sweet potatoes	11.2	129
4	Maize	9.9	115
5	Sorghum	5.7	66
6	Cassava	5.6	64
7	Peas	4.1	47
8	Coffee	2.0	23
9	Other fresh fruit	1.9	22
10	Rice	1.8	21

Source: FAOSTAT (FAO 2010).
Note: All values are based on the three-year average for 2006–2008.

TABLE 3.6 Value of production for leading agricultural commodities in Burundi, 2005–2007 (millions of constant 2000 US$)

Rank	Crop	Percent of total	Value of production
	Total	100.0	1,516.6
1	Bananas	44.7	677.2
2	Cassava	12.4	187.6
3	Sweet potatoes	12.2	184.9
4	Beans	8.9	135.1
5	Other fresh vegetables	3.8	58.1
6	Rice	3.1	47.3
7	Maize	2.5	37.5
8	Coffee	2.2	33.9
9	Other fresh fruit	2.2	33.8
10	Peas	2.0	30.5

Source: FAOSTAT (FAO 2010).
Note: All values are based on the three-year average for 2005–2007.

TABLE 3.7   Consumption of leading food commodities in Burundi, 2003–2005 (thousands of metric tons)

Rank	Crop	Percent of total	Food consumption
	Total	100.0	3,958
1	Bananas	26.2	1,036
2	Sweet potatoes	21.0	830
3	Cassava	17.5	692
4	Fermented beverages	8.0	315
5	Other vegetables	5.7	225
6	Beans	5.4	213
7	Maize	4.5	178
8	Beer	2.3	93
9	Other fruits	2.1	85
10	Other roots and tubers	2.0	79

Source: FAOSTAT (FAO 2010).
Note: All values are based on the three-year average for 2003–2005.

FIGURE 3.8   Yield (metric tons per hectare) and harvest area density (hectares) for rainfed plantains and bananas in Burundi, 2000

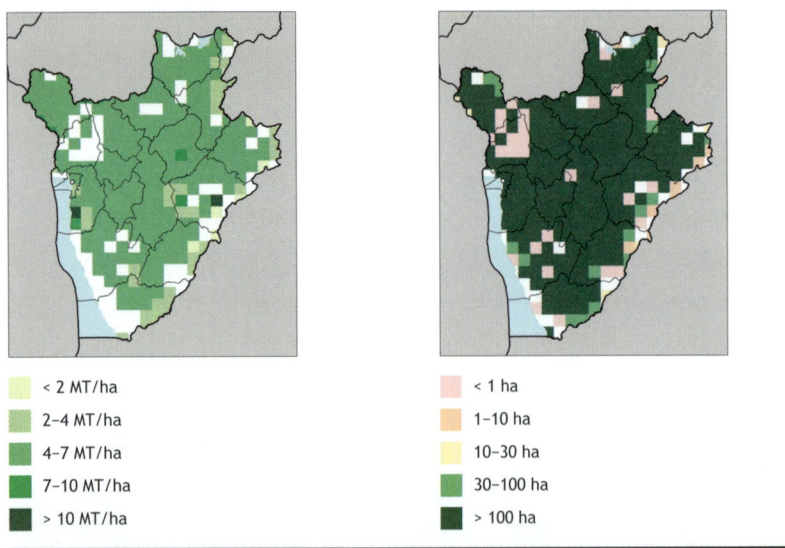

< 2 MT/ha	< 1 ha
2–4 MT/ha	1–10 ha
4–7 MT/ha	10–30 ha
7–10 MT/ha	30–100 ha
> 10 MT/ha	> 100 ha

Source: SPAM (Spatial Production Allocation Model) (You and Wood 2006; You, Wood, and Wood-Sichra 2006, 2009).
Note: ha = hectare; MT/ha = metric tons per hectare.

FIGURE 3.9 Yield (metric tons per hectare) and harvest area density (hectares) for rainfed beans in Burundi, 2000

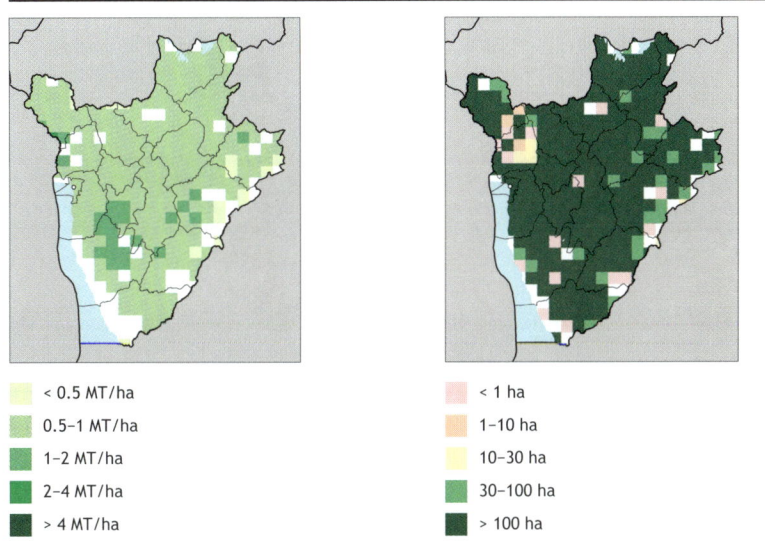

< 0.5 MT/ha	< 1 ha
0.5–1 MT/ha	1–10 ha
1–2 MT/ha	10–30 ha
2–4 MT/ha	30–100 ha
> 4 MT/ha	> 100 ha

Source: SPAM (Spatial Production Allocation Model) (You and Wood 2006; You, Wood, and Wood-Sichra 2006, 2009).
Note: ha = hectare; MT/ha = metric tons per hectare.

FIGURE 3.10 Yield (metric tons per hectare) and harvest area density (hectares) for rainfed sweet potatoes and yams in Burundi, 2000

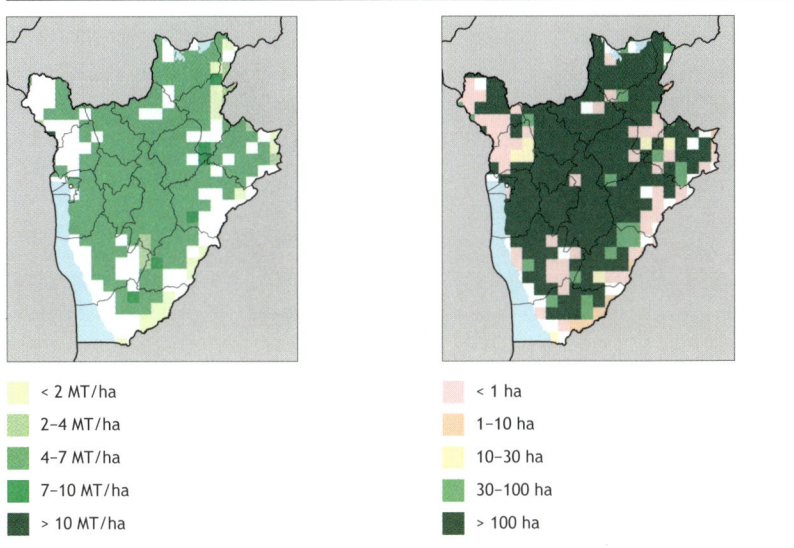

< 2 MT/ha	< 1 ha
2–4 MT/ha	1–10 ha
4–7 MT/ha	10–30 ha
7–10 MT/ha	30–100 ha
> 10 MT/ha	> 100 ha

Source: SPAM (Spatial Production Allocation Model) (You and Wood 2006; You, Wood, and Wood-Sichra 2006, 2009).
Note: ha = hectare; MT/ha = metric tons per hectare.

FIGURE 3.11   Yield (metric tons per hectare) and harvest area density (hectares) for rainfed cassava in Burundi, 2000

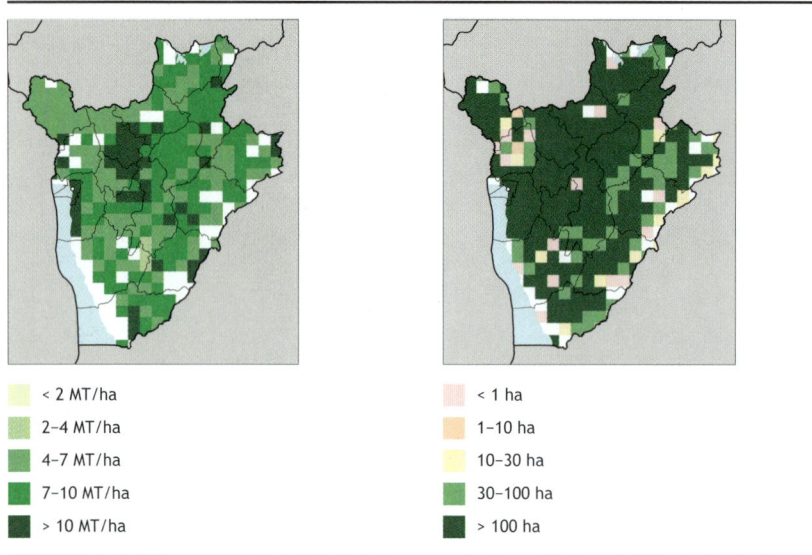

☐ < 2 MT/ha	☐ < 1 ha
☐ 2–4 MT/ha	☐ 1–10 ha
☐ 4–7 MT/ha	☐ 10–30 ha
☐ 7–10 MT/ha	☐ 30–100 ha
☐ > 10 MT/ha	☐ > 100 ha

Source: SPAM (Spatial Production Allocation Model) (You and Wood 2006; You, Wood, and Wood-Sichra 2006, 2009).
Note: ha = hectare; MT/ha = metric tons per hectare.

FIGURE 3.12   Yield (metric tons per hectare) and harvest area density (hectares) for rainfed maize in Burundi, 2000

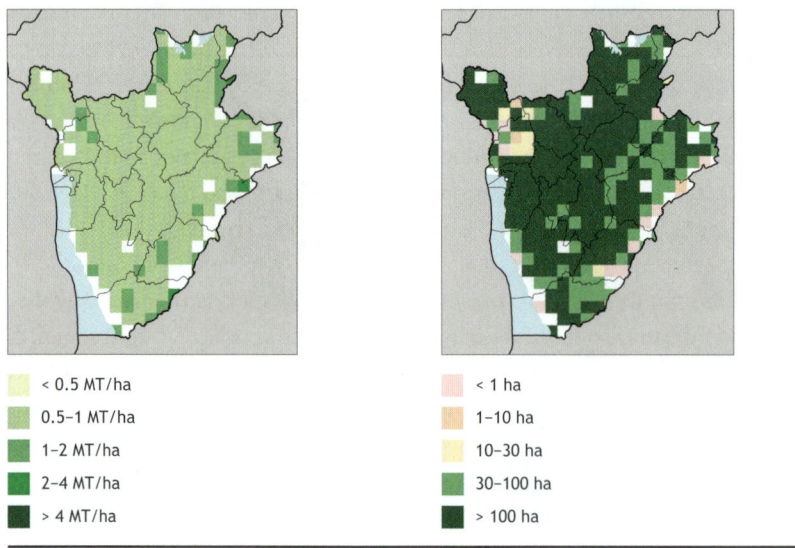

☐ < 0.5 MT/ha	☐ < 1 ha
☐ 0.5–1 MT/ha	☐ 1–10 ha
☐ 1–2 MT/ha	☐ 10–30 ha
☐ 2–4 MT/ha	☐ 30–100 ha
☐ > 4 MT/ha	☐ > 100 ha

Source: SPAM (Spatial Production Allocation Model) (You and Wood 2006; You, Wood, and Wood-Sichra 2006, 2009).
Note: ha = hectare; MT/ha = metric tons per hectare.

main uses of bananas, besides as food, include juice, local brew, flour, dried fruits, and wine. The other key commodities in terms of total value are beans, sweet potatoes and yams, cassava, and maize. The maps show that the crops are produced throughout the country.

Scenarios for the Future

Economic and Demographic Indicators

Population

Figure 3.13 shows a substantial population increase in all scenarios: a doubling of population by 2050 under the high variant, an increase of around 75 percent under the medium variant, and a 50 percent increase for the low variant.

A high rate of population growth will most likely increase the pressure on basic social services and the natural resource base, which are already strained, resulting in limited access to land, on which most Burundians depend for their livelihood. Consequently, food security is likely to be compromised, with adverse effects on malnutrition and other health indicators, especially for net food buyers. An urban population influx could occur as households move to urban centers in search of alternative sources of income.

FIGURE 3.13   Population projections for Burundi, 2010–2050

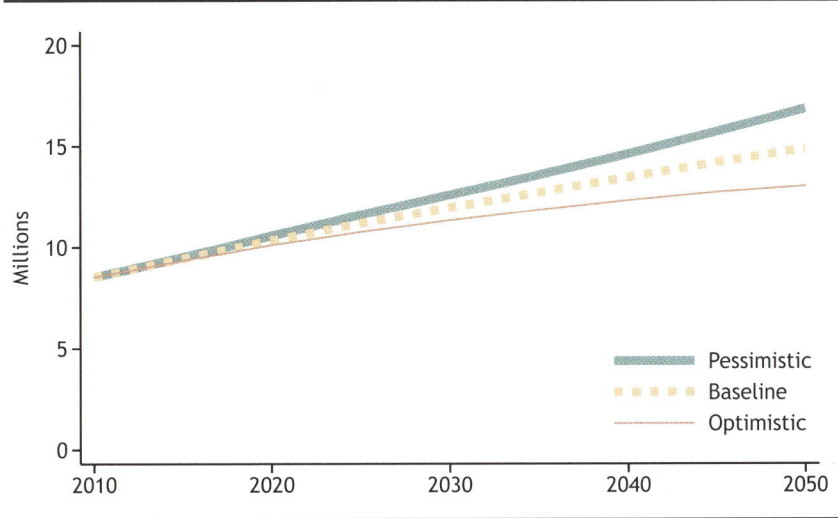

Source: UNPOP (2009).

FIGURE 3.14   Gross domestic product (GDP) per capita future scenarios for Burundi, 2010–2050

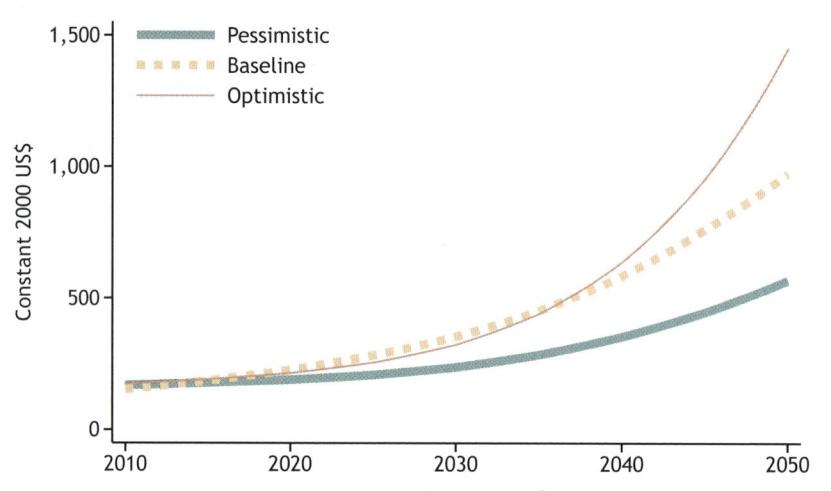

Sources: Computed from GDP data from the World Bank Economic Adaptation to Climate Change project (World Bank 2010), from the Millennium Ecosystem Assessment (2005), and from population data from the United Nations (UNPOP 2009).
Note: US$ = US dollars.

Income

Figure 3.14 shows the three scenarios for GDP per capita used for this study. These are the result of combining three GDP projections with the three population projections of Figure 3.13. The optimistic scenario combines high GDP with a low rate of population growth; the baseline scenario combines the medium GDP projection with a low rate of population growth; and the pessimistic scenario combines the low GDP projection with the a low rate of population growth.

All three scenarios show slow growth in GDP per capita for the first 13 years, perhaps because significant reforms are needed to lift the country out of the effects of the past conflict. Without measures to rein in the high rate of population growth, the pessimistic scenario depicts a per capita GDP of only $569 by 2050. GDP per capita rises to $1,450 in the optimistic scenario, assuming that measures to curb population increase are instituted. Raising GDP per capita would also require investments in productivity for the key agricultural commodities—bananas, beans, cassava, sweet potatoes, and maize—that would lead to broad-based growth, especially if coupled with improved marketing infrastructure and services. Equally important is improved access to education to enhance the productivity of the labor force. Alternative growth sectors such as services and industry need to be encouraged to absorb the population that lacks access to land.

Biophysical Analysis

Climate Models

Figure 3.15 shows projected precipitation changes under the four downscaled climate models we use in this chapter with the A1B scenario.[2] The models show significant differences. The CNRM-CM3 and ECHAM 5 general circulation models (GCMs) show minimal changes in precipitation across the country.[3] CSIRO Mark 3 shows precipitation losses of 50–100 millimeters in the northern and eastern provinces—areas that already have a history of intermittent droughts; this implies a need for drought-tolerant agricultural technologies in these regions to address climate change. However, MIROC 3.2 shows precipitation increasing by up to 200 millimeters in the western provinces and more than 200 millimeters in the rest of the country. This is favorable for crop production, but it could potentially increase the risk of flooding, especially in the Central Highlands and the Congo–Nile Divide, where precipitation levels are already above average.

Figure 3.16 shows changes in maximum temperature for the month with the highest mean daily maximum temperature. All models show Burundi getting warmer by 1°–2.5°C. The CSIRO model projects the coolest future, with increases in the 1°–1.5°C range. Both the CNRM and the ECHAM model predict the warmest future, with changes from 2° to 2.5°C. Given the tropical humid climate of Burundi, this means that evapotranspiration rates would be higher, reducing the water available for plant growth and other productive uses. High temperatures are also likely to improve conditions for the proliferation of disease and vermin, which could raise the costs of prevention or impair overall labor productivity.

Crop Models

The DSSAT (Decision Support Software for Agrotechnology Transfer) crop modeling system was used to compare future yields with the baseline yield (with 2000 climate unchanged). The results show uncertainty in yield outcomes for maize production in Burundi.

---

2 The A1B scenario is a greenhouse gas emissions scenario that assumes fast economic growth, a population that peaks midcentury, and the development of new and efficient technologies, along -with a balanced use of energy sources.

3 CNRM-CM3 is National Meteorological Research Center–Climate Model 3. ECHAM 5 is a fifth-generation climate model developed at the Max Planck Institute for Meteorology in Hamburg. CSIRO Mark 3 is a climate model developed at the Australia Commonwealth Scientific and Industrial Research Organization. MIROC 3.2 is the Model for Interdisciplinary Research on Climate, developed at the University of Tokyo Center for Climate System Research.

FIGURE 3.15   Changes in mean annual precipitation in Burundi, 2000–2050, A1B scenario (millimeters)

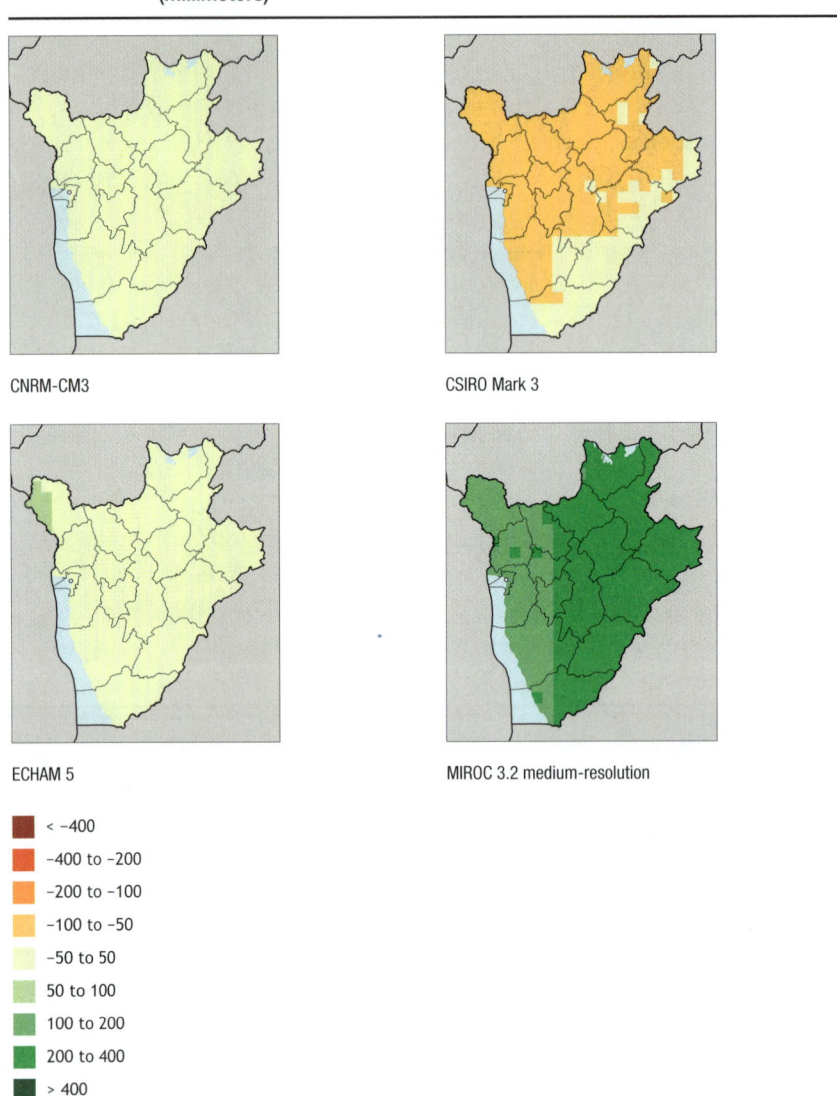

CNRM-CM3

CSIRO Mark 3

ECHAM 5

MIROC 3.2 medium-resolution

- < −400
- −400 to −200
- −200 to −100
- −100 to −50
- −50 to 50
- 50 to 100
- 100 to 200
- 200 to 400
- > 400

Source: Authors' calculations based on Jones, Thornton, and Heinke (2009).

Notes: A1B = greenhouse gas emissions scenario that assumes fast economic growth, a population that peaks midcentury, and the development of new and efficient technologies, along with a balanced use of energy sources; CNRM-CM3 = National Meteorological Research Center–Climate Model 3; CSIRO = climate model developed at the Australia Commonwealth Scientific and Industrial Research Organisation; ECHAM 5 = fifth-generation climate model developed at the Max Planck Institute for Meteorology (Hamburg); GCM = general circulation model; MIROC = Model for Interdisciplinary Research on Climate, developed by the University of Tokyo Center for Climate System Research.

FIGURE 3.16   Change in monthly mean maximum daily temperature in Burundi for the warmest month, 2000–2050, A1B scenario (°C)

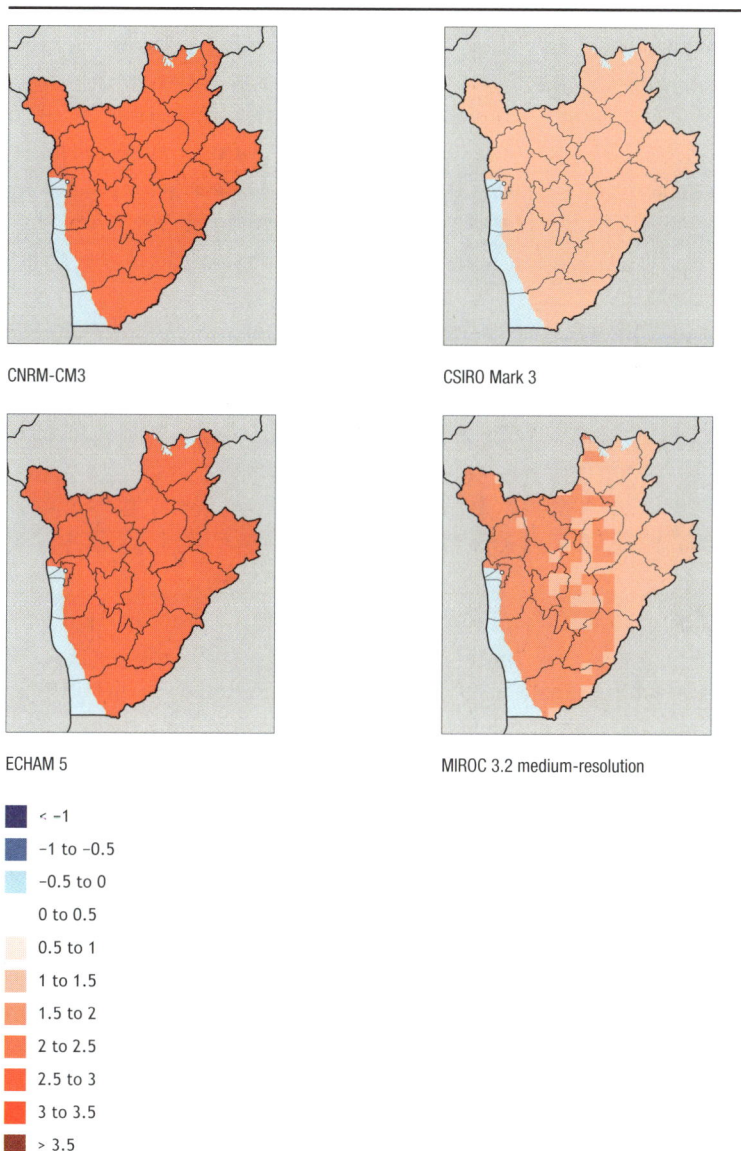

CNRM-CM3

CSIRO Mark 3

ECHAM 5

MIROC 3.2 medium-resolution

■ < −1
■ −1 to −0.5
□ −0.5 to 0
□ 0 to 0.5
□ 0.5 to 1
■ 1 to 1.5
■ 1.5 to 2
■ 2 to 2.5
■ 2.5 to 3
■ 3 to 3.5
■ > 3.5

Source: Authors' calculations based on Jones, Thornton, and Heinke (2009).

Notes: A1B = greenhouse gas emissions scenario that assumes fast economic growth, a population that peaks midcentury, and the development of new and efficient technologies, along with a balanced use of energy sources; CNRM-CM3 = National Meteorological Research Center–Climate Model 3; CSIRO = climate model developed at the Australia Commonwealth Scientific and Industrial Research Organisation; ECHAM 5 = fifth-generation climate model developed at the Max Planck Institute for Meteorology (Hamburg); GCM = general circulation model; MIROC = Model for Interdisciplinary Research on Climate, developed by the University of Tokyo Center for Climate System Research.

Figure 3.17 shows that predictions of yield loss for maize differ greatly between the GCMs. In CNRM-CM3 we see that most of the country is predicted to have yield losses exceeding 25 percent, whereas in ECHAM 5 we see that most of the country is predicted to have yield gains of between 5 percent and 25 percent, with a number of areas showing even greater increases. CSIRO Mark 3 and MIROC 3.2 are in between, with both showing that most of the area will have yield declines of between 5 percent and 25 percent, with a few areas showing yield increases of the same amount. Given that maize is one of the five most important foods in Burundi, yield losses would lead to a worsening of food security.

These results do not seem to relate in any consistent way to the climate change maps just presented, for both CNRM-CM3 and ECHAM 5 showed very similar temperature and precipitation changes. It is important to keep in mind that the climate maps showed annual changes, whereas the crop models respond to changes only during the months in which a crop is being grown.

Vulnerability

Figure 3.18 shows the impact of future GDP and population scenarios on the number of malnourished children under age five. Figure 3.19 shows the share of children who are malnourished. Only the optimistic scenario shows a decline in numbers of malnourished children under age five, from the baseline level of 780,000 to 630,000 in 2050. However, all scenarios show downward trends after 2040. With population increasing, however, the malnutrition rates are slightly lower in 2050 than in 2010. A higher rate of GDP growth, if sufficiently broad based, would mean that more income would be available for food purchase even for net food buyers.

Figure 3.20 shows modeled trends for available calories per capita. The daily per capita calorie intake rises to the required standard of 2,000–2,250 calories (recommended by the World Health Organization) only in the optimistic scenario, and then only in 2050. A lower rate of population growth may reduce demand for food and hence consumer food prices, thus enhancing access. As long as a large proportion of the population is dependent on agriculture for employment, attaining food sufficiency for everyone in the country in the short term will require more than just growth in GDP; other factors, such as improved productivity and market access, will be needed to fast-track progress.

With the huge rise in GDP per capita between 2010 and 2050, even in the pessimistic scenario, we might expect a larger increase in mean calorie

FIGURE 3.17   Yield change under climate change: Rainfed maize in Burundi, 2000–2050, A1B scenario

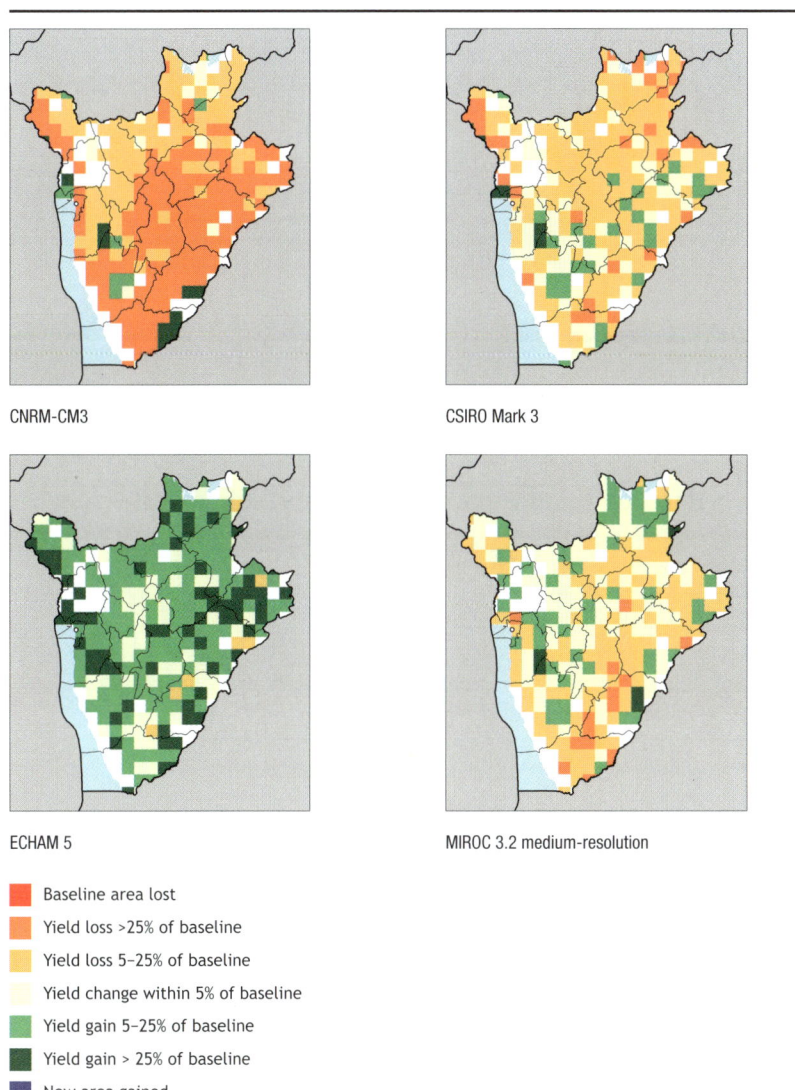

CNRM-CM3

CSIRO Mark 3

ECHAM 5

MIROC 3.2 medium-resolution

■ Baseline area lost
■ Yield loss >25% of baseline
■ Yield loss 5–25% of baseline
■ Yield change within 5% of baseline
■ Yield gain 5–25% of baseline
■ Yield gain > 25% of baseline
■ New area gained

Source: Authors' calculations.

Notes: A1B = greenhouse gas emissions scenario that assumes fast economic growth, a population that peaks midcentury, and the development of new and efficient technologies, along with a balanced use of energy sources; CNRM-CM3 = National Meteorological Research Center–Climate Model 3; CSIRO = climate model developed at the Australia Commonwealth Scientific and Industrial Research Organisation; ECHAM 5 = fifth-generation climate model developed at the Max Planck Institute for Meteorology (Hamburg); GCM = general circulation model; MIROC = Model for Interdisciplinary Research on Climate, developed by the University of Tokyo Center for Climate System Research.

FIGURE 3.18   Number of malnourished children under five years of age in Burundi in multiple income and climate scenarios, 2010–2050

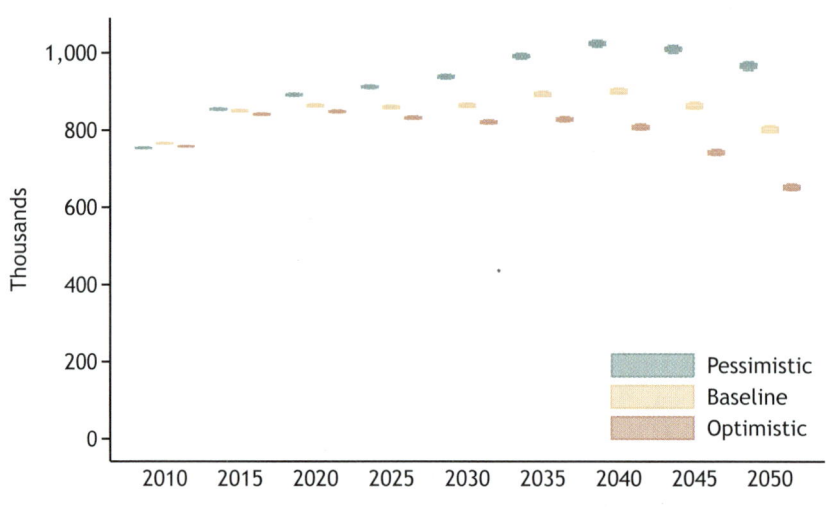

Source: Based on analysis conducted for Nelson et al. (2010).

Note: The box and whiskers plot for each socioeconomic scenario shows the range of effects from the four future climate scenarios.

FIGURE 3.19   Share of malnourished children under five years of age in Burundi in multiple income and climate scenarios, 2010–2050

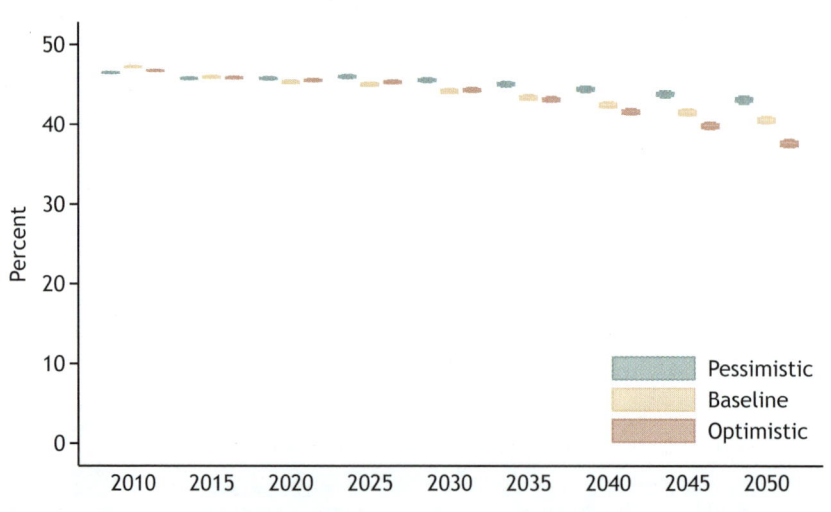

Source: Based on analysis conducted for Nelson et al. (2010).

Note: The box and whiskers plot for each socioeconomic scenario shows the range of effects from the four future climate scenarios.

FIGURE 3.20   Kilocalories per capita in Burundi in multiple income and climate scenarios, 2010–2050

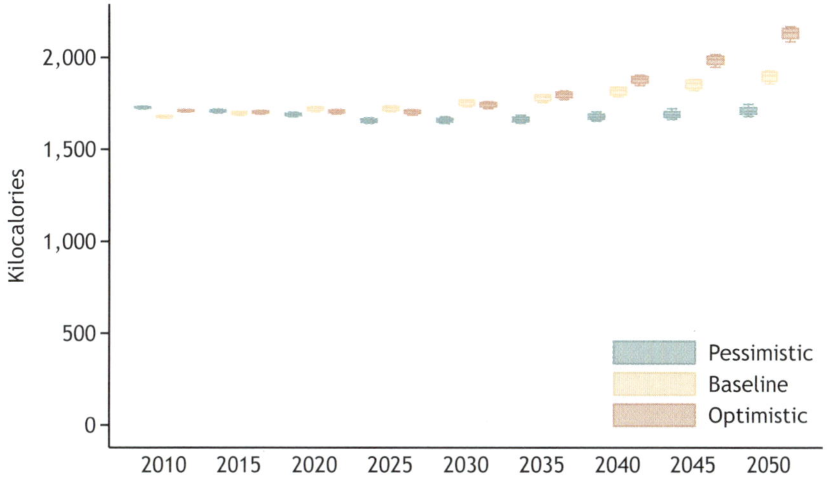

Source: Based on analysis conducted for Nelson et al. (2010).
Note: The box and whiskers plot for each socioeconomic scenario shows the range of effects from the four future climate scenarios.

consumption in the pessimistic scenario. Staple food prices also experience large increases over the same period, though not as much as the GDP per capita. We explain the low response in calorie consumption in part by noting that many of the positive income effects are negated by the negative price effects. But we also note that in the IMPACT model (International Model for Policy Analysis of Agricultural Commodities and Trade) used in this analysis, the own-price elasticities were probably set too high; therefore, we expect a better consumption response if income and prices rise as predicted in this scenario.

Agricultural Outcomes

Figures 3.21–3.23 show simulation results from the IMPACT model for key agricultural crops in Burundi. Each featured crop has five graphs: production, yield, area, net exports, and world prices.

For sweet potatoes and yams, the three scenarios show similar trends for all the variables analyzed. Production of sweet potatoes is shown to be rising by about 300,000 tons, reflecting growth in yield from 7.5 tons per hectare to 20.0 tons per hectare. This outcome is comparable to yields currently obtainable under good management practices (Okori and Wathum 2007), implying

FIGURE 3.21   Impact of changes in GDP and population on sweet potatoes and yams in Burundi, 2010–2050

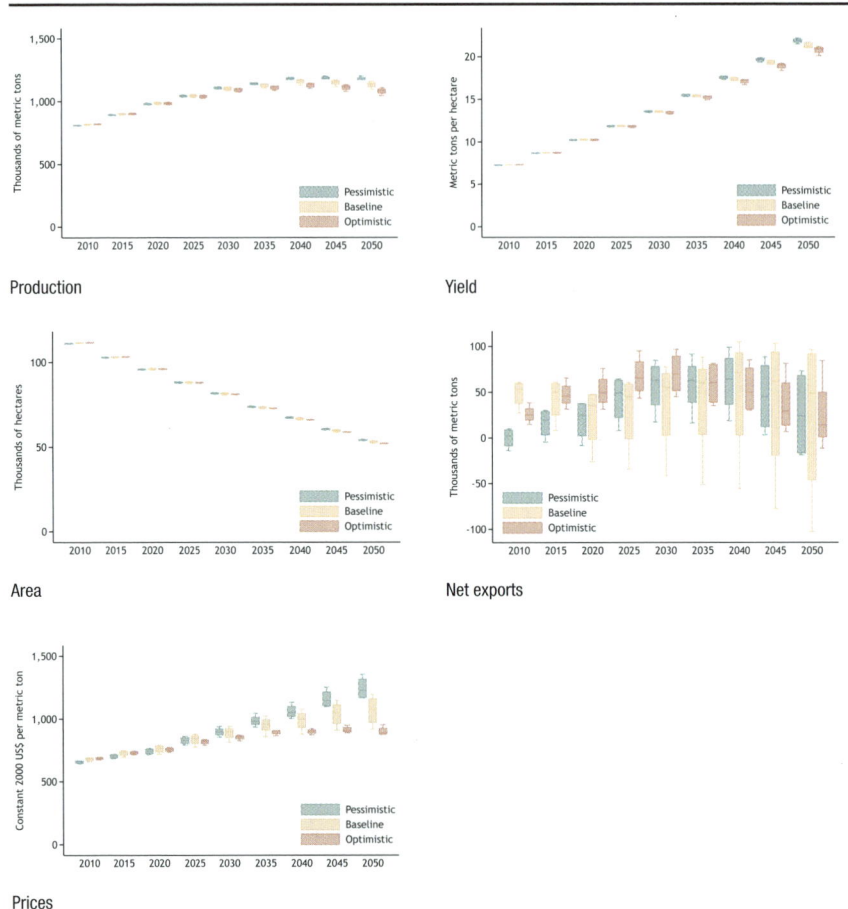

Source: Based on analysis conducted for Nelson et al. (2010).

Notes: The box and whiskers plot for each socioeconomic scenario shows the range of effects from the future climate scenarios. GDP = gross domestic product; US$ = US dollars.

an intensification of production. Harvested area declines from 110,000 hectares to less than 50,000 hectares. It is difficult to generalize what happens to net exports, because the variance of the outcomes is so large that they range from increasing to decreasing over time. It does appear, however, that in most climate models and most GDP and population scenarios, net exports increase.

Cassava production does not change much in spite of a doubling of yield because the area sown with cassava falls. With production mostly unchanged, Burundi will turn to imports to meet its domestic demand.

FIGURE 3.22   Impact of changes in GDP and population on cassava in Burundi, 2010–2050

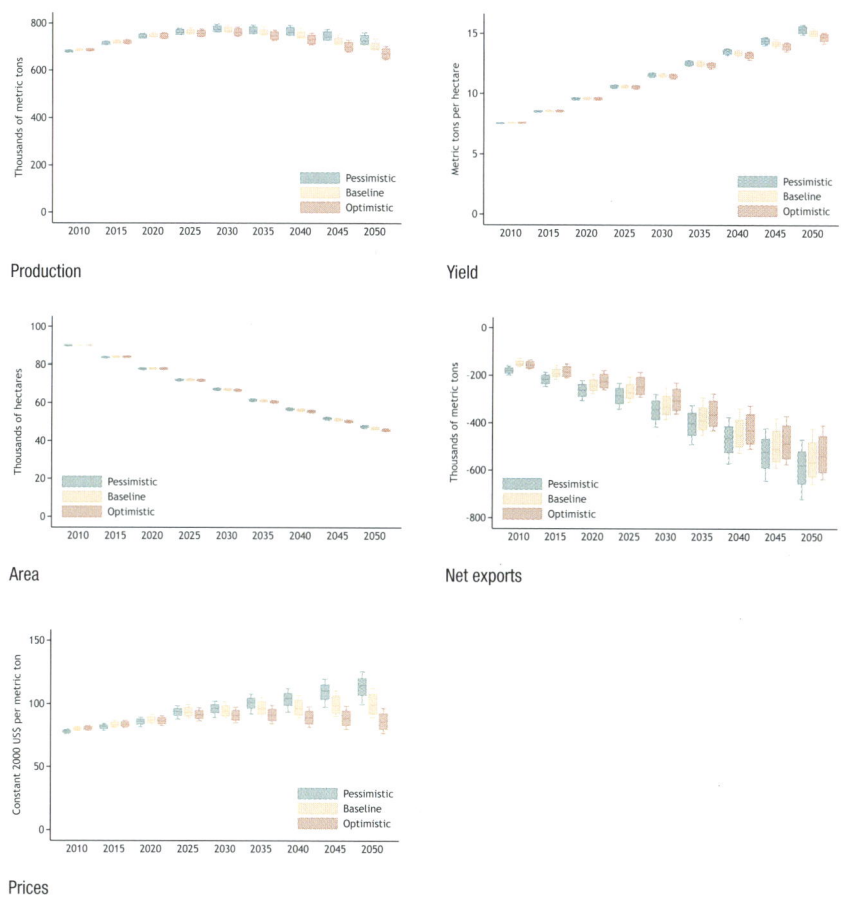

Production

Yield

Area

Net exports

Prices

Source: Based on analysis conducted for Nelson et al. (2010).

Notes: The box and whiskers plot for each socioeconomic scenario shows the range of effects from the future climate scenarios. GDP = gross domestic product; US$ = US dollars.

Production of maize rises due to an increase of roughly 80 percent in yield against a 25 percent reduction in area. Although the variance for net exports is rather large by 2050, it appears that all scenarios and climate models project that Burundi will be an importer of maize in 2050.

Although such results suggest that Burundi will not be able to meet its food requirements by itself, projected increases in GDP per capita will allow Burundi to import sufficient foods, as reflected in trends toward lower malnutrition rates and greater calorie consumption per person (except in the pessimistic scenario).

FIGURE 3.23   Impact of changes in GDP and population on maize in Burundi, 2010–2050

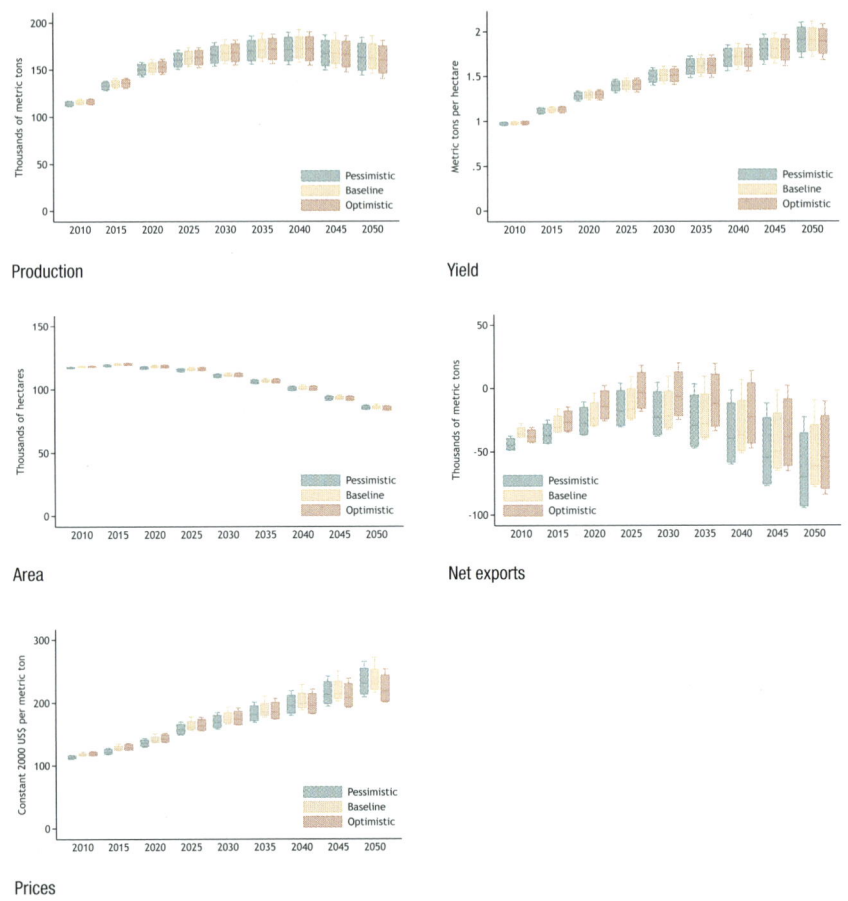

Source: Based on analysis conducted for Nelson et al. (2010).

Notes: The box and whiskers plot for each socioeconomic scenario shows the range of effects from the future climate scenarios. GDP = gross domestic product; US$ = US dollars.

Conclusions and Policy Recommendations

The four models used in this book differ regarding the extent and even the direction of future changes in precipitation in Burundi. They are in accord, however, in showing a general rise in temperatures within a range of 1°–2.5°C, though the models differ as to how much. Considering Burundi's current population growth rates, limited access to land, low level of GDP per capita, low education level, and other variables, the country is judged to have limited resources to cope with climate change shocks. Strategies for adaptation and mitigation will be needed to ensure increased agricultural productivity and food security.

Agricultural outcomes are also uncertain. Two models show significant yield losses for maize, while the other two predict some yield gain in most maize-producing areas. Given that maize is among the five most important foods in Burundi, food security challenges could arise from any yield loss. Even with some increase in agricultural production, and in an optimistic scenario of population and GDP growth, the available calories per capita meet the level recommended by the World Health Organization only in 2050.

Key strategies for addressing climate change will encompass measures for slowing population growth, enhancing agricultural productivity, and investing in the nonagricultural sector to expand livelihood options. These will include measures for

- supporting family planning facilities;

- increasing agriculture productivity through appropriate policies, technologies, and practices that should probably include increasing funding to national agricultural research and extension services;

- promoting accessible nonagriculture livelihood options;

- investing in education;

- implementing land reforms;

- promoting sustainable land management;

- prioritizing the agriculture sector by increasing its budget to at least 10 percent of the national budget; and

- implementing sensitization and awareness campaigns on the potential effects of climate change on agriculture.

References

Banderembako, D. 2006. "The Link between Land, Environment, Employment, and Conflict in Burundi." Produced by Nathan Associates Inc. for U.S. Agency for International Development (USAID), Washington, DC.

Bartholome, E., and A. S. Belward. 2005. "GLC2000: A New Approach to Global Land Cover Mapping from Earth Observation Data." *International Journal of Remote Sensing* 26 (9): 1959–1977.

Beck, J., G. Citegetse, J. Ko, and S. Sieber, S. 2010. "Burundi Environmental Threats and Opportunities Assessment." Report prepared for U.S. Agency for International Development (USAID), Washington, DC.

Burundi, Ministry for Land Management, Tourism, and Environment. 2007. *National Adaptation Plan of Action to Climate Change.* Bujumbura. 85 pp.

Burundi, Ministry of Agriculture and Livestock. 2008. *National Agriculture Strategy Plan of the Ministry of Agriculture and Livestock 2008–2015.* Bujumbura.

Burundi, Ministry of Finance. 2007. "Poverty Reduction Strategy Paper." Bujumbura.

Burundi, Ministry of Water, Environment, Land Management, and Urban Planning. 2009. *National Communication on Climate Change and Adaptation (NCCCA).* Bujumbura.

Burundi, Republic du, Ministère de l'Eau, de l'Environnement, de l'Aménagement du Territoire et de l'Urbanisme. 2009. "Deuxième Communication Nationale sur les Changements Climatiques." Bujumbura.

CIA (Central Intelligence Agency). 2011. *The World Fact Book.* Washington, DC.

CIESIN (Center for International Earth Science Information Network), Columbia University, IFPRI (International Food Policy Research Institute), World Bank, and CIAT (Centro Internacional de Agricultura Tropical). 2004. *Global Rural–Urban Mapping Project (GRUMP), Alpha Version: Population Density Grids.* Palisades, NY, US: Socioeconomic Data and Applications Center (SEDAC), Columbia University. Accessed October 18, 2005. at http://sedac.ciesin.columbia.edu/gpw.

FAO (Food and Agriculture Organization of the United Nations). 2010. FAOSTAT Database on Agriculture. Rome.

IFAD. 2011. Programme de developpement des filieres (PRODEFI). Accessed June 17, 2012. www.fidafrique.net/IMG/pdf/DAO_VEHICULES_PRODEFI.pdf.

Jones, P. G., P. K. Thornton, and J. Heinke. 2009. *Generating Characteristic Daily Weather Data Using Downscaled Climate Model Data from the IPCC's Fourth Assessment.* Project report for the International Institute for Land Reclamation and Improvement, Wageningen, the Netherlands. Accessed May 7, 2010. www.ccafs-climate.org/pattern_scaling/.

Lehner, B., and P. Döll. 2004. "Development and Validation of a Global Database of Lakes, Reservoirs, and Wetlands." *Journal of Hydrology* 296 (1–4): 1–22.

Millennium Ecosystem Assessment. 2005. *Ecosystems and Human Well-being: Synthesis.* Washington, DC: Island Press. www.maweb.org/en/Global.aspx.

Nelson, G. C., M. W. Rosegrant, A. Palazzo, I. Gray, C. Ingersoll, R. Robertson, S. Tokgoz, et al. 2010. *Food Security, Farming, and Climate Change to 2050: Scenarios, Results, Policy Options.* Washington, DC: International Food Policy Research Institute.

Nkurunziza H., A. Gebhardt, and J. Pilz, J. 2010. "Bayesian Modeling of the Effect of Climate on Malaria in Burundi." *Malaria Journal* 9: 114. doi:10.1186/1475-2875-9-114. www .malariajournal.com/content/9/1/114.

Okori, P., and P. Wathum. 2007. "Situational Analysis of Staple Crops in ECA: Background Paper for Strategy Planning and Priority Setting for the Staple Crops Program of the Association for Strengthening Agricultural Research in Eastern and Central Africa (ASARECA)." Entebbe, Uganda.

Omamo, S. W., X. Diao, S. Wood, J. Chamberlin, L. You, S. Benin, U. Wood-Sichra, and A. Tatwangire, A. 2006. *Strategic Priorities for Agricultural Development in Eastern and Central Africa.* IFPRI Research Report 150. Washington, DC: International Food Policy Research Institute.

UNEP (United Nations Environment Programme) and IUCN (International Union for Conservation of Nature). 2009. World Database on Protected Areas (WDPA). Annual Release. No longer available online.

UNPOP (United Nations Department of Economic and Social Affairs–Population Division). 2009. *World Population Prospects: The 2008 Revision.* New York. http://esa.un.org/unpd/wpp/.

Wood, S., G. Hyman, U. Deichmann, E. Barona, R. Tenorio, Z. Guo, et al. 2010. *Sub-national Poverty Maps for the Developing World Using International Poverty Lines: Preliminary Data Release.* Washington, DC: Harvest Choice and International Food Policy Research Institute. http:// labs.harvestchoice.org/2010/08/poverty-maps/.

World Bank. 2009. *World Development Indicators 2009.* Washington, DC.

You, L., and S. Wood. 2006. "An Entropy Approach to Spatial Disaggregation of Agricultural Production." *Agricultural Systems* 90 (1–3): 329–347.

You, L., S. Wood, and U. Wood-Sichra. 2006. "Generating Global Crop Distribution Maps: From Census to Grid." Paper presented at the International Association of Agricultural Economists Conference, Brisbane, Australia, August 11–18.

———. 2009. "Generating Plausible Crop Distribution and Performance Maps for Sub-Saharan Africa Using a Spatially Disaggregated Data Fusion and Optimization Approach." *Agricultural Systems* 99 (2–3): 126–140.

Chapter 4

DEMOCRATIC REPUBLIC OF CONGO

Blandine M. Nsombo, Timothy S. Thomas, Miriam Kyotalimye,
and Michael Waithaka

The Democratic Republic of Congo (DRC) committed to the mitigation of the effects of climate change by signing the Kyoto Protocol for climate change and other related environmental management protocols. Since 1994, DRC has produced two national climate change communication documents (RDC, Ministère de l'Environnement, Conservation de la Nature, Eaux, et Forêts 2001; RDC, Ministère de l'Environnement, Conservation de la Nature et Tourisme 2009). These documents give the strategic orientation for the application of the climate change convention with a focus on sectors vulnerable to climate change, the extent of greenhouse gas emissions, and the potential for carbon sequestration, adaptation, and mitigation strategies. This research monograph focuses on agriculture. It looks at indicators of economic growth among the sectors most vulnerable to climate change and gives future scenarios for the 10 major crops in the country according to four climate models.

Review of the Current Situation and Trends

Economic and Demographic Indicators

This section discusses the population trends and income indicators in DRC. Different projections show about the same tendency and show that the urban population of DRC has grown at a high rate compared to the total and rural populations, even if evidence shows that the agricultural sector is the most important source of income for the total population (Bhaduri et al. 2007; UNPOP 2009; World Bank 2009).

Population

Figure 4.1 shows the country's total and rural population counts (left axis) and the share of its urban population (right axis) from 1960 to 2008. Table 4.1 provides additional information concerning rates of population

FIGURE 4.1   Population trends in Democratic Republic of Congo: Total population, rural population, and percent urban, 1960–2008

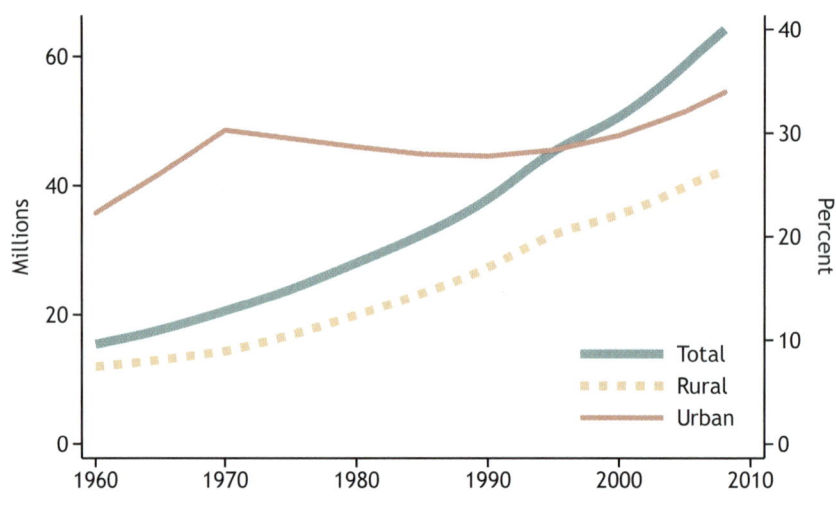

Source: *World Development Indicators* (World Bank 2009).

growth. The geographic distribution of the population within DRC is further shown in Figure 4.2. All figures are estimates based on census data and other sources.

The population of DRC has more than tripled over the past five decades, from 16 million people in 1960 to 64 million people in 2010. Figure 4.1 shows a gradual but slow rise in total and rural population from 1960 to the mid-1990s. Thereafter, there was a slight decline in the rural population, while the total population continued to rise. The share of urban population rose rapidly

TABLE 4.1   Population growth rates in Democratic Republic of Congo, 1960–2008 (percent)

Decade	Total growth rate	Rural growth rate	Urban growth rate
1960–1969	2.80	1.80	5.90
1970–1979	3.10	3.30	2.60
1980–1989	2.90	3.10	2.60
1990–1999	3.00	2.70	3.60
2000–2008	3.00	2.30	4.60

Source: Authors' calculations based on *World Development Indicators* (World Bank 2009).

FIGURE 4.2 Population distribution in Democratic Republic of Congo, 2000 (persons per square kilometer)

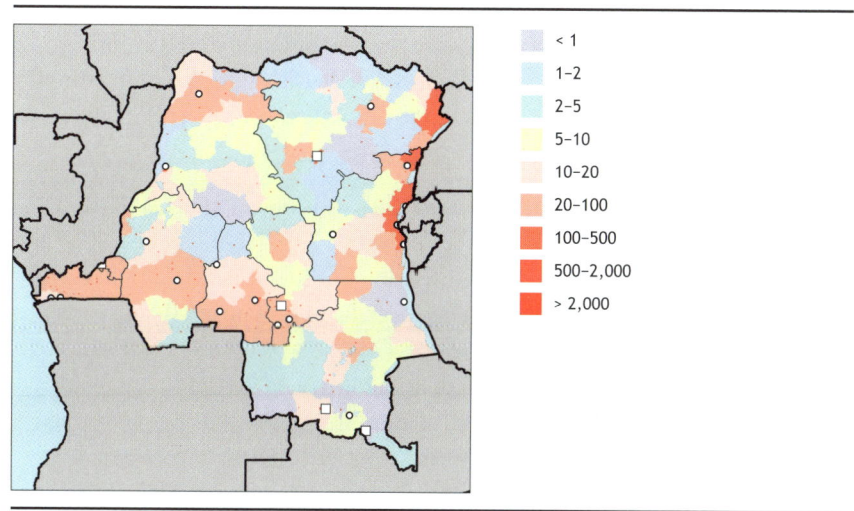

	< 1
	1–2
	2–5
	5–10
	10–20
	20–100
	100–500
	500–2,000
	> 2,000

Source: CIESIN et al. (2004).

to about 30 percent in 1970, and then it experienced a slight decline until 2000, when there was resumption in urban population growth.

Table 4.1 shows that total population growth rate has been around 3 percent per year. The growth rate of the urban population has always been higher than the rural rate except from the 1970s to the end of 1980s. Precisely during these periods, the urban growth rate fell by 44 percent of its level in the previous decade. Within the same period, the rural population growth rate rose by nearly similar proportions. The rural growth rate has been declining since then, reaching half the urban growth rate in the past decade. A high rate of urban population growth, caused when farmers abandon agriculture and migrate to cities in search of better services and job opportunities, can lead to a reduction in agricultural production. This can be a threat to national food security if those who have migrated to cities are unable to find jobs and the nation is unable to produce goods and services to trade with other nations for food commodities. However, in many cases the migration may improve food security if people are more productive in the urban sector than in the rural sector.

Rapid urbanization can also cause environmental and climatic hazards such as air and water pollution, disease, flooding, and land degradation, which can extend to surrounding rural areas. Considering the population distribution (see Figure 4.2), the savanna regions are more populated because they are more accessible and urbanized than are forested ones, which are humid and hence less accessible. The Uganda–Rwanda border towns of Ituri, Bunia, Butembo, Goma, and Bukavu have areas with the highest population densities, 500–2,000 persons per square kilometer. Population densities are also high around the western seaport and Kinshasa and in the Northern Equateur Province.

Income

Figure 4.3 shows trends in GDP per capita and the proportion of GDP from agriculture. GDP per capita declined from the mid-1970s until around 2000 and has risen since then. However, the contribution of agriculture to GDP has tended to trend in the opposite direction, peaking in the late 1990s and early 2000 but declining steeply in recent years. With the agricultural sector still leading in importance, small changes in climate patterns are likely to have a major impact on agricultural GDP and overall economic growth. The fact is that agricultural production relies heavily on weather patterns. These impacts will be enhanced by lack of appropriate technology and climate change adaptation strategies common in the rural agricultural areas. However, the country is endowed with some $24 trillion in untapped raw mineral deposits, including tungsten,

FIGURE 4.3   Per capita GDP in Democratic Republic of Congo (constant 2000 US$) and share of GDP from agriculture (percent), 1960–2008

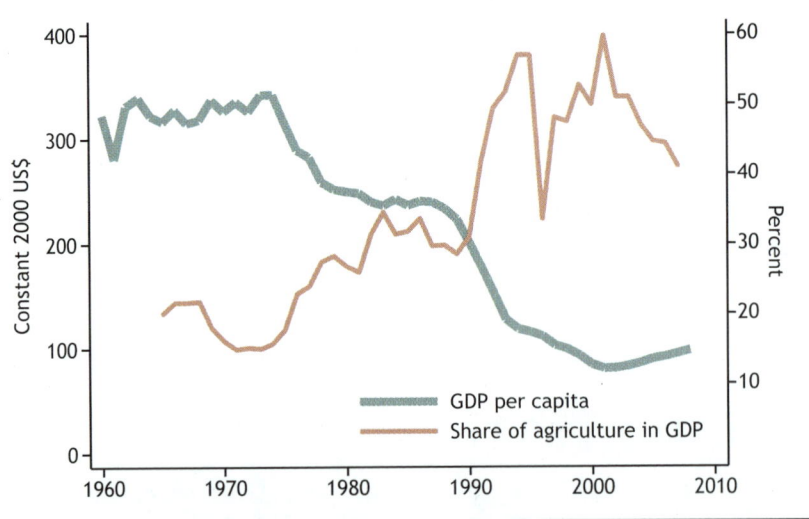

Source: *World Development Indicators* (World Bank 2009).
Note: GDP = gross domestic product; US$ = US dollars.

tantalum, tin, gold, diamonds, copper, cobalt, zinc, and, recently, oil in the Albertine rift. Assuming that there is an end to civil conflict, the exploitation of these should lead to overall GDP growth through growth in the nonagriculture sector. Unfortunately, growth in mineral exports does not always translate into a higher standard of living for large segments of the population.

Vulnerability to Climate Change

Table 4.2 provides some data on additional indicators of vulnerability and resiliency to economic shocks: the population's level of education and literacy and the concentration of labor in poorer or less dynamic sectors. From Table 4.2 it appears that between the time of students' enrollment in primary and secondary school, their withdrawal rate is more than 50 percent. Generalized poverty keeps many pupils out of school and girls stop studying early due to marriage and the need for farm labor. However, the relatively high adult literacy rate is encouraging given that it shows a substantial population base with a potential for the uptake of knowledge and skills on strategies for adaptation to and mitigation of climate change, which could lessen the adverse impacts of the same. Table 4.2 also shows that malnutrition among those under age five in the country is still high, at 33.6 percent.

Figure 4.4 shows two noneconomic correlates of poverty, life expectancy and under-five mortality. A quick review of Figure 4.4 shows a slow increase in life expectancy at birth since 1997 coupled with decreases in the under-five mortality rate around the same years. Life expectancy is still well below the average for Africa south of the Sahara, with estimates showing that about 540,000 Congolese die annually before their prime due to war, starvation, and disease (Nzongola-Ntalaja 2004). Despite the rural exodus accentuated by civil war, global efforts in terms of mother-and-child health and nutrition improvements have brought a certain stability, which has translated into an improvement in general well-being. Maintenance and further gains in these indicators

TABLE 4.2   Education and nutrition statistics for Democratic Republic of Congo, 2000s

Indicator	Year	Percent
Primary school enrollment: Percent gross (three-year average)	2007	85.1
Secondary school enrollment: Percent gross (three-year average)	2007	33.4
Adult literacy rate	2001	67.2
Under-five malnutrition (weight for age)	2001	33.6

Source: *World Development Indicators* (World Bank 2009).

FIGURE 4.4   Well-being indicators in Democratic Republic of Congo, 1960–2008

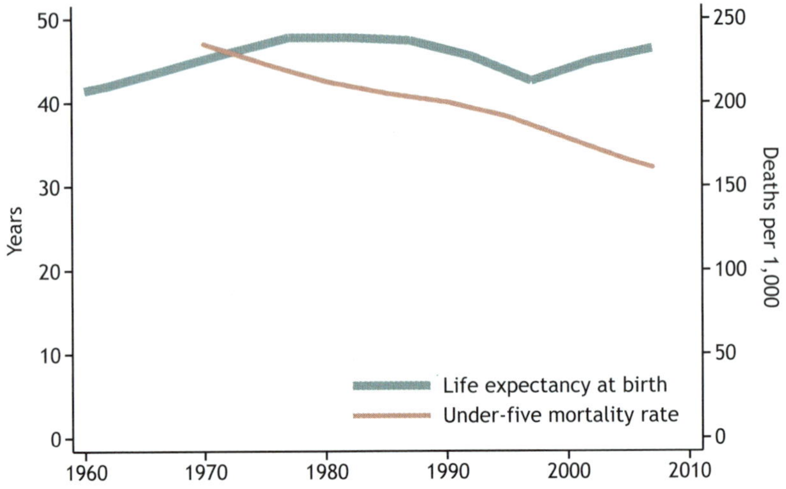

Source: *World Development Indicators* (World Bank 2009).

is threatened by ongoing climate change, which will affect agriculture production and alter the abundance and distribution of vector hosts, promoting the spread of diseases such as malaria, diarrheal illnesses, and cholera in the country (Griffith, Kelly-Hope, and Miller 2006; Paaijmans et al. 2008). The effects of increased incidences of these diseases would be increased nutrition deficiencies and underproductivity due to sickness and death.

Review of Land Use, Potential, and Limitations

The following section highlights the area occupied by each major crop in DRC to show, on the one hand, the proportion of coverage through the country and, on the other hand, how harvests reach markets. The longer smallholder farmers' produce takes to reach the market, the more vulnerable they are, because most of their products are perishable (Bartholome and Belward 2005; You and Wood, 2006; You, Wood, and Wood-Sichra 2006, 2009; UNEP and IUCN 2009).

Land Use Overview

Figure 4.5 shows land cover and use in DRC as of 2000. Although about 70 percent of the total population depends on agriculture for their survival, only about 7 percent of the country's land area is used for cropping and

FIGURE 4.5   Land cover and land use in Democratic Republic of Congo, 2000

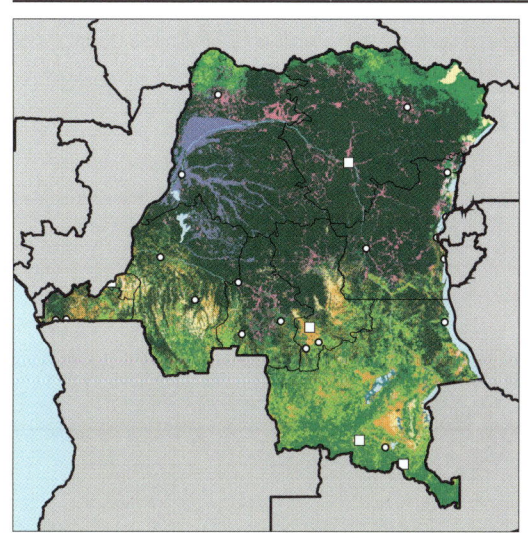

■ Tree cover, broadleaved, evergreen

■ Tree cover, broadleaved, deciduous, closed

■ Tree cover, broadleaved, open

■ Tree cover, broadleaved, needle-leaved, evergreen

■ Tree cover, broadleaved, needle-leaved, deciduous

■ Tree cover, broadleaved, mixed leaf type

■ Tree cover, broadleaved, regularly flooded, fresh water

■ Tree cover, broadleaved, regularly flooded, saline water

■ Mosaic of tree cover/other natural vegetation

■ Tree cover, burnt

■ Shrub cover, closed–open, evergreen

■ Shrub cover, closed–open, deciduous

☐ Herbaceous cover, closed–open

■ Sparse herbaceous or sparse shrub cover

■ Regularly flooded shrub or herbaceous cover

☐ Cultivated and managed areas

■ Mosaic of cropland/tree cover/other natural vegetation

■ Mosaic of cropland/shrub/grass cover

■ Bare areas

■ Water bodies

☐ Snow and ice

■ Artificial surfaces and associated areas

☐ No data

Source: GLC2000 (Bartholome and Belward 2005).

FIGURE 4.6   Protected areas in Democratic Republic of Congo, 2009

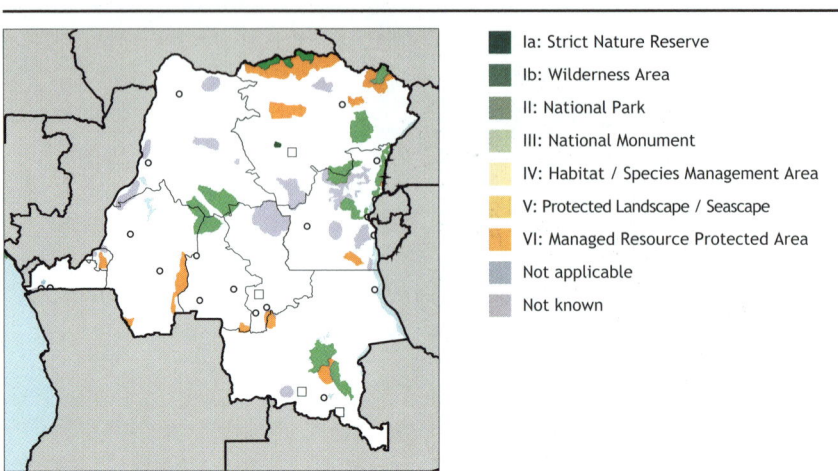

- Ia: Strict Nature Reserve
- Ib: Wilderness Area
- II: National Park
- III: National Monument
- IV: Habitat / Species Management Area
- V: Protected Landscape / Seascape
- VI: Managed Resource Protected Area
- Not applicable
- Not known

Sources: Protected areas are from the World Database on Protected Areas (UNEP and IUCN 2009). Water bodies are from the World Wildlife Federation's Global Lakes and Wetlands Database (Lehner and Döll 2004).

livestock activities. These areas are located around the main cities and along major roads, especially in the eastern part of the country. Forests occupy the northern part of the country, whereas the southern part is covered by savanna. Protected areas including parks and reserves cover about 10 percent of the country's total area. The country also has one of the most extensive networks of water resources in Africa, served by the Congo River and its tributaries, which have a regular flow fed by rivers and streams from both sides of the equator due to the complementary alteration of rainy and dry seasons.

Figure 4.6 shows the location of protected areas in DRC, including parks and reserves. These locations provide important protection for fragile environmental areas, which may also be important for the tourism industry.

Dispersed all over the country, protected areas occupy about 15 percent of the total land area and are situated in all the different ecosystems. These areas represent important sites for biodiversity conservation and mitigation of the impacts of climate change through carbon sequestration and other environmental services. In addition they act as safety nets for the poor by serving as sources of subsistence food and cash incomes in seasons of crop failure, illness, job loss, and shortage in energy supplies. However, emerging evidence shows that there is increasing deforestation in forested areas and reserves through activities such as selective logging, agriculture, mining, infrastructure construction, and fuel burning, which contribute to reductions in biodiversity, increases in carbon dioxide emissions, and other environmental impacts such as flooding (Nkem, Idinoba, and Sendanshonga 2008). Although the effects of these actions might be negligible now, their impact is likely to be magnified as the population grows and war's conflict dies down. The sustenance of forested and protected areas in DRC is vital given that they help regulate weather patterns in a region highly dependent on rainfed agriculture. The potential of these areas for tourism is also relatively high. DRC's most important potential tourist resources include the Massif du Ruwenzori (Ruwenzori Mountains or Mountains of the Moon) between Lakes Albert and Edward, which are also the highest range in Africa. The height of these mountains and their location on the equator make for a varied and spectacular flora. Other sites include the Virunga Mountains north of Lac Kivu, site of several active volcanoes and a game park.

Figure 4.7 shows travel times to urban areas of various sizes, which provide potential markets for agricultural products as well as potentially serving as places where farmers can acquire agricultural inputs and consumer goods. Overall, the maps show that the markets that are fastest to reach are urban centers toward the borders of the country, especially on the eastern, southern, and southwestern sides of the county.

Most of DRC is served by the Congo River and its tributaries, which provide the country with the most extensive and evenly distributed network of navigable waterways in Africa (about 16,238 kilometers). Other forms of the national transport network include four international airports and about 145,000 kilometers of roads, 7,400 kilometers of which are well-paved urban roads. However, most of the transport infrastructure is poor, and access to agricultural markets is low (DRSP 2006).

Improving DRC's infrastructure is likely to enhance agricultural growth and reduce poverty in the country. A good and diverse transportation network can reduce time and costs. Research has indicated that reductions in travel costs are linked to increases in producer prices, agricultural production, and market access (Ulimwengu et al. 2009). It has also been found that poor-quality roads lead to higher transportation costs, market prices, and trader margins in DRC (Minten and Kyle 1999). The unfavorable effects are likely to be felt more in times of flooding and will be experienced most by agricultural communities in the central regions that appear remote.

Agriculture

Tables 4.3 and 4.4 show key agricultural commodities in terms of area harvested and consumption, ranked by weight. Table 4.4 shows that cassava is the most important staple crop in terms of harvest area and consumption, followed by maize and, in relatively small proportion, groundnuts. Production of

FIGURE 4.7   Travel time to urban areas of various sizes in Democratic Republic of Congo, circa 2000

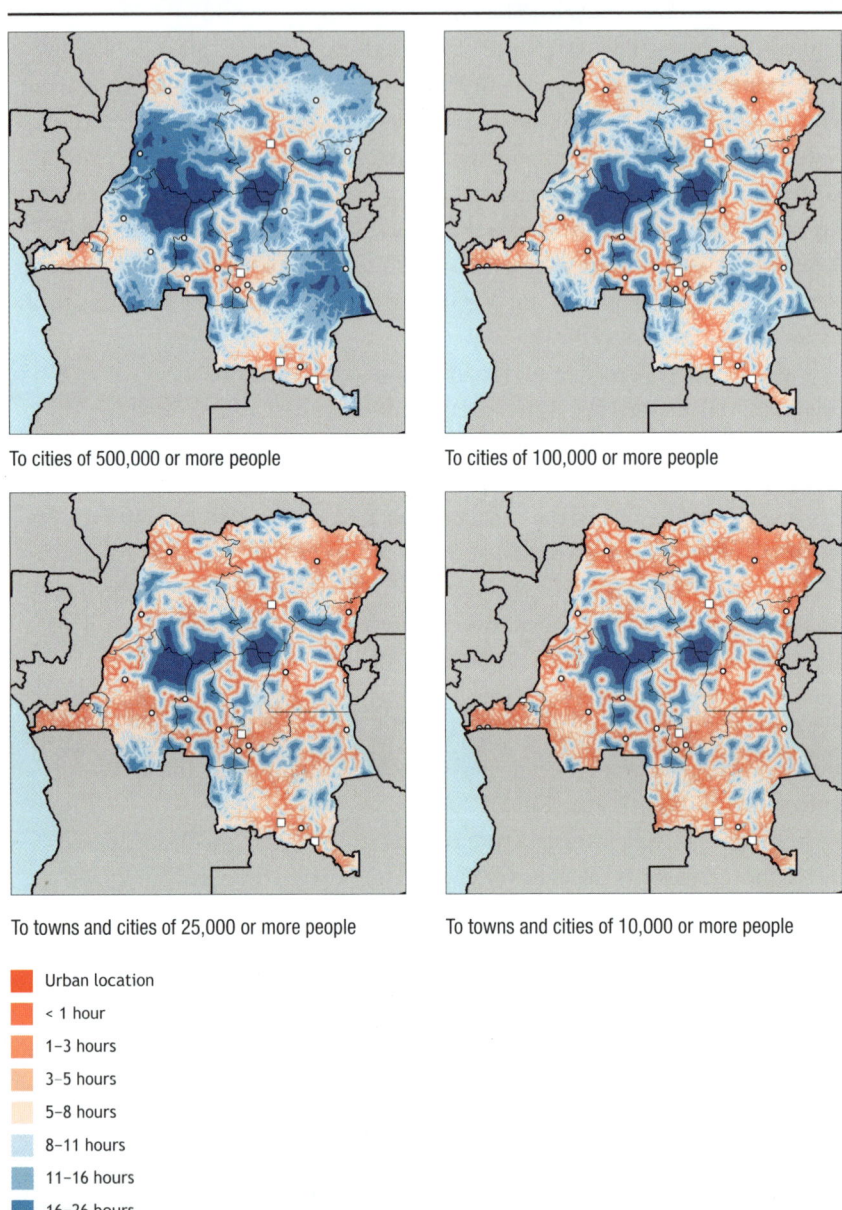

To cities of 500,000 or more people

To cities of 100,000 or more people

To towns and cities of 25,000 or more people

To towns and cities of 10,000 or more people

Urban location
< 1 hour
1–3 hours
3–5 hours
5–8 hours
8–11 hours
11–16 hours
16–26 hours
> 26 hours

Source: Authors' calculations.

TABLE 4.3   Harvest area of leading agricultural commodities in Democratic Republic of Congo, 2006–2008 (thousands of hectares)

Rank	Crop	Percent of total	Harvest area
	Total	100.0	5,862
1	Cassava	31.7	1,859
2	Maize	25.3	1,484
3	Groundnuts	8.1	475
4	Rice	7.1	419
5	Plantains	4.6	268
6	Beans	3.5	207
7	Oil palm fruit	2.9	172
8	Cowpeas	2.0	116
9	Melon seeds	1.5	90
10	Bananas	1.4	84

Source: FAOSTAT (FAO 2010).
Note: All values are based on the three-year average for 2006–2008.

TABLE 4.4   Consumption of leading food commodities in Democratic Republic of Congo, 2003–2005 (thousands of metric tons)

Rank	Crop	Percent of total	Food consumption
	Total	100.0	23,036
1	Cassava	67.9	15,643
2	Maize	5.1	1,178
3	Fermented beverages	4.4	1,008
4	Other fruits	2.1	487
5	Wheat	2.0	455
6	Plantains	1.8	425
7	Other vegetables	1.6	377
8	Sugarcane	1.3	303
9	Bananas	1.2	282
10	Rice	1.2	275

Source: FAOSTAT (FAO 2010).
Note: All values are based on the three-year average for 2003–2005.

the first two crops has been found to be sensitive to precipitation changes that occur due to climate change (Van-Minnen, Alcamo, and Haupt 2000). Priority should be given to cassava and maize when designing adaption and mitigation strategies geared toward improving crop production, given the significance of these foods in the household diet. The low productivity of itinerant sub-sistent agriculture can be improved through strategies that encourage the set-tlement of agriculture, improvement of land productivity through biological approaches based on agroforestry systems, use of improved crop varieties, and adoption of postharvest handling and storage technologies.

Figures 4.8–4.11 show the estimated yield as well as the harvest and grow-ing areas for key crops in DRC in 2000. In Figure 4.8 we see that cassava is mostly produced in the southern half of the country, with potential yields exceeding 10 tons per hectare but with actual reported yields averaging around 8 tons per hectare. DRC is the greatest producer of cassava in eastern and cen-tral Africa. However, cassava yields have been suffering due to changes in farm-ing techniques (farmers cultivate land before it has undergone a fallow period), pests and diseases (especially the mosaic disease), and competition from other crops (especially maize that is grown year round). The crop offers a means of food security and incomes for the poor.

FIGURE 4.8   Yield (metric tons per hectare) and harvest area density (hectares) for rainfed cassava in Democratic Republic of Congo, 2000

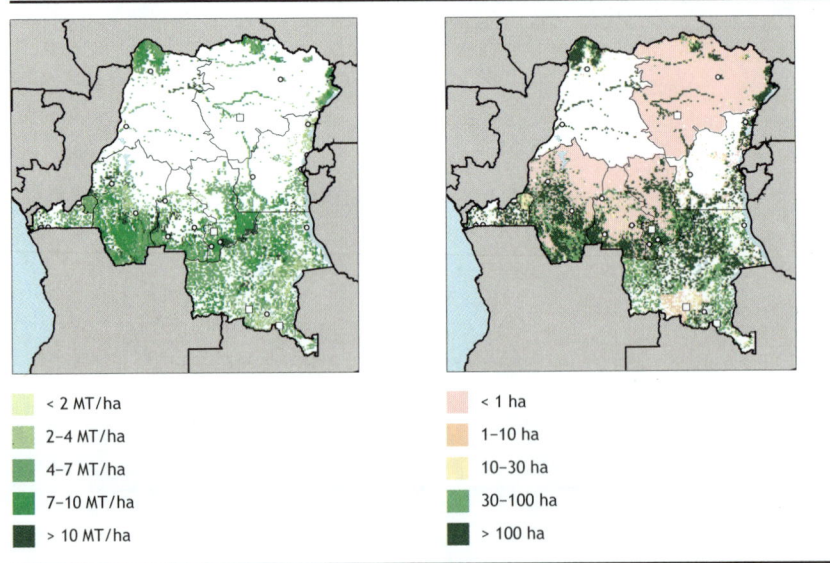

< 2 MT/ha	< 1 ha
2–4 MT/ha	1–10 ha
4–7 MT/ha	10–30 ha
7–10 MT/ha	30–100 ha
> 10 MT/ha	> 100 ha

Source: SPAM (Spatial Production Allocation Model) (You and Wood 2006; You, Wood, and Wood-Sichra 2006, 2009).
Note: ha = hectare; MT/ha = metric tons per hectare.

FIGURE 4.9 Yield (metric tons per hectare) and harvest area density (hectares) for rainfed maize in Democratic Republic of Congo, 2000

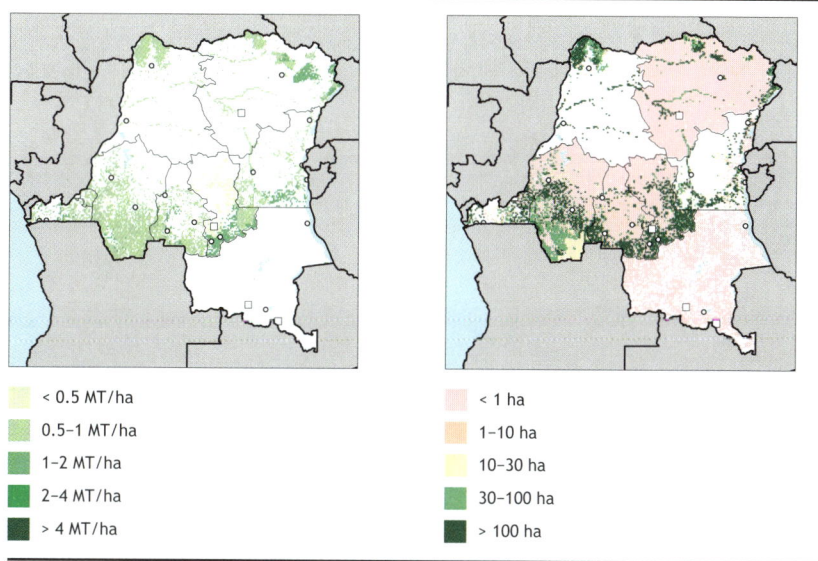

< 0.5 MT/ha	< 1 ha
0.5–1 MT/ha	1–10 ha
1–2 MT/ha	10–30 ha
2–4 MT/ha	30–100 ha
> 4 MT/ha	> 100 ha

Source: SPAM (Spatial Production Allocation Model) (You and Wood 2006; You, Wood, and Wood-Sichra 2006, 2009).
Note: ha = hectare; MT/ha = metric tons per hectare.

FIGURE 4.10 Yield (metric tons per hectare) and harvest area density (hectares) for rainfed groundnuts in Democratic Republic of Congo, 2000

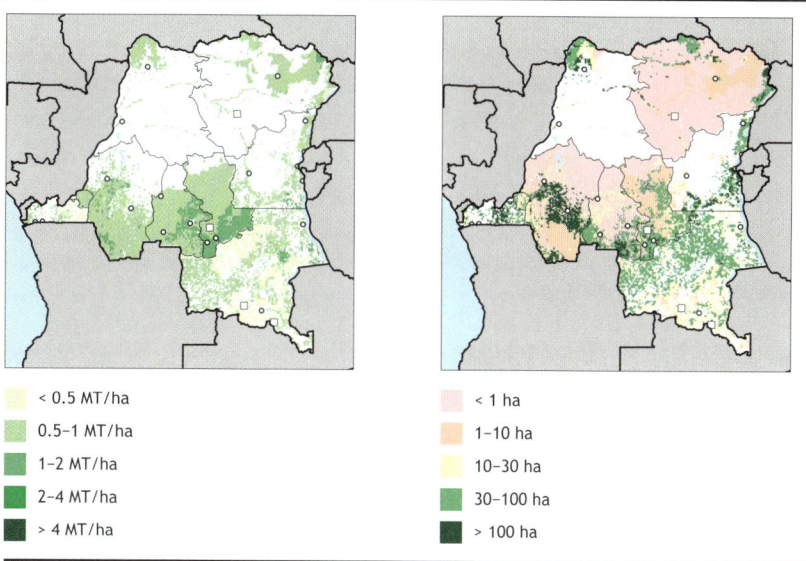

< 0.5 MT/ha	< 1 ha
0.5–1 MT/ha	1–10 ha
1–2 MT/ha	10–30 ha
2–4 MT/ha	30–100 ha
> 4 MT/ha	> 100 ha

Source: SPAM (Spatial Production Allocation Model) (You and Wood 2006; You, Wood, and Wood-Sichra 2006, 2009).
Note: ha = hectare; MT/ha = metric tons per hectare.

FIGURE 4.11   Yield (metric tons per hectare) and harvest area density (hectares) for rainfed rice in Democratic Republic of Congo, 2000

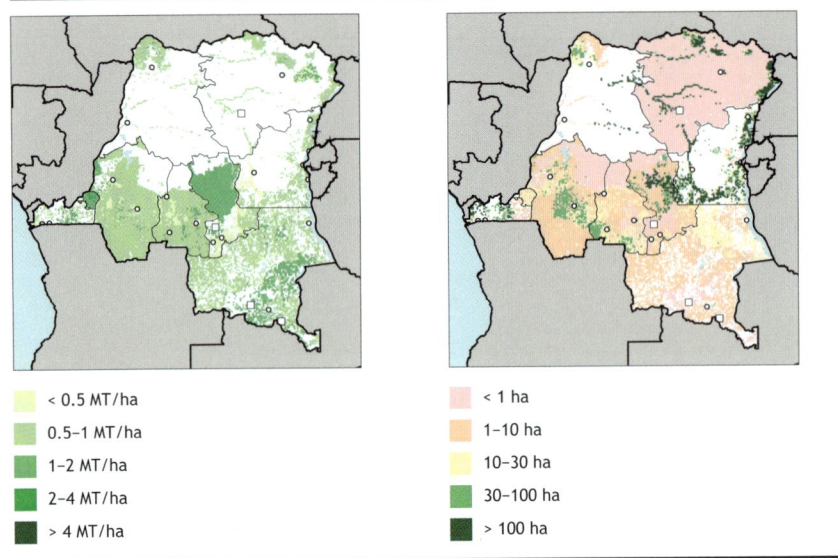

< 0.5 MT/ha	< 1 ha
0.5–1 MT/ha	1–10 ha
1–2 MT/ha	10–30 ha
2–4 MT/ha	30–100 ha
> 4 MT/ha	> 100 ha

Source: SPAM (Spatial Production Allocation Model) (You and Wood 2006; You, Wood, and Wood-Sichra 2006, 2009).
Note: ha = hectare; MT/ha = metric tons per hectare.

The maps show that maize is produced mainly in the central regions of the country. The average national maize yield is around 1.2 tons per hectare. In a few instances where inorganic fertilizers and improved seeds are used, yields of up to 10 tons per hectare are common (RDC, Ministère de l'Agriculture 2009). Groundnuts and rice are produced in much smaller quantities than the first two crops. Higher yields occur in central regions of Kinshasa, Kasai-Oriental, Kasai-Occidental, and Bandundu Provinces. Despite high production potential, more than 90 percent of the rice consumed in DRC is imported.

Scenarios for the Future

Economic and Demographic Indicators

Population
Figure 4.12 shows population projections by the UN Population Division through 2050. DRC's population is growing and will continue to grow for the foreseeable future. Its major cities are likely to become overpopulated, posing challenges of food security and potential damage to natural resources as settlements, infrastructure, and food production expand.

FIGURE 4.12   Population projections for Democratic Republic of Congo, 2010–2050

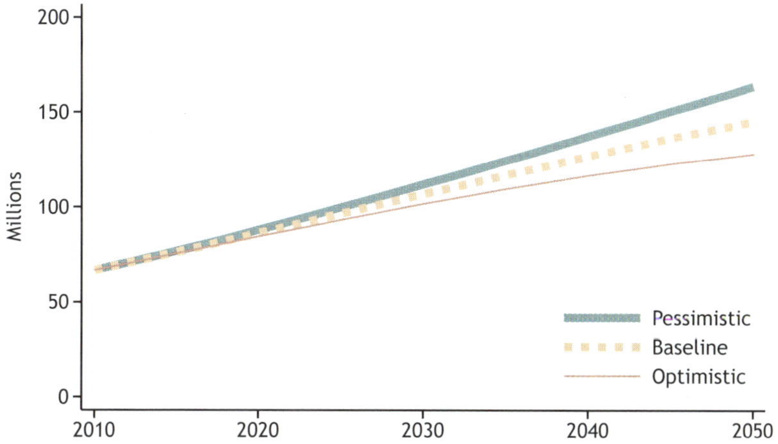

Source: UNPOP (2009).

Income

Figure 4.13 shows three GDP per capita scenarios used for this study. These results combine three GDP projections with the three population projections of Figure 4.12 from the UN Population Division. The optimistic scenario combines high GDP with a low rate of population growth. The baseline scenario combines the medium GDP projection with the medium population projection. Finally, the pessimistic scenario combines the low GDP projection with the high population projection.

The differences between the scenarios are striking. In the pessimistic scenario, by 2050 GDP per capita is $277; in the baseline scenario it is $440; in the optimistic scenario it is $715.

Biophysical Analysis

Climate Models

Figure 4.14 shows projected precipitation changes under the four downscaled climate models used in this book with the A1B scenario.[1] Model predictions of mean annual precipitation changes for DRC to 2050 show differences between the general circulation models (GCMs), although three of the models show

---

1 The A1B scenario is a greenhouse gas emissions scenario that assumes fast economic growth, a population that peaks midcentury, and the development of new and efficient technologies, along with a balanced use of energy sources.

FIGURE 4.13   Gross domestic product (GDP) per capita in Democratic Republic of Congo, future scenarios, 2010–2050

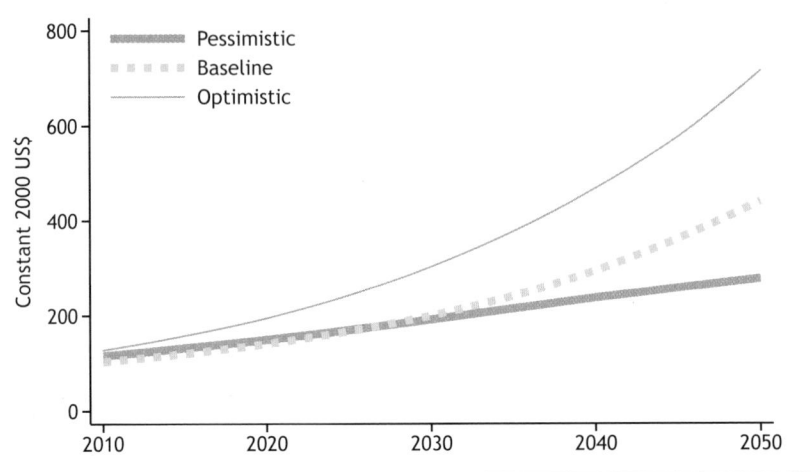

Sources: Computed from GDP data from the World Bank Economic Adaptation to Climate Change project (World Bank 2010), from the Millennium Ecosystem Assessment (2005) reports, and from population data from the United Nations (UNPOP 2009).
Note: US$ = US dollars.

increases of up 200 millimeters but variations in location. CNRM-CM3 suggests precipitation increases concentrated in the west, ECHAM 5 suggests increases concentrated in the north, and MIROC 3.2 suggests precipitation increases concentrated in the east. The ECHAM 5 GCM shows similar precipitation patterns to those of SCNCC (RDC, Ministère de l'Environnement, Conservation de la Nature et Tourisme 2009) and Ntombi and Pangu (2011).[2] The CSIRO model shows patches of rainfall reduction of up to 150 millimeters annually, with a large patch in the upper eastern portion of the country, as well as in the west. Although the MIROC 3.2 model predicts rainfall increasing in the west, around Kinshasa, the rainfall is predicted to decrease by around 200 millimeters or more per year.

The model predictions of temperature changes to 2050 are more consistent with each other than were the precipitation changes, although there are differences in the amount of change in each model. Temperatures in DRC will rise by 1° to a maximum of 3°C. CSIRO Mark 3 seems to be the appropriate model to simulate the normal annual maximum temperature for DRC, because

---

2 CNRM-CM3 is National Meteorological Research Center–Climate Model 3. MIROC 3.2 is the Model for Interdisciplinary Research on Climate, developed at the University of Tokyo Center for Climate System Research. CSIRO Mark 3 is a climate model developed at the Australia Commonwealth Scientific and Industrial Research Organisation. ECHAM 5 is a fifth-generation climate model developed at the Max Planck Institute for Meteorology in Hamburg.

FIGURE 4.14   Changes in mean annual precipitation in Democratic Republic of Congo, 2000–2050, A1B scenario (millimeters)

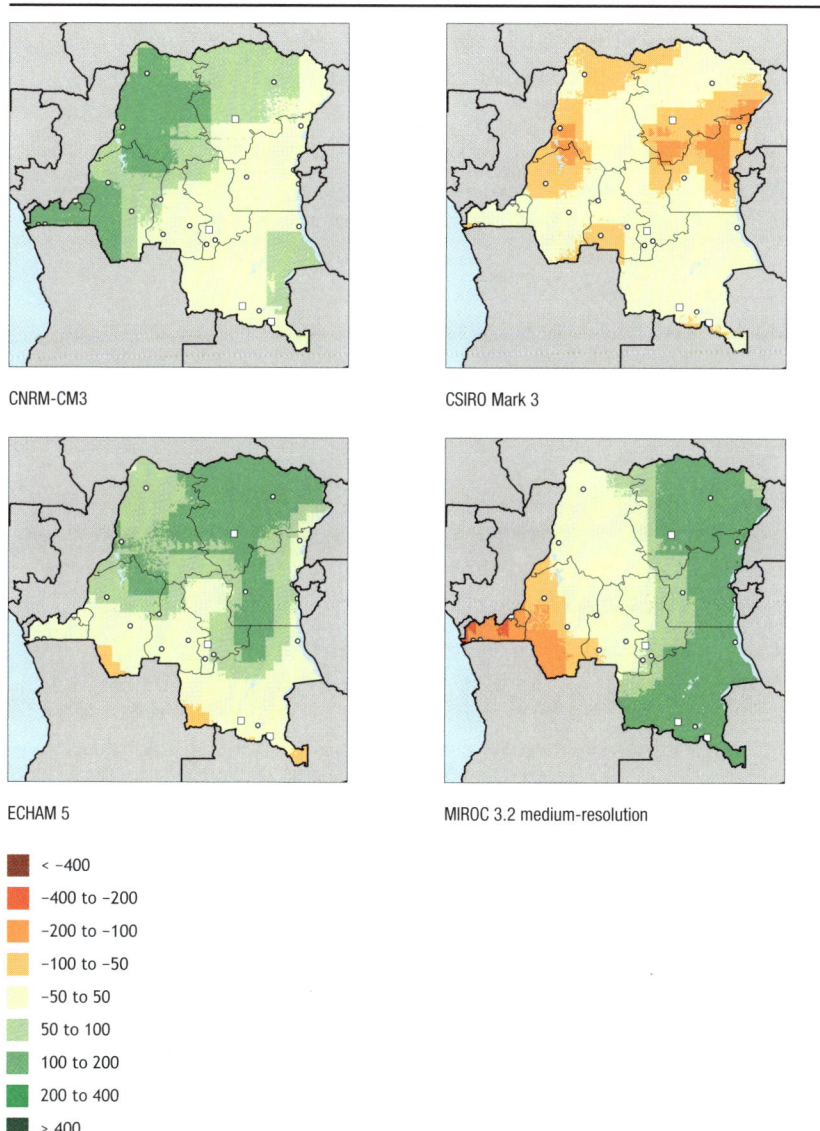

CNRM-CM3

CSIRO Mark 3

ECHAM 5

MIROC 3.2 medium-resolution

- < –400
- –400 to –200
- –200 to –100
- –100 to –50
- –50 to 50
- 50 to 100
- 100 to 200
- 200 to 400
- > 400

Source: Authors' calculations based on Jones, Thornton, and Heinke (2009).

Notes: A1B = greenhouse gas emissions scenario that assumes fast economic growth, a population that peaks midcentury, and the development of new and efficient technologies, along with a balanced use of energy sources; CNRM-CM3 = National Meteorological Research Center–Climate Model 3; CSIRO = climate model developed at the Australia Commonwealth Scientific and Industrial Research Organisation; ECHAM 5 = fifth-generation climate model developed at the Max Planck Institute for Meteorology (Hamburg); GCM = general circulation model; MIROC = Model for Interdisciplinary Research on Climate, developed by the University of Tokyo Center for Climate System Research.

its simulations match better with others in the country (RDC, Ministère de l'Environnement, Conservation de la Nature, Eaux, et Forêts 2001).

Figure 4.15 shows changes in mean daily maximum temperature for the warmest month. The model predictions of temperature changes to 2050 are more consistent, although with varying intensities. Temperatures in DRC will rise by 1° to a maximum of 3°C. The CSIRO model appears to predict the coolest future, with the temperature for most of the country rising by only 1.5°C. MIROC 3.2 appears to predict only slightly warmer temperatures. However, in the CNRM-CM3 and ECHAM 5 models most areas have changes greater than 2°C, and in the MIROC 3.2 model some areas in the extreme south approach 3°C warmer. The projections of CSIRO Mark 3 seems to be most similar to the projections in other studies of the country (RDC, Ministère de l'Environnement, Conservation de la Nature, Eaux, et Forêts 2001).

Crop Models

The Decision Support Software for Agrotechnology Transfer (DSSAT) crop model was used to analyze current and future crop yields under temperature and precipitation changes in the climate of DRC from 2000 to 2050. The results from comparing the mean yields for the two years are presented in Figures 4.16 and 4.17. Considering the four scenarios for maize and rice, it appears that with the changes in the climate, there will be yield losses in some locales; however, these seem to be offset by increases in yields in other parts of the country. It is expected that with the increase in temperatures, incidences of pests and diseases will intensify. The DSSAT model did not take that into consideration; on the other hand, we intentionally ignored the impact of increased CO_2 in the atmosphere on crop productivity in order to compensate for increases in pests and diseases. To that extent, the results still hold. In Figure 4.16, we see surprising agreement between the GCMs that maize yields will generally increase in Bandundu Province and parts of Kasai-Occidental but will mostly decrease elsewhere. In Figure 4.17 we also see surprising agreement between the models in terms of yield increases for rice in Bandundu. We see further agreement in terms of yield increases in the eastern portion of Kivu Province.

Vulnerability

Figure 4.18 shows the impact of future GDP and population scenarios on the number of malnourished children under age five. Figure 4.19 shows the share of children who are malnourished. The wars over the past 15–20 years in DRC and neighboring countries have contributed to an increase in incidences of abandoned children, refugee invasion, and insecurity, especially among women who abandoned farming, thereby exacerbating nutritional deficiencies in their

FIGURE 4.15   Changes in monthly mean maximum daily temperature in Democratic Republic of Congo for the warmest month, 2000–2050, A1B scenario (°C)

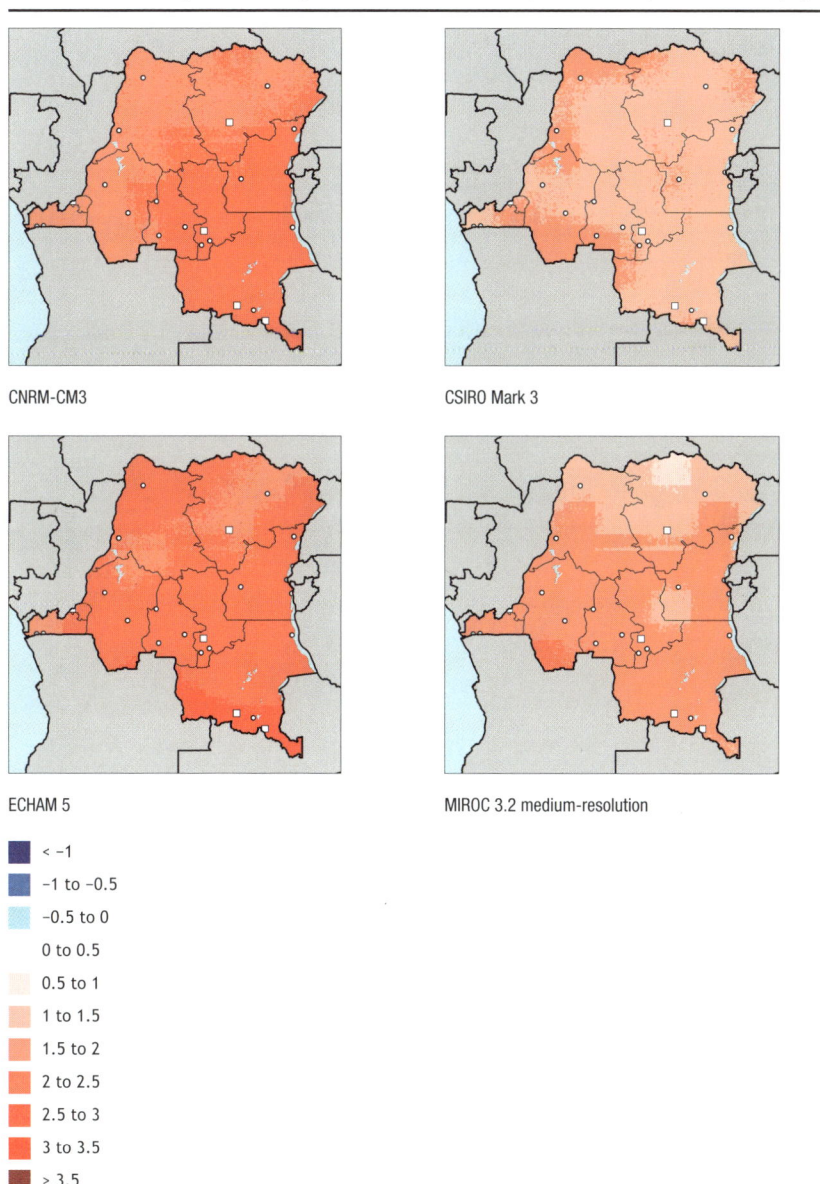

CNRM-CM3

CSIRO Mark 3

ECHAM 5

MIROC 3.2 medium-resolution

■	< −1
■	−1 to −0.5
■	−0.5 to 0
	0 to 0.5
	0.5 to 1
	1 to 1.5
	1.5 to 2
	2 to 2.5
	2.5 to 3
	3 to 3.5
■	> 3.5

Source: Authors' calculations based on Jones, Thornton, and Heinke (2009).

Notes: A1B = greenhouse gas emissions scenario that assumes fast economic growth, a population that peaks midcentury, and the development of new and efficient technologies, along with a balanced use of energy sources; CNRM-CM3 = National Meteorological Research Center–Climate Model 3; CSIRO = climate model developed at the Australia Commonwealth Scientific and Industrial Research Organisation; ECHAM 5 = fifth-generation climate model developed at the Max Planck Institute for Meteorology (Hamburg); GCM = general circulation model; MIROC = Model for Interdisciplinary Research on Climate, developed by the University of Tokyo Center for Climate System Research.

FIGURE 4.16   Yield change under climate change: Rainfed maize in Democratic Republic of Congo, 2000–2050, A1B scenario

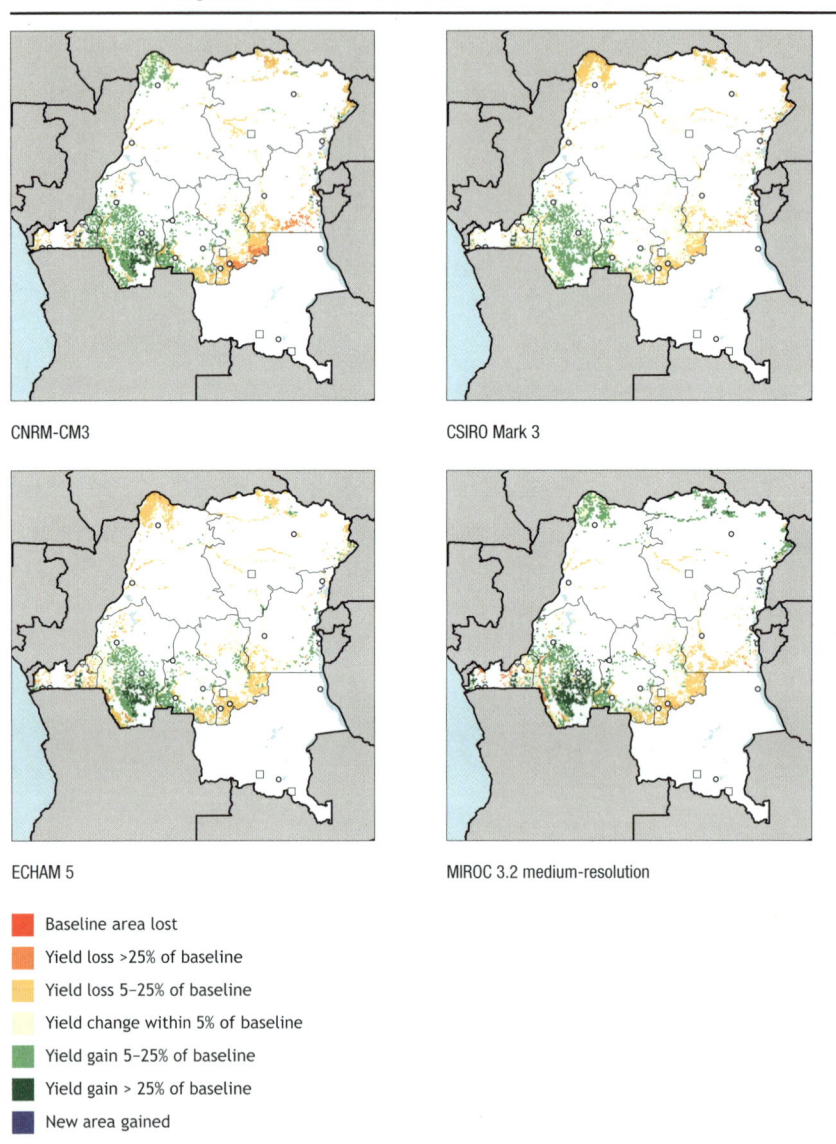

CNRM-CM3

CSIRO Mark 3

ECHAM 5

MIROC 3.2 medium-resolution

■ Baseline area lost
■ Yield loss >25% of baseline
■ Yield loss 5–25% of baseline
□ Yield change within 5% of baseline
■ Yield gain 5–25% of baseline
■ Yield gain > 25% of baseline
■ New area gained

Source: Authors' calculations.

Notes: A1B = greenhouse gas emissions scenario that assumes fast economic growth, a population that peaks midcentury, and the development of new and efficient technologies, along with a balanced use of energy sources; CNRM-CM3 = National Meteorological Research Center–Climate Model 3; CSIRO = climate model developed at the Australia Commonwealth Scientific and Industrial Research Organisation; ECHAM 5 = fifth-generation climate model developed at the Max Planck Institute for Meteorology (Hamburg); GCM = general circulation model; MIROC = Model for Interdisciplinary Research on Climate, developed by the University of Tokyo Center for Climate System Research.

FIGURE 4.17   Yield change under climate change: Rainfed rice in Democratic Republic of Congo, 2000–2050, A1B scenario

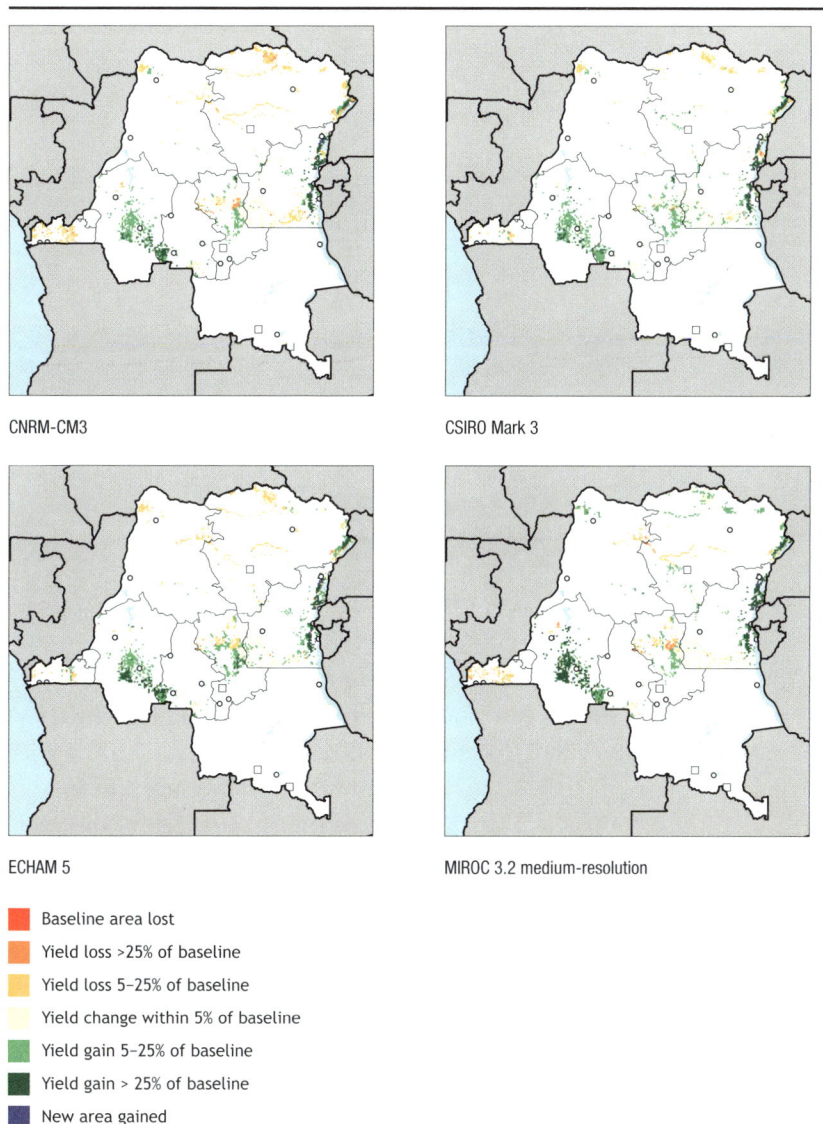

CNRM-CM3

CSIRO Mark 3

ECHAM 5

MIROC 3.2 medium-resolution

- Baseline area lost
- Yield loss >25% of baseline
- Yield loss 5–25% of baseline
- Yield change within 5% of baseline
- Yield gain 5–25% of baseline
- Yield gain > 25% of baseline
- New area gained

Source: Authors' calculations.

Notes: A1B = greenhouse gas emissions scenario that assumes fast economic growth, a population that peaks midcentury, and the development of new and efficient technologies, along with a balanced use of energy sources; CNRM-CM3 = National Meteorological Research Center–Climate Model 3; CSIRO = climate model developed at the Australia Commonwealth Scientific and Industrial Research Organisation; ECHAM 5 = fifth-generation climate model developed at the Max Planck Institute for Meteorology (Hamburg); GCM = general circulation model; MIROC = Model for Interdisciplinary Research on Climate, developed by the University of Tokyo Center for Climate System Research.

FIGURE 4.18   Number of malnourished children under five years of age in Democratic Republic of Congo in multiple income and climate scenarios, 2010–2050

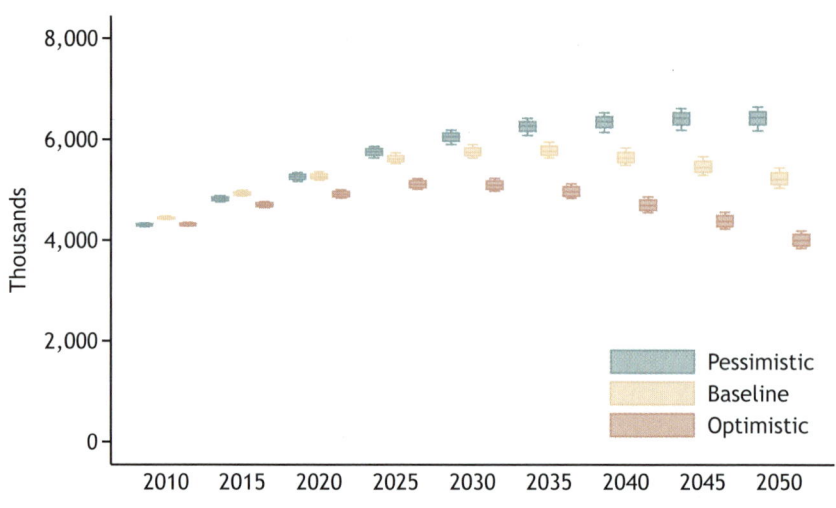

Source: Based on analysis conducted for Nelson et al. (2010).

Note: The box and whiskers plot for each socioeconomic scenario shows the range of effects from the four future climate scenarios.

children under age five. All scenarios show an increasing trend until around 2025–2030. The percentage of malnourished children increased from 12 to 16 percent around 2006 (RDC, Ministère du Plan, 2006); if there is no improvement in food security, this tendency will confirm the pessimistic scenario. It may be helpful to keep in mind that although the number of malnourished children is projected to be larger in 2050 than in 2010 in the baseline scenario, with population growth this larger figure nevertheless represents a slight decline in the rate of malnutrition among children under age five except in the pessimistic scenario.

Figure 4.20 shows the kilocalories per capita available to each person in DRC. We see a relatively large improvement in calorie consumption in the optimistic scenario, though the baseline remains largely unchanged, and in the pessimistic scenario we note a decline. Although incomes rise in all scenarios, so do food prices, especially those for staple foods in the pessimistic scenario—though they do not rise as much as income. The negative price effects on consumption offset some of the positive income effects in all models and scenarios. However, the decline in calorie consumption noted in the pessimistic scenario likely also reflects the fact that in the International Model for Policy Analysis of Agricultural Commodities and Trade (IMPACT), own-price elasticities of staples were set a little too high relative to the income elasticities, which caused the price effect to exceed the income effect, a result we would not normally expect.

FIGURE 4.19 Share of malnourished children under five years of age in Democratic Republic of Congo in multiple income and climate scenarios, 2010–2050

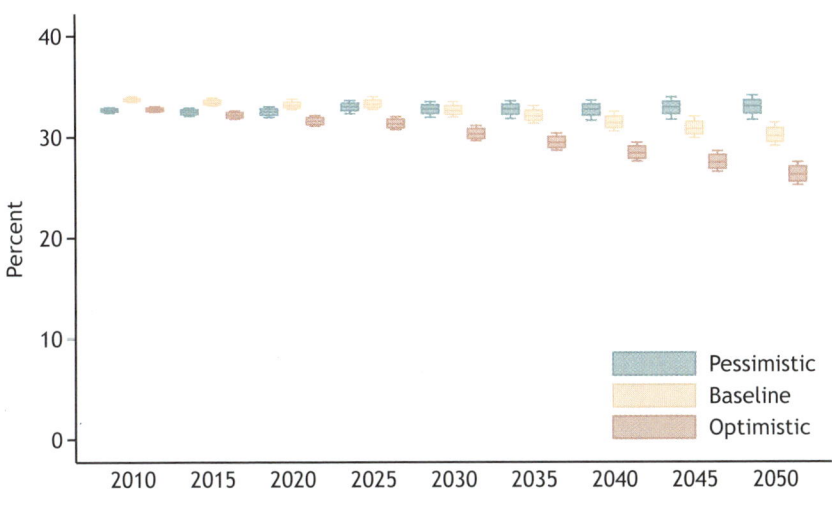

Source: Based on analysis conducted for Nelson et al. (2010).

Note: The box and whiskers plot for each socioeconomic scenario shows the range of effects from the four future climate scenarios.

FIGURE 4.20 Kilocalories per capita in Democratic Republic of Congo in multiple income and climate scenarios, 2010–2050

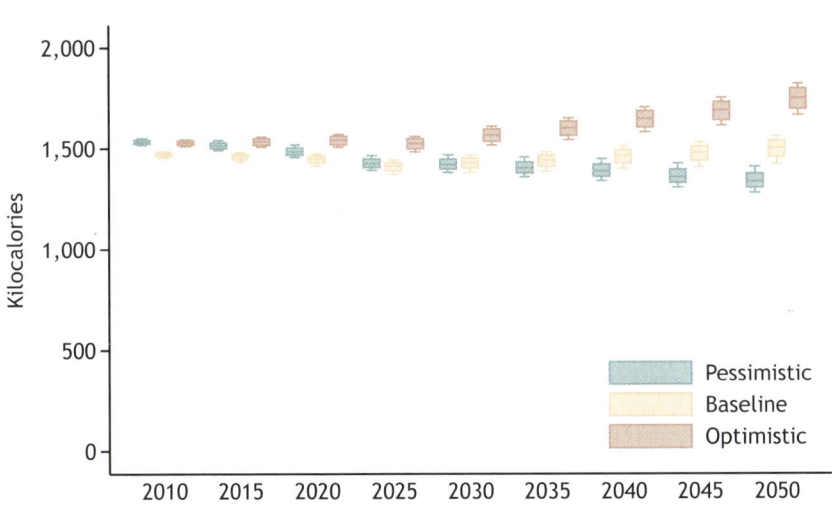

Source: Based on analysis conducted for Nelson et al. (2010).

Note: The box and whiskers plot for each socioeconomic scenario shows the range of effects from the four future climate scenarios.

FIGURE 4.21   Impact of changes in GDP and population on cassava in Democratic Republic of Congo, 2010–2050

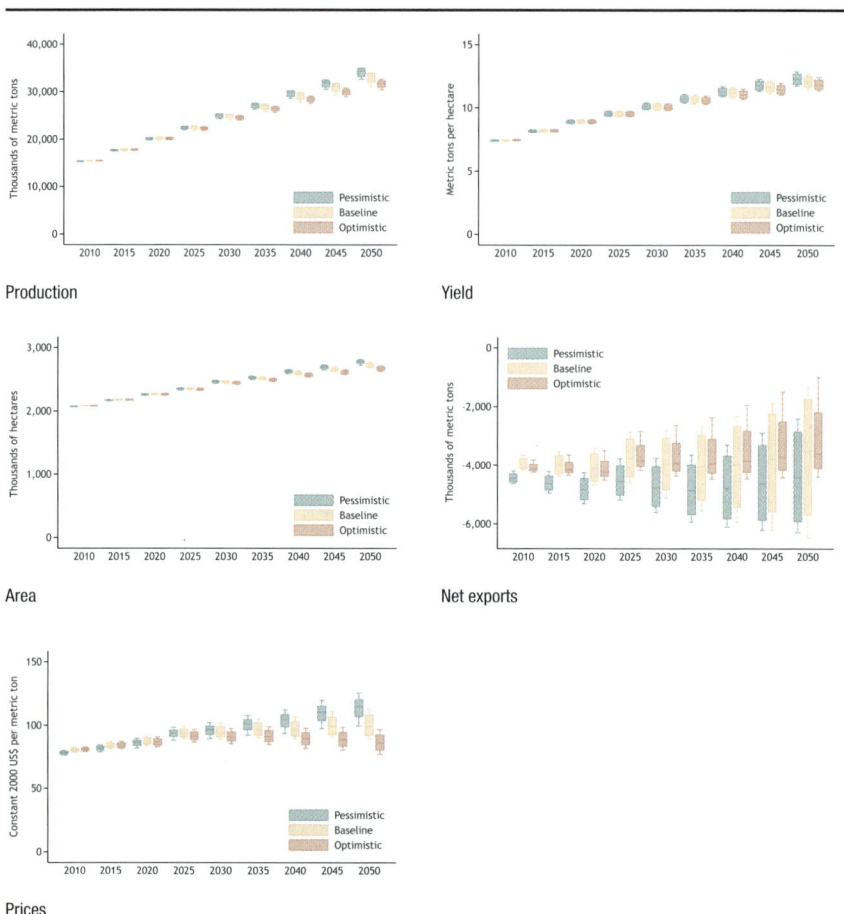

Production

Yield

Area

Net exports

Prices

Source: Based on analysis conducted for Nelson et al. (2010).

Notes: The box and whiskers plot for each socioeconomic scenario shows the range of effects from the four future climate scenarios. GDP = gross domestic product; US$ = US dollars.

Agricultural Outcomes

Figures 4.21–4.24 show simulation results from the IMPACT model associated with key agricultural crops in DRC. Each featured crop has five graphs, one each showing production, yield, area, net exports, and world price.

Cassava is the leading crop in the country and is the pillar of its food security. Cassava production is projected to increase due to yield gains and the allocation of more land to its production (see Figure 4.21). However, the model predicts that the country will need to continue to import cassava through

FIGURE 4.22   Impact of changes in GDP and population on maize in Democratic Republic of Congo, 2010–2050

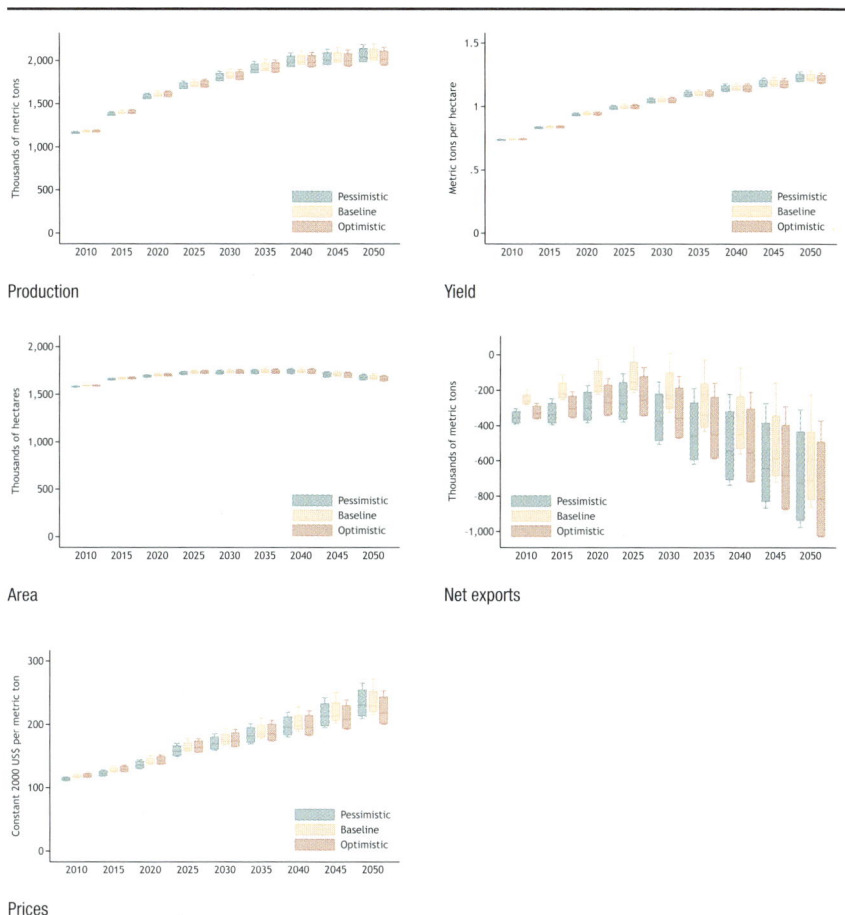

Production

Yield

Area

Net exports

Prices

Source: Based on analysis conducted for Nelson et al. (2010).

Notes: The box and whiskers plot for each socioeconomic scenario shows the range of effects from the four future climate scenarios. GDP = gross domestic product; US$ = US dollars.

2050 to meet the demand. The rising demand for cassava is mostly due to a high rate of population growth, and it may be that domestic supply will continue to be constrained by poor postharvest technologies.

Maize is among the major cereals consumed in the country. It is produced mainly by small farmers in mixed cropping systems. The IMPACT model projects that yields will increase by around 70 percent, reaching around 1.2 tons per hectare, which is still relatively low (see Figure 4.22). As in the case of cassava, domestic demand for maize is projected to outstrip domestic

FIGURE 4.23 Impact of changes in GDP and population on groundnuts in Democratic Republic of Congo, 2010–2050

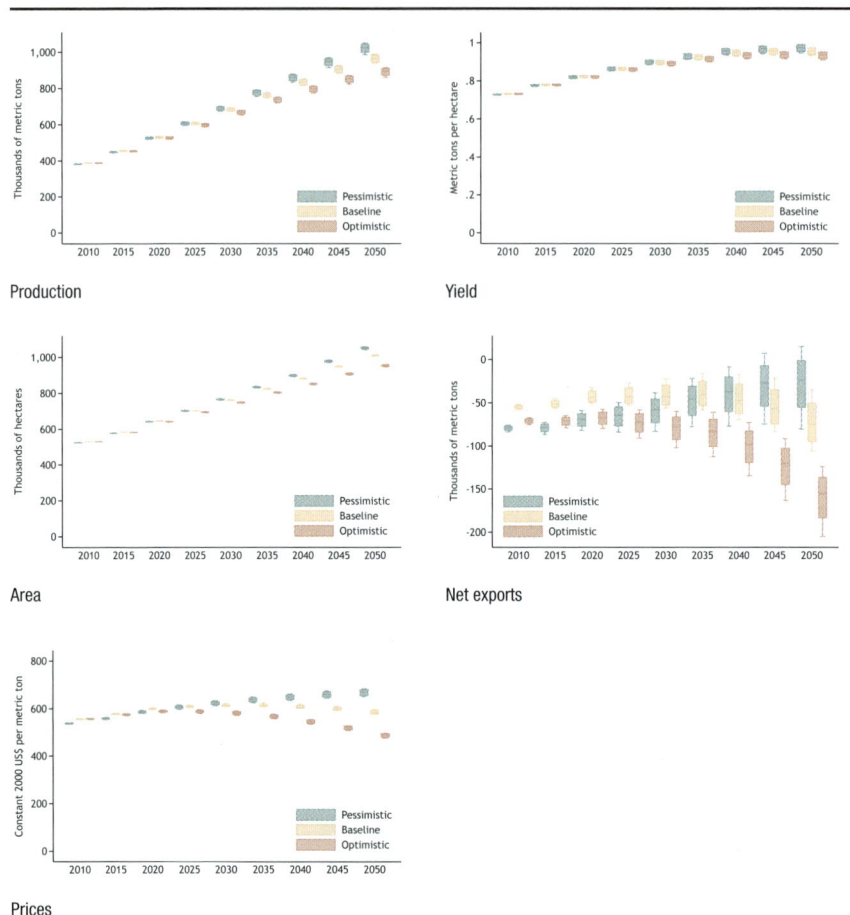

Source: Based on analysis conducted for Nelson et al. (2010).

Notes: The box and whiskers plot for each socioeconomic scenario shows the range of effects from the four future climate scenarios. GDP = gross domestic product; US$ = US dollars.

production, though the model predicts an improvement until 2025, after which imports will grow again.

The rate of production of groundnuts is currently low because of low soil fertility, the particular varieties used, and lack of fertilizers. The IMPACT model suggests that there will be moderate improvements in yield through 2050, but with a near-doubling of harvested area, production should more than double (see Figure 4.23). There is uncertainty between GDP and population scenarios and between climate models as to the direction in which trade

FIGURE 4.24   Impact of changes in GDP and population on rice in Democratic Republic of Congo, 2010–2050

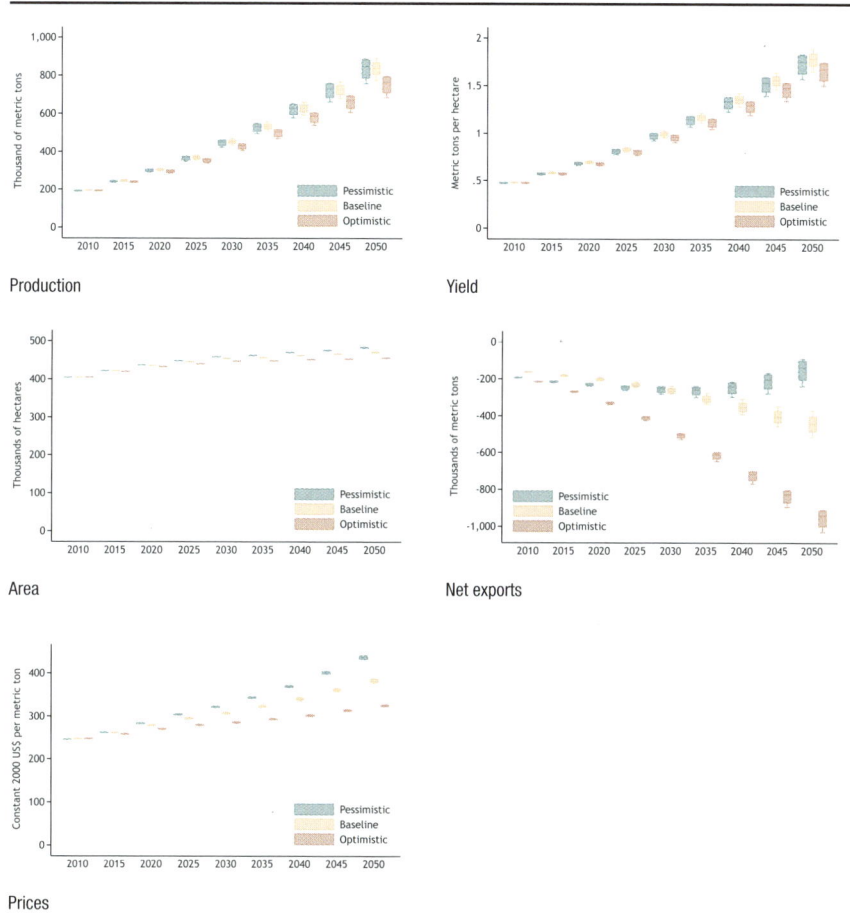

Source: Based on analysis conducted for Nelson et al. (2010).

Notes: The box and whiskers plot for each socioeconomic scenario shows the range of effects from the four future climate scenarios. GDP = gross domestic product; US$ = US dollars.

will change, but it appears that under almost every case tested, DRC will likely be an importer of groundnuts through 2050.

Although the potential of rice production is high, all rice consumed in DRC is currently imported. The IMPACT model used older FAOSTAT data that reported lower yields than are currently shown in FAOSTAT (FAO 2010)—which suggests that around 800 kilograms per hectare is the average yield—and lower than those reported by the national agricultural statistics service (RDC, Ministère de l'Agriculture 2009), which gives yields of between 625 and

1,500 kilograms per hectare for smallholder farmers. In any case, IMPACT suggests that rice yields could more than triple between now and 2050 (see Figure 4.24). Although it projects that the area sown with rice will grow only slightly, we see that the total production of rice could quadruple by 2050. Such an increase still may fall short of keeping pace with domestic demand. Efforts in opening access to roads and markets and making new technologies and better management practices available to small producers will be needed to satisfy local demand and leave surpluses for exports.

Given that yields are projected to improve in response to climate change for all key crops, significant investments are required in intensification and knowledge creation for better agricultural production to feed the country. Also required is a strategy for opening up more land to sustainable agricultural production to ensure adequate supplies.

Conclusions and Policy Recommendations

This chapter assessed the vulnerability of agriculture to climate change in DRC. The purpose of the chapter is to help policymakers and researchers better understand and anticipate the likely impacts of climate change on agriculture and on vulnerable households in DRC. The chapter reviews current data on agriculture and economic development, models anticipated changes in climate between now and 2050, applies crop modeling to assess the impact of climate changes on agricultural production, and models global supply and demand for food to predict relevant food price trends.

Based on existing data and future projections, temperatures will increase by 1°–3°C by 2050. The direction of precipitation change is uncertain given that three of our four models show increases of 50–200 millimeters per year but in varying locations, while the other model shows less rainfall in large portions of the country. But the models are not geographically consistent: even the three models that generally predict increased rainfall in portions of the country disagree on which portions. Each indicates that many areas will not experience significant change. Even one of the models that shows increased rainfall also suggests that part of the country will have much less rainfall.

Concerning the response of major crops, the crop models show mixed responses, though probably more areas will have higher yields of maize and rice than will show lower yields. The IMPACT model predicts increased yields; however, these will generally not be enough to match the domestic demands.

The IMPACT model shows expansion of cultivated area for some crops. Expansion of cultivated land could be detrimental to natural resources and, more specifically, could lead to the cutting down tropical forest. This points to the need for policymakers to decide how much of the forests, if any, should be allowed to be converted to agricultural land. One option would be for the extensive forest resource to be conserved and used for carbon sequestration and carbon trade under the carbon fund initiative.

Because the majority of DRC's citizens rely on rainfed agriculture and are likely to continue to do so in the future, in the face of climate variability and change, decisionmakers and researchers need to work together to ameliorate the challenges ahead. Major opportunities to be considered include encouraging settlement agriculture, improving the productivity of arable land by adopting high-yielding crop varieties, providing crop protection, and improving postharvest techniques. It is also important to consider more environmentally friendly agricultural techniques such as agroforestry.

Further, there is a need for attention to interactions between soils, livestock, and crop production and agroecosystems, not only because of their important roles in agriculture but also to minimize the potential contribution to climate change of methane emissions.

All of these options seem to suggest a need for increased funding to strengthen the agricultural research and extension system so that technologies can be tested and adapted to the local environment and this knowledge passed on to farmers.

References

Bartholome, E., and A. S. Belward. 2005. "GLC2000: A New Approach to Global Land Cover Mapping from Earth Observation Data." *International Journal of Remote Sensing* 26 (9–10): 1959–1977.

Bhaduri, B., E. Bright, P. Coleman, and M. Urban. 2007. "LandScan USA: A High-Resolution Geospatial and Temporal Modeling Approach for Population Distribution and Dynamics." *GeoJournal* 69: 103–117.

CIESIN (Center for International Earth Science Information Network), Columbia University, IFPRI (International Food Policy Research Institute), World Bank, and CIAT (Centro Internacional de Agricultura Tropical). 2004. *Global Rural–Urban Mapping Project (GRUMP), Alpha Version: Population Density Grids.* Palisades, NY, US: Socioeconomic Data and Applications Center (SEDAC), Columbia University. http://sedac.ciesin.columbia.edu/gpw.

FAO (Food and Agriculture Organization of the United Nations). 2010. FAOSTAT. Rome. http://faostat.fao.org.

Griffith, D., L. Kelly-Hope, and M. Miller. 2006. "Review of Reported Cholera Outbreaks Worldwide, 1995–2005." *American Journal of Tropical Medicine and Hygiene* 75 (5): 973–977.

Jones, P. G., P. K. Thornton, and J. Heinke. 2009. *Generating Characteristic Daily Weather Data Using Downscaled Climate Model Data from the IPCC's Fourth Assessment.* Project report for the International Livestock Research Institute. Geneva: International Panel on Climate Change.

Lehner, B., and P. Döll. 2004. "Development and Validation of a Global Database of Lakes, Reservoirs, and Wetlands." *Journal of Hydrology* 296 (1–4): 1–22.

Millennium Ecosystem Assessment. 2005. *Ecosystems and Human Well-being: Synthesis.* Washington, DC: Island Press. www.maweb.org/en/Global.aspx.

Minten, B., and S. Kyle. 1999. "The Effect of Distance and Road Quality on Food Collection, Marketing Margins, and Traders' Wages: Evidence from the Former Zaire." *Journal of Development Economics* 60 (2): 467–495.

Nelson, G. C., M. W. Rosegrant, A. Palazzo, I. Gray, C. Ingersoll, R. Robertson, S. Tokgoz, et. al. 2010. *Food Security, Farming, and Climate Change to 2050: Scenarios, Results, Policy Options.* IFPRI Research Monograph. Washington, DC: International Food Policy Research Institute.

Nkem, J., M. Idinoba, and C. Sendashonga. 2008. *Forests for Climate Change Adaptation in the Congo Basin: Responding to an Urgent Need with Sustainable Practice.* CIFOR Environmental Brief 2. Bogor, Indonesia: Center for International Forestry Research.

Ntombi, K., and S. Pangu. 2011. "Evolution de la pluviosité en république démocratique du Congo et stratégies d'adaptation agro climatiques." Paper presented at the 2011 Colloque Climatologie intertropicale, Libreville/Gabon.

Nzongola-Ntalaja, G. 2004. "The International Dimensions of the Congo Crisis." *Global Dialogue* 6 (3–4): 116–126.

Paaijmans, K. P., S. Blanford, A. S. Bell, J. Blanford, A. F. Read, and B. T. Thomas. 2008. "Influence of Climate on Malaria Transmission Depends on Daily Temperature Variation." *Proceedings of the National Academy of Sciences U.S.A.* 107 (34): 15135–15139.

RDC (République Démocratique du Congo), Ministère de l'Agriculture. 2009. *Service national des statistiques agricole.* Rapport 2009. Kinshasa.

RDC, Ministère de l'Environnement, Conservation de la Nature et Tourisme. 2009. *Seconde Communication Nationale à la Convention Cadre sur les Changements Climatiques/RDC.* Kinshasa.

RDC, Ministère de l'Environnement, Conservation de la Nature, Eaux, et Forêts. 2001. *La communication nationale initiale sur les changements climatiques/RDC.* Kinshasa.

RDC, Ministère du Plan. 2006. *DSRP (Document Stratégique pour la Réduction de la Pauvreté).* Kinshasa.

Ulimwengu, J., J. Funes, D. Headey, and L. You. 2009. "Paving the Way for Development: The Impact of Road Infrastructure on Agricultural Production and Household Wealth in the Democratic Republic of Congo." Paper presented at the Annual Meeting of the Agriculture and Applied Economics Association, Milwaukee, WI, US, July 26–28. No longer available online.

UNEP (United Nations Environment Programme) and IUCN (International Union for Conservation of Nature). 2009. World Database on Protected Areas (WDPA) Annual Release 2009. No longer available online.

UNPOP (United Nations Secretariat, Department of Economic and Social Affairs, Population Division). 2009. *World Population Prospects: The 2008 Revision.* New York. http://esa.un.org/unpp.

Van Minnen, J. G., J. Alcamo, and W. Haupt. 2000. "Deriving and Applying Response Surface Diagrams for Evaluating Climate Change Impacts on Crop Production." *Climatic Change* 46 (3): 317–338.

Wingqvis, G. Ö. 2008. *Democratic Republic of Congo.* Environmental and Climate Change Policy Brief. Gothenburg, Germany: University of Gothenburg, School of Business and Law.

World Bank. 2009. *World Development Indicators.* Accessed May 2011. http://data.worldbank.org/data-catalog/world-development-indicators.

———. 2010. *Economics of Adaptation to Climate Change: Synthesis Report.* Washington, DC. http://climatechange.worldbank.org/content/economics-adaptation-climate-change-study-homepage.

You, L., and S. Wood. 2006. "An Entropy Approach to Spatial Disaggregation of Agricultural Production." *Agricultural Systems* 90 (1–3): 329–347.

You, L., S. Wood, and U. Wood-Sichra. 2006. "Generating Global Crop Distribution Maps: From Census to Grid." Paper presented at the International Atomic Energy Agency Conference, Brisbane, Australia, August 11–18.

———. 2009. "Generating Plausible Crop Distribution and Performance Maps for Sub-Saharan Africa Using a Spatially Disaggregated Data Fusion and Optimization Approach." *Agricultural Systems* 99 (2–3): 126–140.

Chapter 5

ERITREA

Bissrat Ghebru, Woldeamlak Araia, Woldeselassie Ogbazghi,
Menghisteab Gebreselassie, and Timothy S. Thomas

Eritrea is located in the Horn of Africa, with a long coastline of about 1,200 kilometers along the Red Sea. The country has varied topography, rainfall, and climate, with altitudes ranging from 60 meters to more than 3,000 meters above sea level (Eritrea 2004b). As one of the arid or semiarid Sahelian countries of Africa, Eritrea faced serious droughts in 1975, 1984, 1985, 1989, and 1991.

Eritrea has a large reserve of gold as well as substantial reserves of potash, zinc, copper, silver, marble, barite, feldspar, kaolin, and rock salt. With the recent licensing of mining concessions, the mining sector is expected to become a significant factor driving the economy. The Bisha mining site, for example, is expected to yield 1.06 million ounces of gold, 10 million ounces of silver, 747 million pounds of copper, and more than 1 billion pounds of zinc over a three-year period beginning in 2011 (ADB 2009).

Eritrea has a number of agricultural systems: rainfed cereal and pulses, semicommercial and periurban agriculture, small-scale irrigated horticulture, commercial farming, agropastoral rainfed farming, and agropastoral spate irrigation systems. The National Development Plan identifies three priority areas in the short to medium term: food security and development of cash crops, physical and social infrastructure, and human capital development.

The climate of Eritrea ranges from hot arid (adjacent to the Red Sea) to temperate subhumid in isolated microcatchments in the eastern escarpment. The Central Highlands part of the country enjoys a semiarid climate, with the lowest temperatures during December and January. The Western Lowlands of the country are relatively hot, especially between April and June, with temperatures reaching 38°C. For the coastal areas, the highest temperatures are between 25° and 40°C (June to August). The country's rainfall intensity is very high—with substantial precipitation within a limited period—resulting in soil erosion and runoff. The annual rainfall in the subhumid zone in the eastern escarpment may reach 1,000 millimeters, while the southern Red Sea region and the northwestern parts of the country receive less than 200 millimeters.

The total rainfall seems to be sufficient for crop production in most parts of the country (400–600 millimeters annually) (Eritrea, Ministry of National Development 2006).

Eritrea has adopted various international environmental laws and policy guidelines. The major policies and programs related to climate change are the Environmental Management Plan for Eritrea, the National Action Program to Combat Desertification and Mitigate Effects of Drought, the National Biodiversity Strategy and Action Plan for Eritrea, Eritrea's Initial National Communication on Climate Change, a Country Assessment Report on Sustainable Development, an Interim Poverty Reduction Strategy Paper, National Environmental Assessment Procedures and Guidelines, Fisheries Legislation and a Strategic Environmental Assessment, a Food Security Strategy for Eritrea, a land use policy, and a coastal areas management policy.

The government has indicated that agriculture is a top priority for economic development (ADB 2009). This will require sustainable land and water management, human resource development, and an integrated rural development package.

Review of the Current Situation and Trends

Economic and Demographic Indicators

Population
The total population of Eritrea has grown from 750,000 in 1943 to a current population of about 5.27 million. Geographically, 50–60 percent of the people live in the highlands, which comprise about 10 percent of the country's total area (Eritrea, Ministry of National Development 2006).

Figure 5.1 shows the country's total and rural population (left axis) as well as the share of its urban population (right axis). The urban population is increasing due to migration from rural areas as well as due to refugees returning to Eritrea. Table 5.1 provides additional information concerning rates of population growth. In general, all population growth rates have been increasing throughout the decades except for the 1980s to 1999, which saw a slowing in population growth. This drop was associated with the war, which led to the deaths and the migration of people away from the country during that period.

Figure 5.2 shows the geographic distribution of the population within Eritrea. The population is concentrated mainly in the central and southern

FIGURE 5.1   Population trends in Eritrea: Total population, rural population, and percent urban, 1960–2008

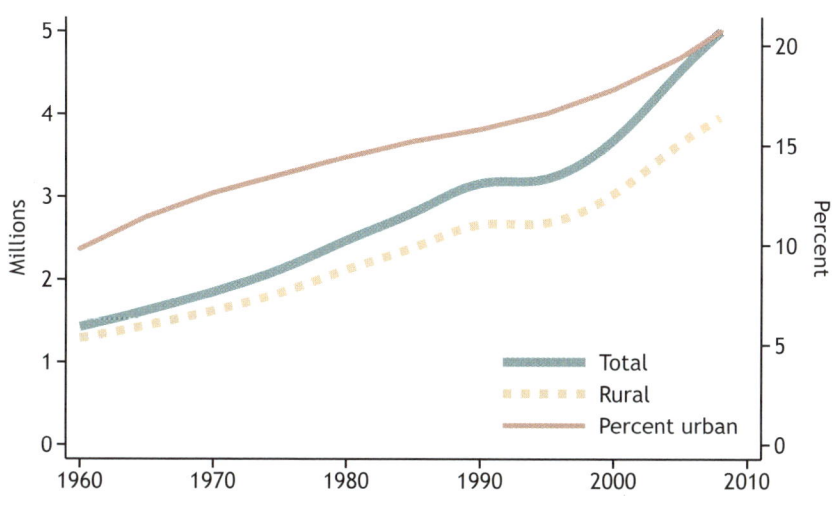

Source: *World Development Indicators* (World Bank 2009).

provinces. The Western Lowlands, a potentially productive agricultural region, has a growing population as well. The population appears to be more densely distributed in areas where the amount of rainfall allows rainfed agriculture (the Highlands and Southwestern Lowlands).

Income

Eritrea is a low-income country. More than 70 percent of the population depends on traditional subsistence agriculture for their livelihood. The main sources of income for rural households are the sale of crops, livestock, and

TABLE 5.1   Population growth rates in Eritrea, 1960–2008 (percent)

Decade	Total growth rate	Rural growth rate	Urban growth rate
1960–1969	2.6	2.3	5.2
1970–1979	2.9	2.7	4.2
1980–1989	2.6	2.4	3.5
1990–1999	1.2	0.9	2.3
2000–2008	3.9	3.4	5.7

Source: Authors' calculations based on *World Development Indicators* (World Bank 2009).

FIGURE 5.2   Population distribution in Eritrea, 2000 (persons per square kilometer)

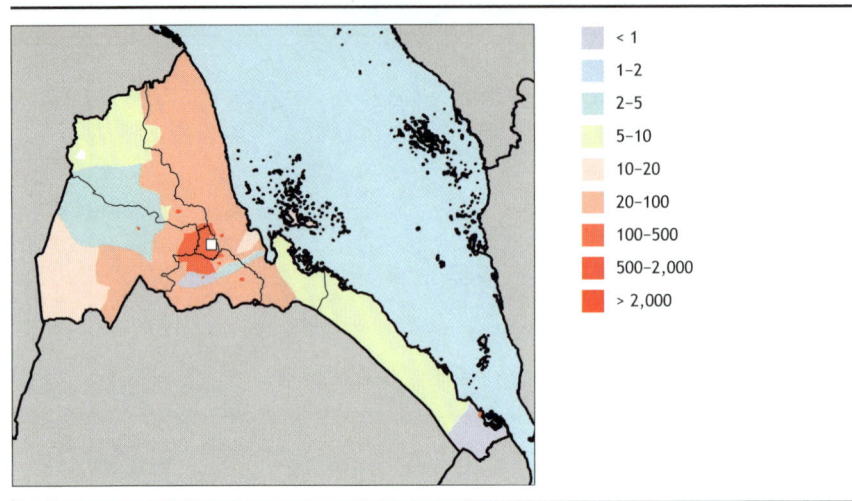

	< 1
	1–2
	2–5
	5–10
	10–20
	20–100
	100–500
	500–2,000
	> 2,000

Source: CIESIN et al. (2004).

livestock products; wages for daily labor; and remittances. In urban areas, people generate income from wage labor, small businesses, petty trade, and poultry farming.

About 66.4 percent of the population is poor, and 37 percent are classified as extremely poor. Urban areas have slightly higher poverty rates (70.3 percent) than rural areas (64.6 percent) but lower rates of extreme poverty (32.7 percent as compared to 38.9 percent) (Eritrea 2004a). In 2004, on average, about 40 percent of household income was obtained from wages and 15 percent from agricultural produce. On the average, nonagricultural activities provided only 5 percent of income. Loans and sale of household assets comprised about 17 percent (Eritrea 2004b). This indicates that the percentage of income obtained from agriculture is still very low.

Figure 5.3 shows a dramatic increase in gross domestic product (GDP) from 1993 to around 1998, when the country registered an annual growth rate of 7 percent, driven mainly by fiscal consolidation following independence from Ethiopia. Thereafter, the trend shows an equally dramatic decrease due to the border conflict with Ethiopia, which lasted from 1998 to 2000. Currently the contribution of the agricultural sector to the country's GDP is not more than 24 percent, low compared to other sectors. Agricultural development has been hampered by drought, poor farm inputs, and inadequate crop and livestock management systems.

FIGURE 5.3   Per capita GDP in Eritrea (constant 2000 US$) and share of GDP from agriculture (percent), 1992–2008

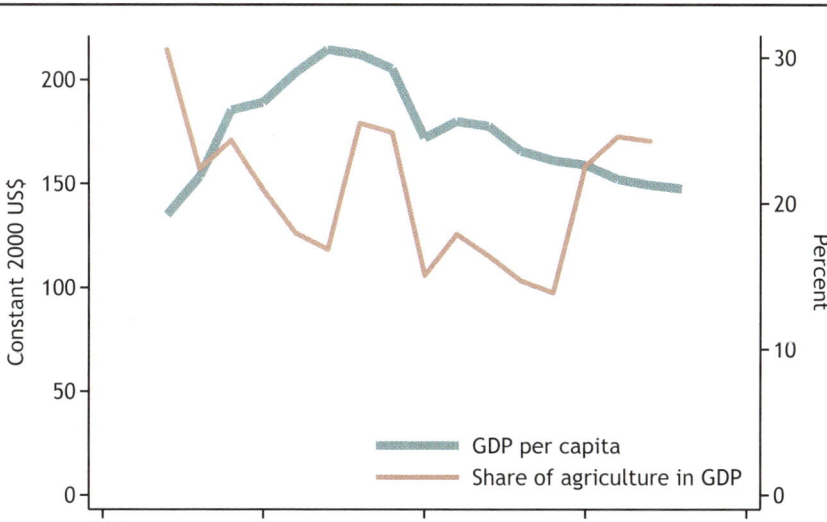

Source: *World Development Indicators* (World Bank 2009).
Note: GDP = gross domestic product; US$ = US dollars.

Vulnerability to Climate Change

Table 5.2 provides some data for Eritrea on indicators of the population's vulnerability and resiliency to economic and environmental shocks: the level of education and literacy and the concentration of labor in poorer or less dynamic sectors. Enrollment in secondary school increased from 12.2 percent in 1991/1992 to 29.2 percent in 2007. Enrollment in primary school similarly

TABLE 5.2   Education and labor statistics for Eritrea, 1990s and 2000s

Indicator	Year	Percent
Primary school enrollment: Percent gross (three-year average)	2007	54.9
Secondary school enrollment: Percent gross (three-year average)	2007	29.2
Adult literacy rate	2002	52.5
Percentage employed in agriculture	2005	52.5
Percentage with vulnerable employment (employed in agriculture on own farms or as day laborers)	1996	7.7
Under-five malnutrition (weight for age)	2002	34.5

Source: *World Development Indicators* (World Bank 2009).

FIGURE 5.4   Well-being indicators in Eritrea, 1960–2008

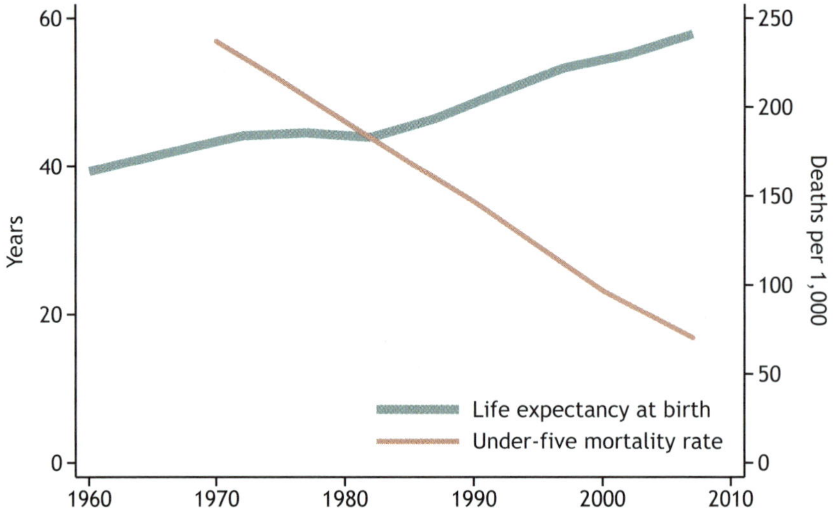

Source: *World Development Indicators* (World Bank 2009).

increased, from 36.3 percent in 1991/1992 to 54.9 percent in 2007. The adult literacy rates also increased, from 38.6 to 52.5 percent (World Bank 2009).

Figure 5.4 shows two noneconomic correlates of poverty, life expectancy and under-five mortality. Life expectancy has increased moderately in Eritrea, from 40 years in the 1960s to about 52 in 2010. Child mortality under age five is decreasing, owing to improved mother and child care. With such improvements in health services, effects of climate change on these correlates are likely to be minimal. Climate change alters several environmental conditions that cause morbidity and mortality; for instance, it increases the transmission of malaria. Recently it has been recorded that malaria morbidity has declined by more than 86 percent and mortality due to malaria by more than 82 percent, making Eritrea one of the few countries of Africa south of the Sahara to have met the Abuja targets (ADB 2009).

Eritrea's geographical location, in the arid and semiarid region of Sahelian Africa, makes it vulnerable to the adverse effects of climate change, such as drought, pest infestation, and degradation of natural resources, which can affect food security, and if it does, can eventually lead to malnutrition or undernutrition in children and adults, increasing the probability of child mortality and reduced life expectancy. For instance, during the drought years of 2002 and 2003, the overall food deficit in Eritrea was estimated at 476,000 tons of grain equivalent. The drought also affected the availability of drinking water, which became increasingly scarce (Eritrea 2004b).

The most vulnerable groups are those in high-rainfall zones who depend on rainfed agriculture as their primary livelihood (Eritrea 2004a). Unfortunately, poor households resort to selling productive assets to mitigate the effects of food shortage. Female-headed households are more vulnerable due to their much lower income levels (ADB 2009).

The country has had to rely on imports to meet the food shortfalls. Currently commercial food imports cover 15–20 percent of domestic demand. The maximum import capacity for grains is around 80,000 tons. In addition, cross-border trade with neighboring countries brings an estimated 70,000 tons. International food assistance is used to fill the remaining food deficit (ADB 2009).

Review of Land Use, Potential, and Limitations

Land Use Overview

A comprehensive land use map of Eritrea is under preparation. The limitations of land use in the country are (1) lack of collaborative implementation of the land use policy among various sectors or users, (2) shortage of institutional and professional capacity in land use, and (3) lack of effective implementation of land use plans.

The existing land cover map shows land in the following categories: cultivated land, grazing land, forests, and barren land. Cultivated land accounts for about 7 percent of the total land area. This figure is likely to change, because there is potential for the expansion of agriculture in various agroecological zones. In general, fertile lands are never allocated for human settlement, because agriculture is considered the breadbasket of the communities. Owing to its topography and limited moisture, much of the land of Eritrea is suited for livestock production rather than farming. The woody and nonwoody species are significant for the survival of pastoralists or agropastoralists (Eritrea, Ministry of Agriculture 2002).

Figure 5.5 shows land cover and land use in Eritrea as of 2000. The map depicts the actual situation on the ground in terms of forest cover and agricultural activities. Intensive agriculture is concentrated in the south and southwestern regions. Evergreen vegetation and the relics of afromontane forest are found on the eastern escarpment and in the southern and northern highlands.

FIGURE 5.5   Land cover and land use in Eritrea, 2000

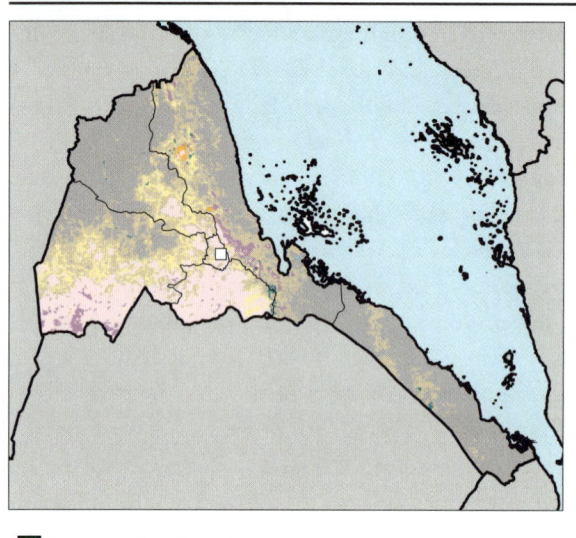

Tree cover, broadleaved, evergreen

Tree cover, broadleaved, deciduous, closed

Tree cover, broadleaved, open

Tree cover, broadleaved, needle-leaved, evergreen

Tree cover, broadleaved, needle-leaved, deciduous

Tree cover, broadleaved, mixed leaf type

Tree cover, broadleaved, regularly flooded, fresh water

Tree cover, broadleaved, regularly flooded, saline water

Mosaic of tree cover/other natural vegetation

Tree cover, burnt

Shrub cover, closed–open, evergreen

Shrub cover, closed–open, deciduous

Herbaceous cover, closed–open

Sparse herbaceous or sparse shrub cover

Regularly flooded shrub or herbaceous cover

Cultivated and managed areas

Mosaic of cropland/tree cover/other natural vegetation

Mosaic of cropland/shrub/grass cover

Bare areas

Water bodies

Snow and ice

Artificial surfaces and associated areas

No data

Source: GLC2000 (Bartholome and Belward 2005).

FIGURE 5.6   Protected areas in Eritrea, 2009

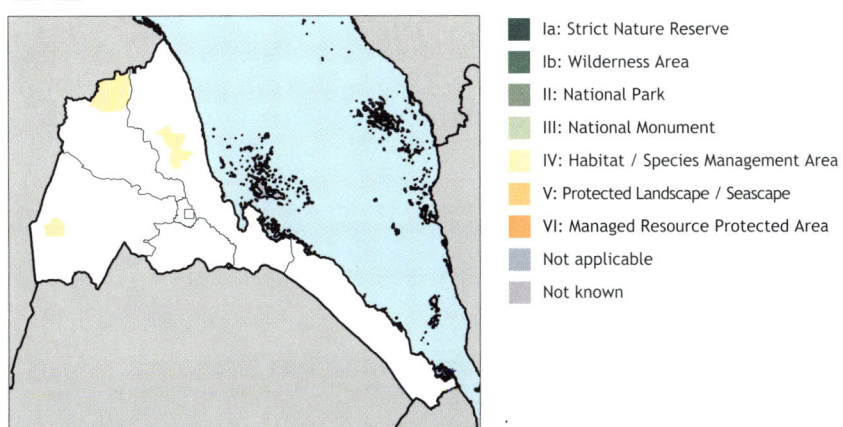

■ Ia: Strict Nature Reserve
■ Ib: Wilderness Area
■ II: National Park
■ III: National Monument
□ IV: Habitat / Species Management Area
■ V: Protected Landscape / Seascape
■ VI: Managed Resource Protected Area
■ Not applicable
■ Not known

Sources: Protected areas are from the World Database on Protected Areas (UNEP and IUCN 2009). Water bodies are from the World Wildlife Federation's Global Lakes and Wetlands Database (Lehner and Döll 2004).

Acacia woodland and shrub are widely distributed in the eastern and western lowlands, with higher temperatures and less rainfall than the highlands. The riverine vegetation makes a significant contribution to the livelihoods of the communities there.

Figure 5.6 shows protected areas in the country. Although these areas are not yet officially gazetted, the proposed protected areas correspond more or less to the categorizations of the International Union for Conservation of Nature. These areas provide important protection for fragile environments that may also be important for emerging ecotourism.

Figure 5.7 shows travel times to the larger cities as potential markets for agricultural products. The major towns in Eritrea are Asmara, Massawa, Assab, Keren, Mendefera, Dekemhare, Adi Quala, Agordet, Barentu, Teseney, Adi Keih, and Segeneity. The border towns are Teseney, Senafe, and Adi Quala. All major towns are connected with asphalt roads.

The maps show that the major cities are within comparatively short travel times from the agriculturally important areas of the south and the western lowlands. Longer travel times are associated with areas that are not agriculturally important, such as the southern Red Sea zone.

As of 2002, the country had about 7,000 kilometers of road network, including primary, secondary, and feeder roads, for an average of 70 kilometers of roads per thousand square kilometers of land. Overall government expenditures on infrastructure (ports, rail, communications, and roads and water

FIGURE 5.7   Travel time to urban areas of various sizes in Eritrea, circa 2000

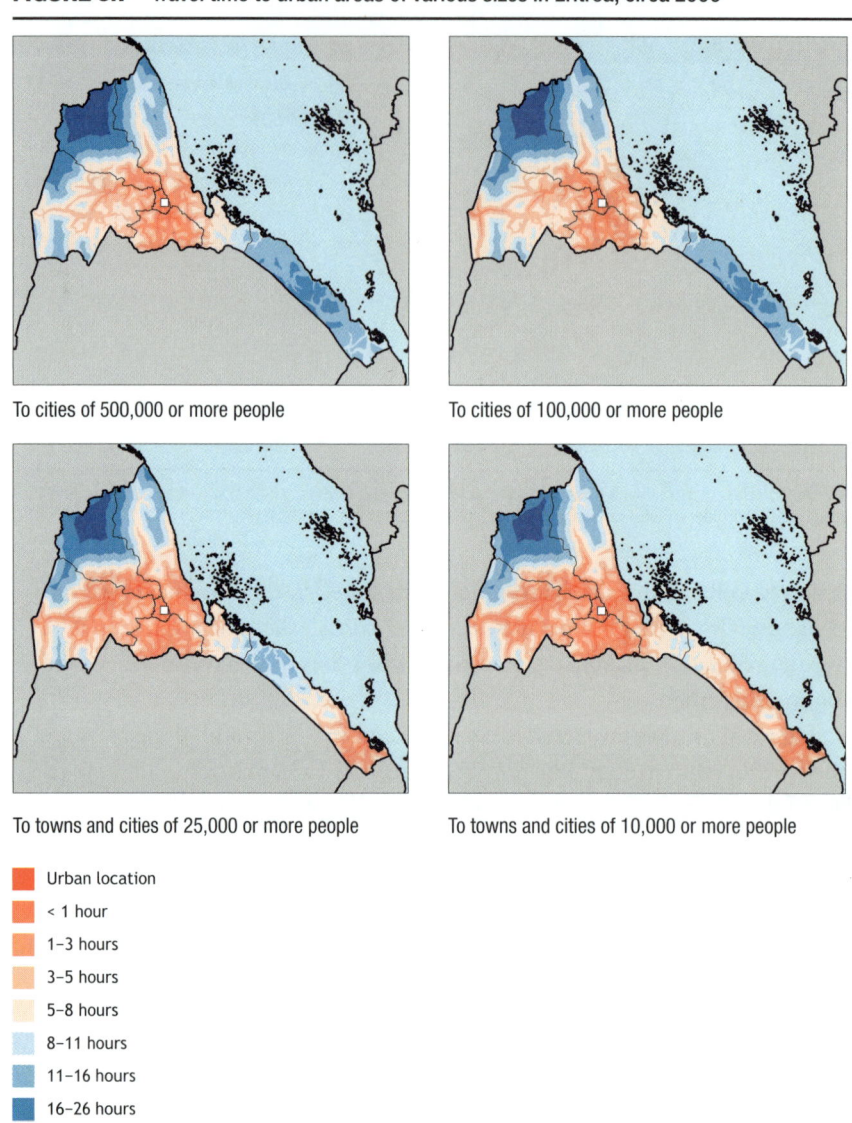

To cities of 500,000 or more people

To cities of 100,000 or more people

To towns and cities of 25,000 or more people

To towns and cities of 10,000 or more people

Urban location

< 1 hour

1–3 hours

3–5 hours

5–8 hours

8–11 hours

11–16 hours

16–26 hours

> 26 hours

Source: Authors' calculations.

supply) increased to 344.3 million nakfa, or 13 percent of total government expenditures for that period.

Agriculture

The major foodcrops grown in Eritrea are sorghum, millet, and barley. Currently sorghum and pearl millet, more drought-tolerant crops, are grown in areas with a well-distributed mean annual rainfall of 300–600 millimeters.

In the southern and northern Red Sea zones, the mean annual rainfall ranges between 50 and 100 millimeters. In this situation, sorghum and pearl millet are produced twice a year under a spate irrigation system. Barley is a drought-tolerant crop that can grow in the highlands, with mean annual rainfall between 400 and 600 millimeters and mean annual temperatures between 10° and 22°C. There is an increased trend toward growing barley under irrigated conditions in the highlands and the lowlands. If temperatures increase, heat stress can bring about desiccation of pollen and hence yield loss in barley.

Tables 5.3–5.5 show the major agricultural commodities in Eritrea in terms of area harvested, production, and food consumption (ranked by weight). Sorghum is by far the most important crop in terms of total area (Table 5.3). Pearl millet and sesame are grown in 9 percent and 8 percent of the lowlands,

TABLE 5.3   Harvest area of leading agricultural commodities in Eritrea, 2006–2008 (thousands of hectares)

Rank	Crop	Percent of total	Harvest area
	Total	100.0	655
1	Sorghum	39.8	261
2	Millet	8.7	57
3	Sesame seeds	7.9	52
4	Barley	7.3	48
5	Other oilseeds	6.1	40
6	Other pulses	6.1	40
7	Other roots and tubers	5.5	36
8	Teff and other cereals	4.7	31
9	Chickpeas	2.9	19
10	Maize	2.8	18

Source: FAOSTAT (FAO 2010).
Note: All values are based on the three-year average for 2006–2008.

TABLE 5.4 Value of production for leading agricultural commodities in Eritrea, 2005–2007 (millions of US$)

Rank	Crop	Percent of total	Value of production
	Total	100.0	229.1
1	Other roots and tubers	34.4	78.9
2	Sorghum	27.1	62.1
3	Sesame seeds	7.1	16.4
4	Millet	7.0	16.1
5	Other fresh vegetables	5.8	13.2
6	Potatoes	4.6	10.6
7	Other pulses	4.4	10.1
8	Barley	2.0	4.6
9	Teff and other cereals	1.9	4.4
10	Chickpeas	1.0	2.3

Source: FAOSTAT (FAO 2010).
Note: All values are based on the three-year average for 2005–2007. US$ = US dollars.

TABLE 5.5 Consumption of leading food commodities in Eritrea, 2003–2005 (thousands of metric tons)

Rank	Crop	Percent of total	Food consumption
	Total	100.0	837
1	Wheat	40.7	341
2	Sorghum	14.3	120
3	Other roots and tubers	11.0	92
4	Other pulses	5.4	45
5	Beer	3.5	30
6	Other vegetables	2.9	25
7	Other oil from oilcrops	2.7	23
8	Sugar	2.7	22
9	Teff and other cereals	2.4	20
10	Beef	2.0	17

Source: FAOSTAT (FAO 2010).
Note: All values are based on the three-year average for 2003–2005.

respectively. Barley, grown in the highlands, takes about 7 percent of the land. Other pulses grown include chickpeas, fava beans, field peas, lentils, and grass peas. Although the total area occupied by vegetable crops is not large compared to that for cereals, the total level of production is higher, with two to three harvests per year.

The leading commodities for human consumption (see Table 5.5) are wheat (40.7 percent), sorghum (14.3 percent), and other roots and tuber crops (11.0 percent). Wheat is used by food industries such as those making pasta, macaroni, bread, cakes, and biscuits. Barley is also a staple crop in the highlands and is used in various forms. The brewing industries use malting barley for beer and local beverages such as *sewa,* accounting for 3.5 percent of its consumption. Sorghum is a staple foodcrop consumed in various forms. Most of the wheat consumed is imported; much remains to be done to produce sufficient maize for domestic consumption.

Figures 5.8 and 5.9 show the estimated yield and harvest area for the two key crops, sorghum and wheat, respectively. In 2000, the major sorghum-growing areas were the southwestern lowlands (53 percent), followed by the southern zone (21 percent), Anseba (13 percent), and the northern Red

FIGURE 5.8   Yield (metric tons per hectare) and harvest area density (hectares) for rainfed sorghum in Eritrea, 2000

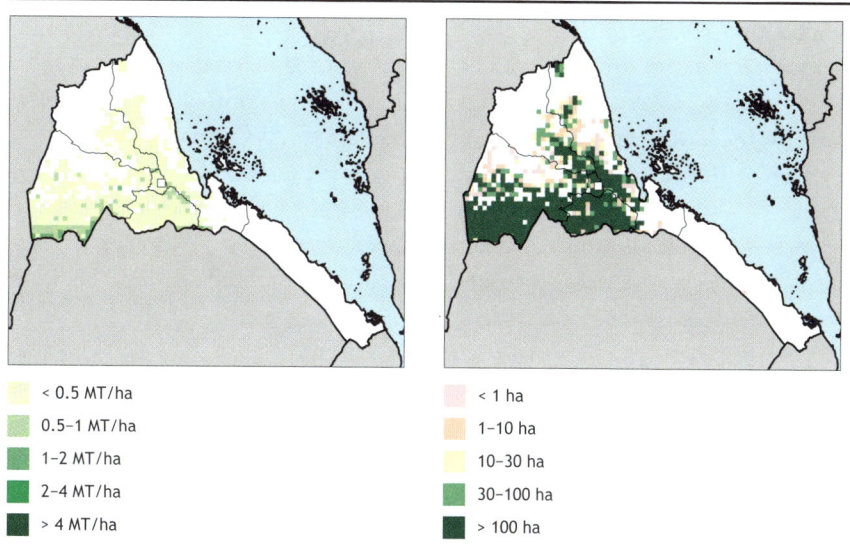

< 0.5 MT/ha
0.5–1 MT/ha
1–2 MT/ha
2–4 MT/ha
> 4 MT/ha

< 1 ha
1–10 ha
10–30 ha
30–100 ha
> 100 ha

Source: SPAM (Spatial Production Allocation Model) (You and Wood 2006; You, Wood, and Wood-Sichra 2006, 2009).
Note: ha = hectare; MT/ha = metric tons per hectare.

FIGURE 5.9 Yield (metric tons per hectare) and harvest area density (hectares) for rainfed wheat in Eritrea, 2000

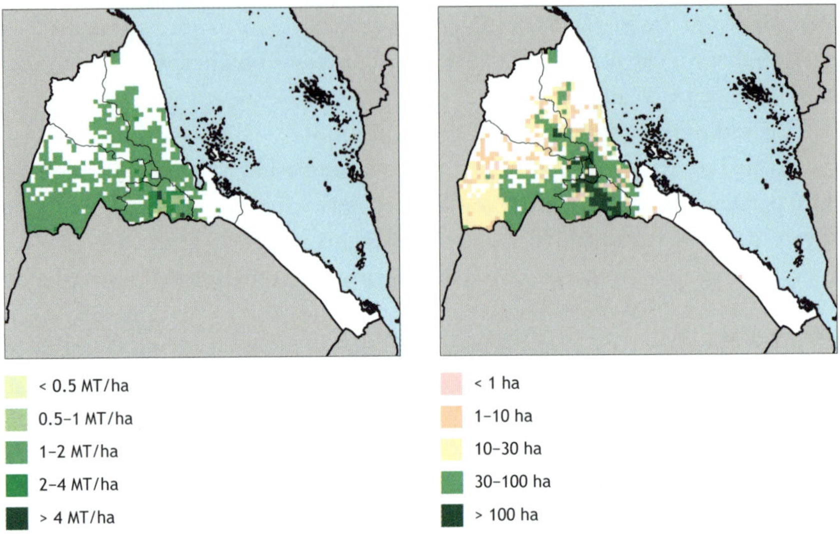

< 0.5 MT/ha	< 1 ha
0.5–1 MT/ha	1–10 ha
1–2 MT/ha	10–30 ha
2–4 MT/ha	30–100 ha
> 4 MT/ha	> 100 ha

Source: SPAM (Spatial Production Allocation Model) (You and Wood 2006; You, Wood, and Wood-Sichra 2006, 2009).
Note: ha = hectare; MT/ha = metric tons per hectare.

Sea area (13 percent). Sorghum has the greatest genetic diversity; it grows throughout the country, adapted to local environmental conditions (moisture, temperature, and so forth). Figure 5.8 shows the geographical distribution and yield of sorghum. The average yield of sorghum for 2000 ranged from less than 0.5 ton per hectare to 1 ton per hectare, but the vast majority of the harvest area is at the lowest end of the range, less than 0.5 ton per hectare.

The major wheat-producing areas, as shown in Figure 5.9, are the southern zone (52 percent) and central zone (39 percent). Other areas of the country have negligible wheat production: the southwestern part (Mekerka, Dekishehay, and other areas), Anseba, and the northern Red Sea area. The average yield for wheat in 2000 was 0.57 tons per hectare.

Scenarios for the Future

Economic and Demographic Indicators

Population

Figure 5.10 shows population projections for Eritrea for 2010–2050 from the UN Population Division. The population is shown reaching 12 million in the high-variant projection, 10.5 million in the medium-variant projection,

FIGURE 5.10   Population projections for Eritrea, 2010–2050

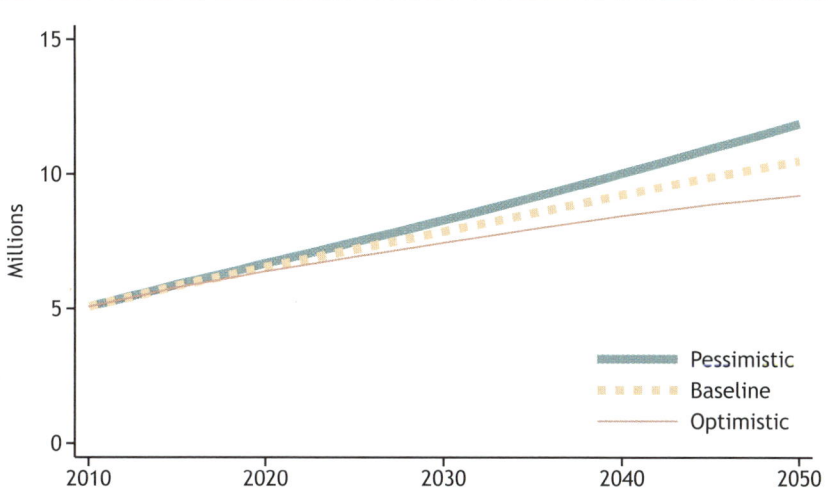

Source: UNPOP (2009).

and 9 million in the low-variant projection. However, estimates by the National Action Program for Eritrea (Eritrea, Ministry of Agriculture 2002) project growth to 6 million by 2015 and an astonishing 22 million by 2050.

Theoretically, the increasing population pressure could aggravate deforestation, overgrazing, and cultivation of crops on steep slopes in response to increased demand for food, shelter, and other resources. However, an educated and well-informed population can also be a vital resource for development. Emphasis must be placed on sustainable development, including protection of the environment, and the promotion of science- and technology-based, environmentally friendly agricultural production systems.

Larger populations can provide the consumer demand needed to generate favorable economies of scale and lower production costs with sufficient domestic investment. (Mining and manufacturing concerns, expected to start soon, are good indicators of foreign direct investment.) The government has indicated that agriculture is a top priority for economic development (ADB 2009). This will require sustainable land and water management, human resource development, and an integrated rural development package.

Income

The three scenarios of GDP per capita shown in Figure 5.11 are derived by combining three GDP projections with the three population projections of Figure 5.10. The optimistic scenario combines the high GDP projection with

FIGURE 5.11   Gross domestic product (GDP) per capita in Eritrea, future scenarios, 2010–2050

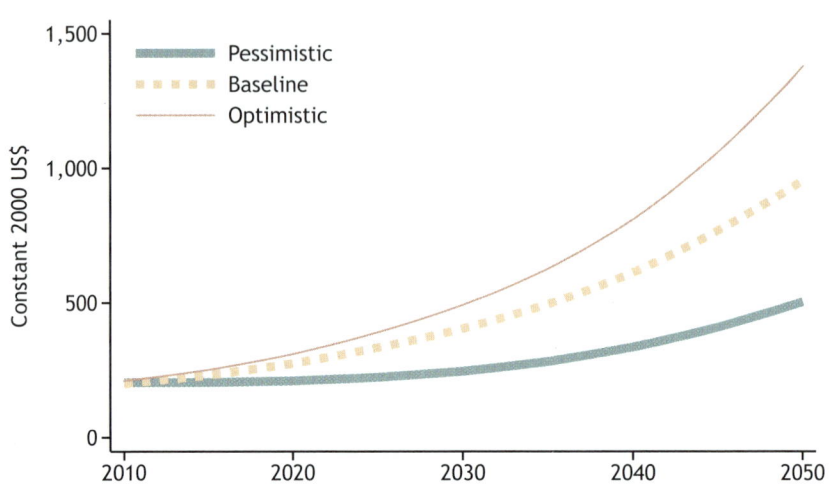

Sources: Computed from GDP data from the World Bank Economic Adaptation to Climate Change project (World Bank 2010), from the Millennium Ecosystem Assessment (2005) reports, and from population data from the United Nations (UNPOP 2009). US$ = US dollars.

low population, the baseline scenario combines the medium GDP projection with the medium population projection, and the pessimistic scenario combines the low GDP projection with the high population projection.

The GDP per capita in Eritrea was $405 in 2010 (World Bank 2011), up from $335 in 2008 and only $150 in 1990. The figures just cited are in current US dollars, whereas the numbers in Figure 5.11 are in constant US dollars with a 2000 base. Following improvement in bilateral relations with the country's neighbors and the exploitation of its largely unexploited resource base, it is expected that growth in investment will drive up per capita GDP and the most likely scenario will be the optimistic scenario.

Biophysical Analysis

Climate Models

Figure 5.12 shows precipitation scenarios for Eritrea for 2000–2050 under the four downscaled climate models we use in this book with the A1B scenario.[1] CNRM-CM3 shows rainfall increasing between 50 and 100 millimeters

---

1 The A1B scenario is a greenhouse gas emissions scenario that assumes fast economic growth, a population that peaks midcentury, and the development of new and efficient technologies, along with a balanced use of energy sources.

FIGURE 5.12   Changes in mean annual precipitation in Eritrea, 2000–2050, A1B scenario (millimeters)

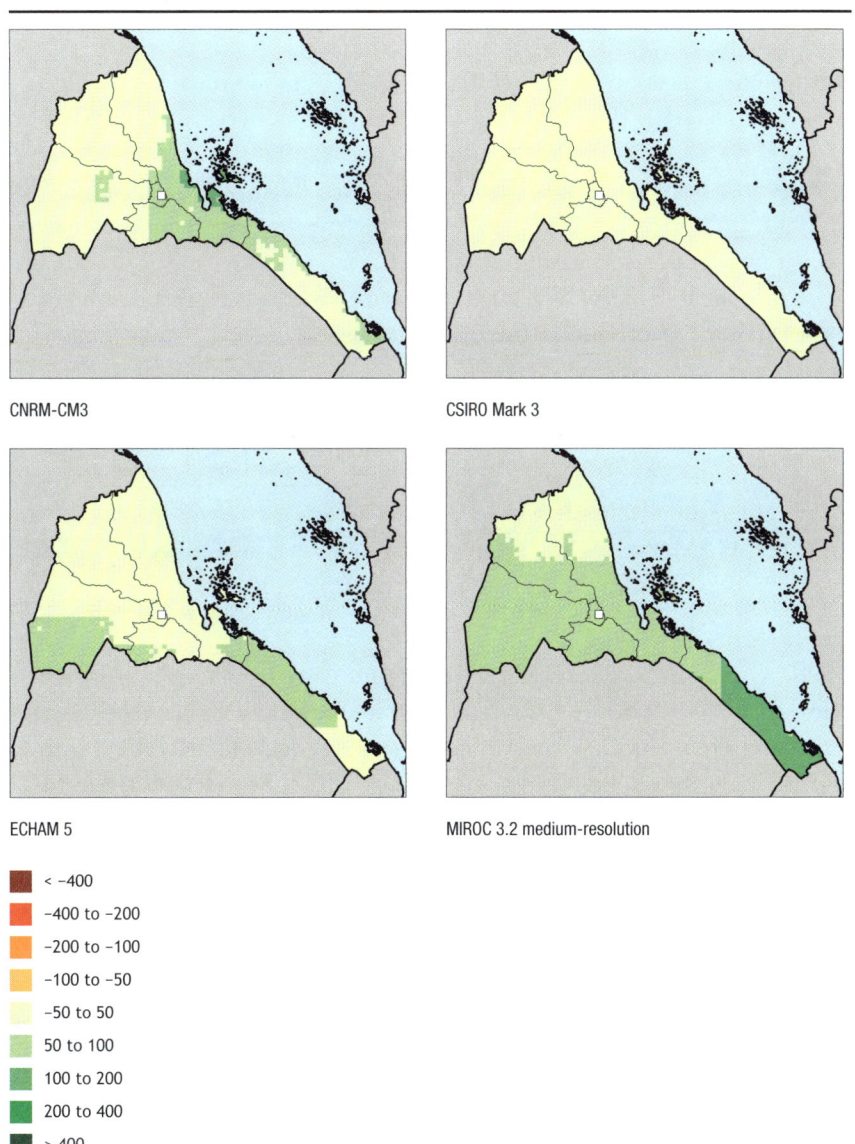

CNRM-CM3

CSIRO Mark 3

ECHAM 5

MIROC 3.2 medium-resolution

- ■ < –400
- ■ –400 to –200
- ■ –200 to –100
- ■ –100 to –50
- ■ –50 to 50
- ■ 50 to 100
- ■ 100 to 200
- ■ 200 to 400
- ■ > 400

Source: Authors' calculations based on Jones, Thornton, and Heinke (2009).

Notes: A1B = greenhouse gas emissions scenario that assumes fast economic growth, a population that peaks midcentury, and the development of new and efficient technologies, along with a balanced use of energy sources; CNRM-CM3 = National Meteorological Research Center–Climate Model 3; CSIRO = climate model developed at the Australia Commonwealth Scientific and Industrial Research Organisation; ECHAM 5 = fifth-generation climate model developed at the Max Planck Institute for Meteorology (Hamburg); GCM = general circulation model; MIROC = Model for Interdisciplinary Research on Climate, developed by the University of Tokyo Center for Climate System Research.

in the central and southern zones and in parts of the northern Red Sea zone but little change over the greater portion of Eritrea, including the south-west (−50 to 50).[2] CSIRO Mark 3 shows minimal change in precipitation throughout the country (−50 to 50). ECHAM 5 shows a small increase in precipitation of 50–100 millimeters in the southwestern part of the country and parts of the southern Red Sea zone—currently an area with very low precipitation. MIROC 3.2 shows areas of the Red Sea zone gaining 100–200 millimeters in precipitation. All models show precipitation remaining the same or even increasing. Similarly, the model done by the Department of Environment (Eritrea, DoE 2001) showed an increase in precipitation in some parts of Eritrea, though the amount is not specified. The most important thing to note is that not a single model used in this analysis shows a reduction in rainfall.

Figure 5.13 shows temperature changes in the mean daily maximum temperature for the warmest month between 2000 and 2050. The CSIRO model is the coolest of the four, showing the entire country getting warmer by 1.0°–1.5°C. ECHAM 5 and CNRM-CM3 are the warmest models, both showing the temperature of most of the country increasing by 2.0°–2.5°C.

Crop Models

The Decision Support Software for Agrotechnology Transfer (DSSAT) software was used to compute yields under current temperature and precipitation regimes in Eritrea for 2000 and for 2050. Yield change results for 2050 with climate change were compared to the current or baseline yield results for the year 2000.

Figure 5.14 shows yield changes for sorghum. Two of the four models show fairly large areas in which sorghum should be able to be grown in the future though it currently cannot be grown. It is not obvious from looking at the temperature and precipitation change maps why these areas would appear. However, DSSAT is concerned only with rainfall and temperature during the growing season, and it may be that rain and temperature shift favorably during the growing months. It could also be that in some of these general circulation models (GCMs) temperatures at higher elevations were too cold for sorghum before climate change. We also note that the ECHAM 5 model shows more

---

2 CNRM-CM3 is National Meteorological Research Center–Climate Model 3. MIROC 3.2 is the Model for Interdisciplinary Research on Climate, developed at the University of Tokyo Center for Climate System Research. CSIRO Mark 3 is a climate model developed at the Australia Commonwealth Scientific and Industrial Research Organisation. ECHAM 5 is a fifth-generation climate model developed at the Max Planck Institute for Meteorology in Hamburg.

FIGURE 5.13   Changes in monthly mean maximum daily temperature in Eritrea for the warmest month, 2000–2050, A1B scenario (°C)

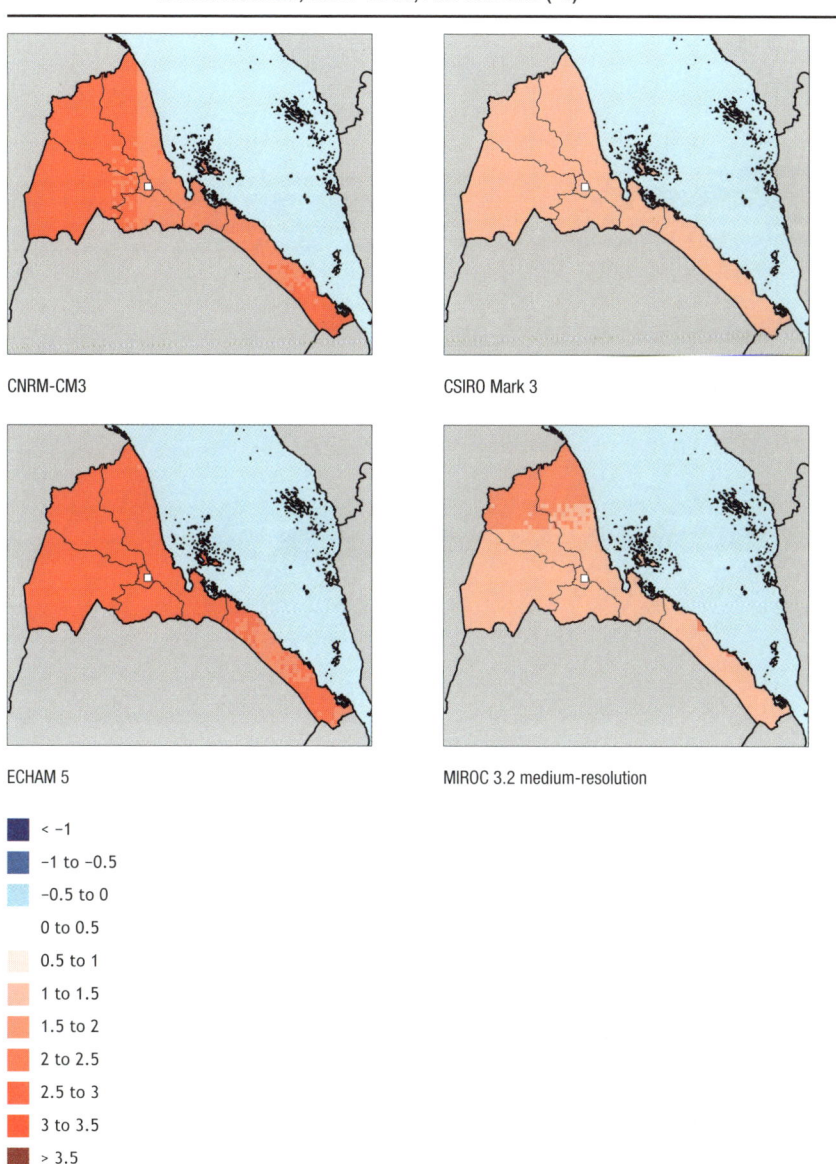

CNRM-CM3

CSIRO Mark 3

ECHAM 5

MIROC 3.2 medium-resolution

- < −1
- −1 to −0.5
- −0.5 to 0
- 0 to 0.5
- 0.5 to 1
- 1 to 1.5
- 1.5 to 2
- 2 to 2.5
- 2.5 to 3
- 3 to 3.5
- > 3.5

Source: Authors' calculations based on Jones, Thornton, and Heinke (2009).

Notes: A1B = greenhouse gas emissions scenario that assumes fast economic growth, a population that peaks midcentury, and the development of new and efficient technologies, along with a balanced use of energy sources; CNRM-CM3 = National Meteorological Research Center–Climate Model 3; CSIRO = climate model developed at the Australia Commonwealth Scientific and Industrial Research Organisation; ECHAM 5 = fifth-generation climate model developed at the Max Planck Institute for Meteorology (Hamburg); GCM = general circulation model; MIROC = Model for Interdisciplinary Research on Climate, developed by the University of Tokyo Center for Climate System Research.

FIGURE 5.14   Yield change under climate change: Rainfed sorghum in Eritrea, 2000–2050, A1B scenario

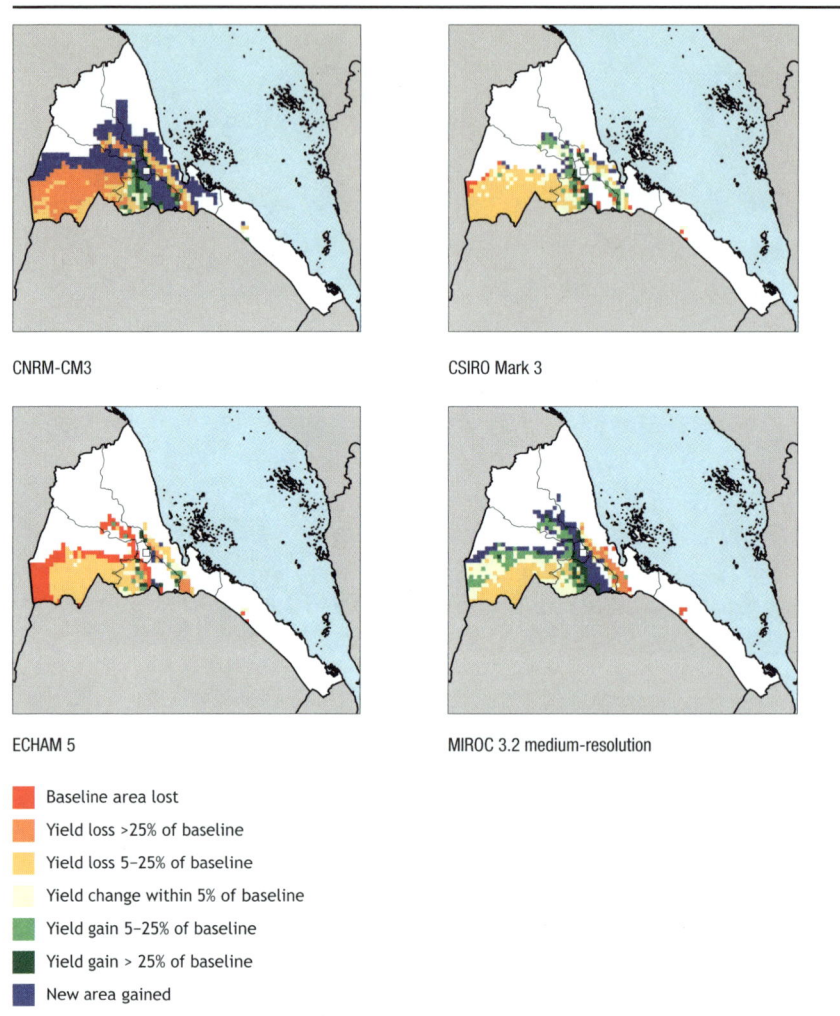

CNRM-CM3

CSIRO Mark 3

ECHAM 5

MIROC 3.2 medium-resolution

■ Baseline area lost
■ Yield loss >25% of baseline
■ Yield loss 5–25% of baseline
■ Yield change within 5% of baseline
■ Yield gain 5–25% of baseline
■ Yield gain > 25% of baseline
■ New area gained

Source: Authors' calculations.

Notes: A1B = greenhouse gas emissions scenario that assumes fast economic growth, a population that peaks midcentury, and the development of new and efficient technologies, along with a balanced use of energy sources; CNRM-CM3 = National Meteorological Research Center–Climate Model 3; CSIRO = climate model developed at the Australia Commonwealth Scientific and Industrial Research Organisation; ECHAM 5 = fifth-generation climate model developed at the Max Planck Institute for Meteorology (Hamburg); GCM = general circulation model; MIROC = Model for Interdisciplinary Research on Climate, developed by the University of Tokyo Center for Climate System Research.

loss of area than the other GCMs. Three of the four GCMs show drops in yield for most areas, but the MIROC 3.2 model is more encouraging, showing areas with no change or actual gains in yield. It will be important to validate any model projections through field-based research.

FIGURE 5.15   Yield change under climate change: Rainfed wheat in Eritrea, 2000–2050, A1B scenario

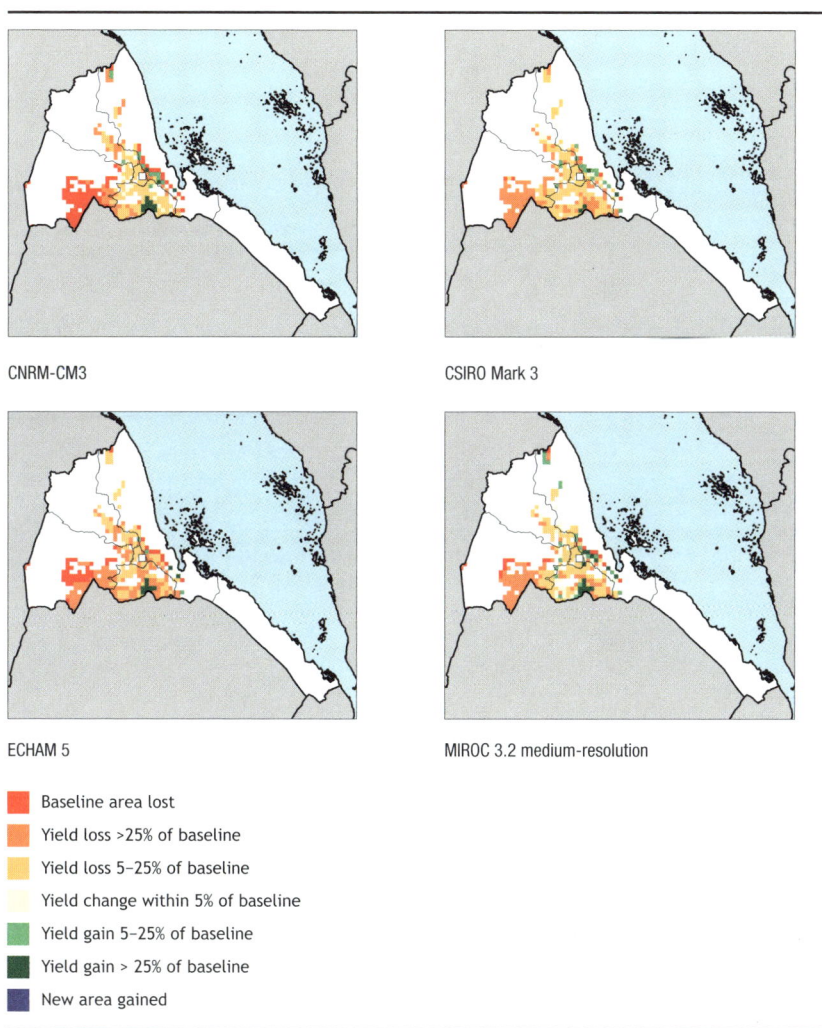

Source: Authors' calculations.

Notes: A1B = greenhouse gas emissions scenario that assumes fast economic growth, a population that peaks midcentury, and the development of new and efficient technologies, along with a balanced use of energy sources; CNRM-CM3 = National Meteorological Research Center–Climate Model 3; CSIRO = climate model developed at the Australia Commonwealth Scientific and Industrial Research Organisation; ECHAM 5 = fifth-generation climate model developed at the Max Planck Institute for Meteorology (Hamburg); GCM = general circulation model; MIROC = Model for Interdisciplinary Research on Climate, developed by the University of Tokyo Center for Climate System Research.

Figure 5.15 shows modeled yield changes for wheat. All the maps show substantial yield losses as high as 25 percent or more, with considerable loss of baseline crop area in some models. Wheat is a crop that is sensitive to hot temperatures, which explains why the losses are so substantial.

FIGURE 5.16   Number of malnourished children under five years of age in Eritrea in multiple income and climate scenarios, 2010–2050

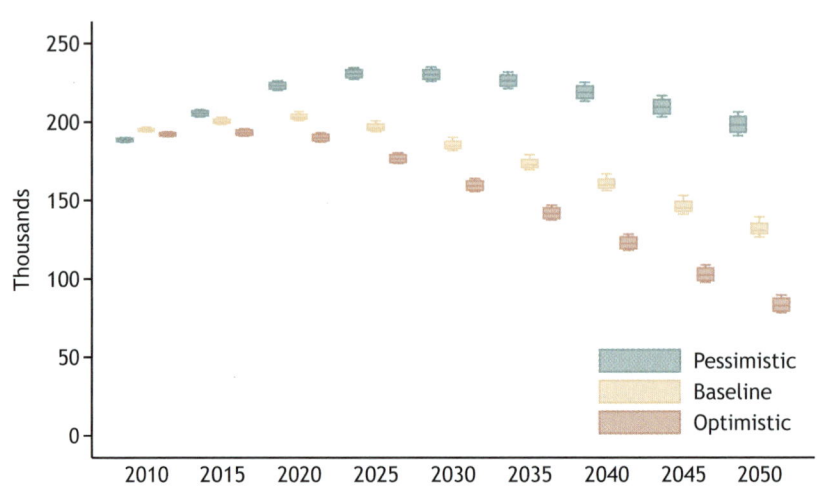

Source: Based on analysis conducted for Nelson et al. (2010).

Note: The box and whiskers plot for each socioeconomic scenario shows the range of effects from the four future climate scenarios.

Vulnerability

Figure 5.16 shows the impact of future GDP and population scenarios on the number of malnourished children in Eritrea under age five. Figure 5.17 shows the share of children who are malnourished. Before 2002, about 38 percent of children suffered from malnutrition or were stunted (short height for age), 15 percent were wasted (low weight for age), and 44 percent were underweight. Since then, Eritrea has shown improvement in these indicators; moreover, infant mortality, under-five mortality, and malaria morbidity have also decreased dramatically due to improved child and mother care and improved nutrition (ADB 2009). In all scenarios, the number of malnourished children rises initially, then falls. For the optimistic scenario, the turning point is 2015, for the baseline 2020, and for the pessimistic scenario sometime between 2025 and 2030. By 2050, only the pessimistic scenario shows more malnourished children than in 2010. With population growth, the malnutrition rate falls steadily.

Figure 5.18 shows the kilocalories per capita available in Eritrea. Both the baseline and the optimistic scenario project rising calorie consumption, with annual increases small at first but becoming large by 2050. In the pessimistic scenario, mean calorie consumption falls until 2025, then begins recovery, ending slightly higher in 2050 than it was in 2010.

FIGURE 5.17   Share of malnourished children under five years of age in Eritrea in multiple income and climate scenarios, 2010–2050

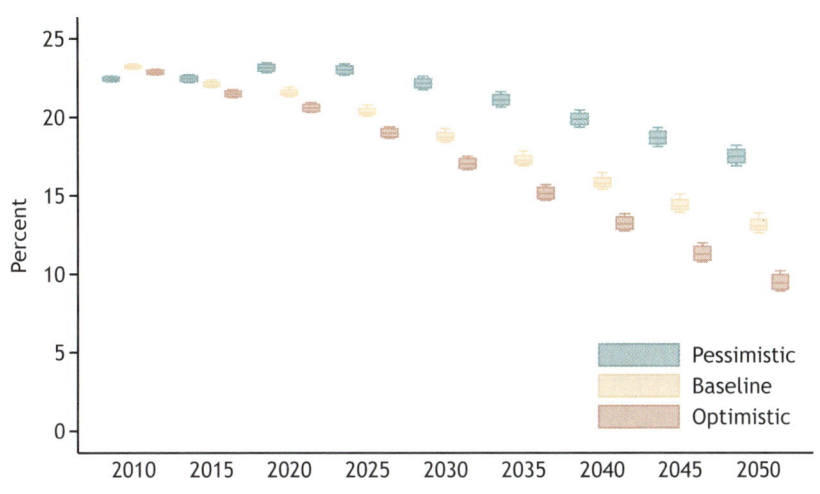

Source: Based on analysis conducted for Nelson et al. (2010).
Note: The box and whiskers plot for each socioeconomic scenario shows the range of effects from the four future climate scenarios.

FIGURE 5.18   Kilocalories per capita in Eritrea in multiple income and climate scenarios, 2010–2050

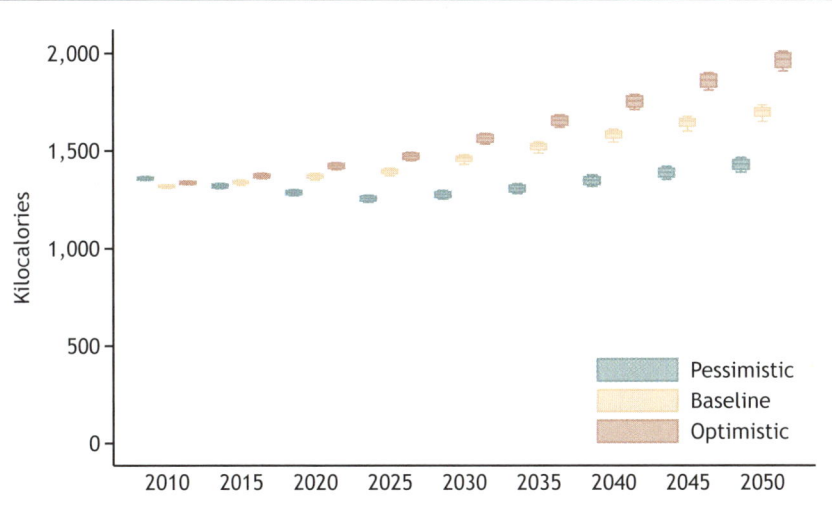

Source: Based on analysis conducted for Nelson et al. (2010).
Note: The box and whiskers plot for each socioeconomic scenario shows the range of effects from the four future climate scenarios.

Agricultural Outcomes

Figures 5.19 and 5.20 show simulation results from the International Model for Policy Analysis of Agricultural Commodities and Trade (IMPACT) associated with key agricultural crops in Eritrea. Each featured crop has five graphs showing production, yield, area, net exports, and world price. The simulation result for sorghum shows a boost in sorghum production in both yield and area of production. The yield is projected to triple, and the area is projected to increase by around 50 percent, resulting in a quadrupling of production (looking at the median results for all scenarios). In all cases, the baseline scenario is very similar to the pessimistic and optimistic scenarios. In the crop models, however, we saw a potential yield loss of sorghum in some parts of the country. We see differences between the crop model and the IMPACT model because the IMPACT model allows for technological change. The net exports have a high variance, making it more difficult to generalize the results. Roughly speaking, however, the model suggests that by 2050, imports of sorghum will be less than in 2010, except possibly in the pessimistic scenario. It seems that high population growth rates can cause consumer demand for sorghum to keep pace with even the most promising gains in production.

Figure 5.20 shows the yield of wheat doubling between 2010 and 2050. But with the area planted increasing only slightly, we note that the yield only doubles as well. In all graphs, the baseline scenario is similar to the optimistic and pessimistic scenarios. Although wheat output will increase, consumer demand for wheat will increase even more, and production levels will not be sufficient to meet domestic demand.

Conclusions and Policy Recommendations

Eritrea is a low-income country with at least 70 percent of the population depending on traditional subsistence agriculture. On average, about 40 percent of its income was obtained from wages and 15 percent from agricultural produce in 2004. The income trends show that the country's GDP increased until around 1998; before that, the country registered an annual growth rate of 7 percent. Since then, GDP growth has been stagnant, and GDP per capita is declining.

The urban population is growing rapidly, partly due to the refugees returning from foreign countries and partly due to high levels of rural-to-urban migration. The population is concentrated mainly in the central and southern provinces. There is a growing population in the Western Lowlands, which is potentially a good agricultural region. The population growth rates in both the rural and the urban centers are high.

FIGURE 5.19   Impact of changes in GDP and population on sorghumt in Eritrea, 2010–2050

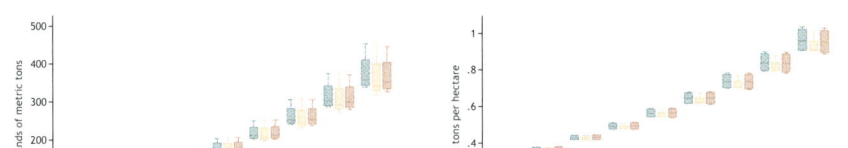

Production

Yield

Area

Net exports

Prices

Source: Based on analysis conducted for Nelson et al. (2010).

Notes: The box and whiskers plot for each socioeconomic scenario shows the range of effects from the four future climate scenarios. GDP = gross domestic product; US$ = US dollars.

The land cover and land use maps show that intensive agriculture is concentrated in the south and southwestern regions of the country. These locations also provide protection for fragile environments considered important for the emerging ecotourism industry.

The GCMs used are in agreement that Eritrea will either have greater precipitation or no change in precipitation in the future. The models differ regarding the likely temperature changes to be experienced. Two suggest that the increases will range from 1°–1.5°C in most parts of the country. However, two others predict a temperature rise of 2°–2.5°C.

FIGURE 5.20   Impact of changes in GDP and population on wheat in Eritrea, 2010–2050

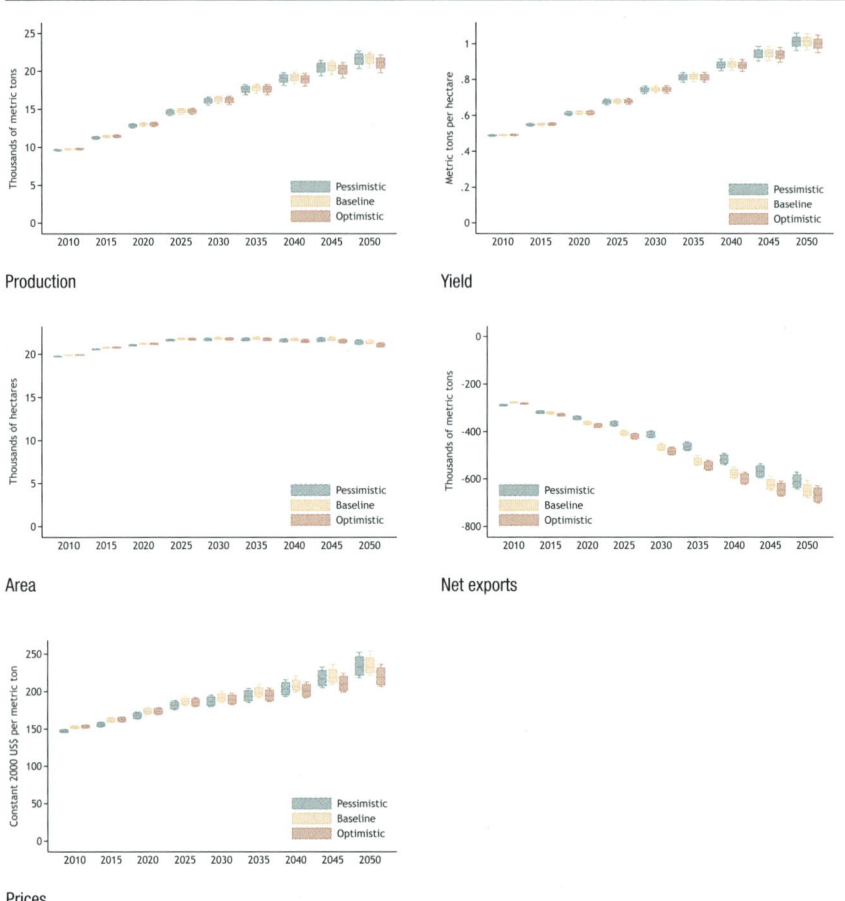

Source: Based on analysis conducted for Nelson et al. (2010).
Notes: The box and whiskers plot for each socioeconomic scenario shows the range of effects from the four future climate scenarios. GDP = gross domestic product; US$ = US dollars.

Most of the crop yield models reveal a potential loss in yields in sorghum growing areas, probably due to temperature increases. The models also suggest that there may be new areas where sorghum can be grown where it currently cannot be grown. This knowledge provides an important opportunity for the country, but it may create policy challenges regarding putting to use land that is currently not used for sorghum and is perhaps not used for agriculture.

Even if the country's rainfall may increase on average, droughts will likely continue to be challenges that farmers face. It is important that appropriate land management practices be introduced, including efficient water harvesting and water management strategies, to cope with water scarcity in drought years and with heavy runoff at other times. Reclamation of degraded lands (sand dunes, gullies, and marginal lands) can be done through the construction of mechanical structures or terraces, and the planting of trees could stabilize these structures. In the coastal areas, drought- and salt-tolerant plant species could be planted both in major urban centers and in rural communities. Widespread tree planting could be encouraged through the promotion of individual tree tenure. A number of environmental policies have been drafted but are yet to be supported with relevant and coherent laws and regulations.

The use of energy-saving stoves will minimize the number of trees to be cut for firewood and charcoal. Other sources of energy, such as wind and solar power, should also be considered.

The introduction of diverse research and development programs is required to address potential risks from climate variability and change affecting both irrigated and rainfed agriculture in Eritrea. Programs at higher levels of learning (MSc and PhD levels) should encourage capacity building in modeling for climate change and in remote sensing.

References

ADB (African Development Bank). 2009. *Interim Country Strategy Paper for Eritrea, 2009–11.* Tunis, Tunisia.

Bartholome, E., and A. S. Belward. 2005. "GLC2000: A New Approach to Global Land Cover Mapping from Earth Observation Data." *International Journal of Remote Sensing* 26 (9–10): 1959–1977.

CIESIN (Center for International Earth Science Information Network), Columbia University, IFPRI (International Food Policy Research Institute), World Bank, and CIAT (Centro Internacional de Agricultura Tropical). 2004. *Global Rural–Urban Mapping Project (GRUMP), Alpha Version: Population Density Grids.* Palisades, NY, US: Socioeconomic Data and Applications Center (SEDAC), Columbia University. http://sedac.ciesin.columbia.edu/gpw.

Eritrea. 2004a. *Food Security Strategy.* Asmara.

———. 2004b. *Interim Poverty Reduction Strategy Paper.* Asmara.

Eritrea, DoE (Department of Environment). 2001. *National Communication under the United Nations Framework Convention on Climate Change (UNFCCC).* Asmara.

———. 2008. *Eritrea Biodiversity Stocktaking Assessment Report.* Asmara, Eritrea.

Eritrea, Ministry of Agriculture. 2002. *Combat Desertification and Mitigate the Effects of Drought.* National Action Program for Eritrea (NAPA). Asmara.

Eritrea, Ministry of National Development. 2006. *Millennium Development Goals Report.* Asmara.

FAO (Food and Agriculture Organization of the United Nations). 2010. FAOSTAT. Rome. http://faostat.fao.org.

Jones, P. G., P. K. Thornton, and J. Heinke. 2009. *Generating Characteristic Daily Weather Data Using Downscaled Climate Model Data from the IPCC's Fourth Assessment.* Project report for the International Institute for Land Reclamation and Improvement, Wageningen, the Netherlands. Accessed May 7, 2010. www.ccafs-climate.org/pattern_scaling/.

Lehner, B., and P. Döll. 2004. "Development and Validation of a Global Database of Lakes, Reservoirs, and Wetlands." *Journal of Hydrology* 296 (1–4): 1–22.

Millennium Ecosystem Assessment. 2005. *Ecosystems and Human Well-being: Synthesis.* Washington, DC: Island Press. http://www.maweb.org/en/Global.aspx.

Nelson, G. C., M. W. Rosegrant, A. Palazzo, I. Gray, C. Ingersoll, R. Robertson, S. Tokgoz, et al. 2010. *Food Security, Farming, and Climate Change to 2050: Scenarios, Results, Policy Options.* Washington, DC: International Food Policy Research Institute.

UNEP (United Nations Environment Programme) and IUCN (International Union for Conservation of Nature). 2009. World Database on Protected Areas (WDPA) Annual Release 2009. No longer available online.

UNPOP (United Nations Department of Economic and Social Affairs–Population Division). 2009. *World Population Prospects: The 2008 Revision.* New York. http://esa.un.org/unpd/wpp/.

World Bank. 2009. *World Development Indicators.* Accessed May 2011. http://data.worldbank.org/data-catalog/world-development-indicators.

———. 2010. *Economics of Adaptation to Climate Change: Synthesis Report.* Washington, DC. http://climatechange.worldbank.org/content/economics-adaptation-climate-change-study-homepage.

———. 2011. *World Development Indicators.* Accessed February 2012. http://data.worldbank.org/data-catalog/world-development-indicators.

You, L., and S. Wood. 2006. "An Entropy Approach to Spatial Disaggregation of Agricultural Production." *Agricultural Systems* 90 (1–3): 329–347.

You, L., S. Wood, and U. Wood-Sichra. 2006. "Generating Global Crop Distribution Maps: From Census to Grid." Paper presented at the International Association of Agricultural Economists Conference, Brisbane, Australia, August 11–18.

———. 2009. "Generating Plausible Crop Distribution and Performance Maps for Sub-Saharan Africa Using a Spatially Disaggregated Data Fusion and Optimization Approach." *Agricultural Systems* 99 (2–3): 126–140.

Chapter 6

ETHIOPIA

Habtamu Admassu, Mezgebu Getinet, Timothy S. Thomas,
Michael Waithaka, and Miriam Kyotalimye

Ethiopia is a country that is particularly exposed to possible adverse impacts of climate change because a large proportion of the population is dependent on agriculture for employment and income and the country has a low gross domestic product (GDP) per capita, offering agricultural households little capacity to compensate for income losses from weather-related shocks.

Understanding the country's socioeconomic and biophysical vulnerability is important in order to devise appropriate adaptation measures that can save the environment and the livelihoods of the people. Knowledge of expected adverse effects of climate change has led to the establishment of the Ethiopian National Forum for Climate Change, established in July 2008, which is playing a significant role in bringing the potential impact of climate change to the attention of political leaders and the public. Several other initiatives are in place: the National Policy on Disaster Prevention and Preparedness, the Plan for Accelerated and Sustainable Development to End Poverty, the National Adaptation Programme of Action, and the Disaster Risk Management Policy. There is a plan to establish a National Commission for Climate Change that will be responsible for the integration of climate issues into development practices and for the coordination of policies.

In this chapter we present our assessments of the current socioeconomic and biophysical vulnerability of the Ethiopian agricultural sector, mainly focusing on key indicators of socioeconomic vulnerability and selected food enterprises. General circulation model (GCM) projections to 2050 are used to uncover what the future holds for the agricultural sector and what should be considered to enhance the sector's adaptive capacity. Crop simulation models are used to assess the changes in prices of major commodities expected with climate change. Finally, a future direction for sustainable planned adaptation is suggested for policymakers to guide their actions.

Review of the Current Situation and Trends

Economic and Demographic Indicators

Population

Figure 6.1 shows the total population and rural population (left axis) of Ethiopia as well as the share of the population that is urban (right axis). The figure shows that both the urban and the rural populations in Ethiopia have been continuously growing. From around 24 million in 1960, the total population reached 34 million in the mid-1970s. The population growth rate has fluctuated, from about 2.5 percent (1960–1969) to 2.0 percent (1970–1979), then increasing to 3.1 percent (1980–1999) and declining again to around 2.6 percent by 2008. In the late 1970s and early 1980s, population growth leveled off owing to high rates of mortality from diseases and famine (Table 6.1). In recent years, growth has picked up, and the population is almost 79 million. Factors in the recent rapid population growth are increasing fertility, reduction in adult mortality, and lack of fertility control. The urban population has grown steadily in recent years due to the expansion of the service sector and industrialization.

FIGURE 6.1   Population trends in Ethiopia: Total population, rural population, and percent urban, 1960–2008

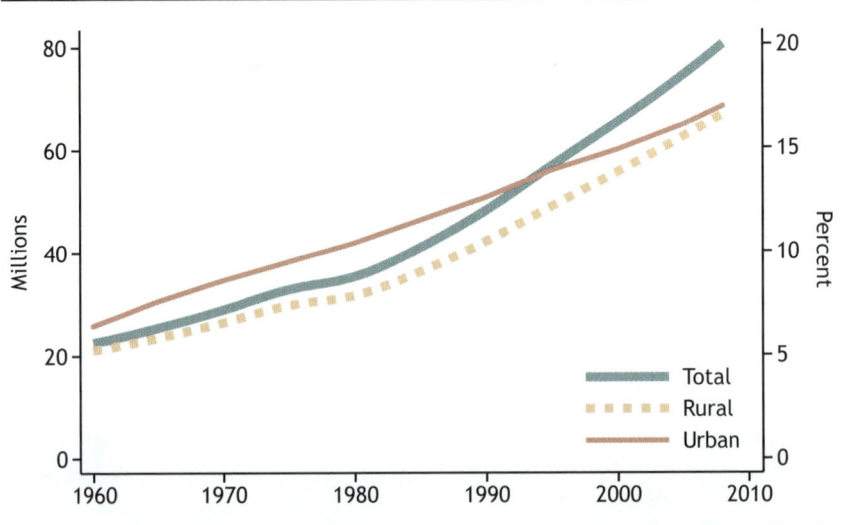

Source: *World Development Indicators* (World Bank 2009).

TABLE 6.1   Population growth rates in Ethiopia, 1960–2008 (percent)

Decade	Total growth rate	Rural growth rate	Urban growth rate
1960–1969	2.5	2.2	5.5
1970–1979	2.0	1.8	3.9
1980–1989	3.1	2.9	5.1
1990–1999	3.1	2.8	4.8
2000–2008	2.6	2.3	4.2

Source: Authors' calculations based on *World Development Indicators* (World Bank 2009).

Figure 6.2 shows the geographic distribution of population in Ethiopia using estimates based on census data. The map for 2000 shows several densely populated areas in the highlands that have recently been settled, including the fertile areas in and around the Rift Valley lakes in the Southern Peoples Nations and Nationalities (SPNN) as well as the eastern highlands of Hararghe. The highest population density—500 people per square kilometer—prevails in the northeastern areas of SPNN and the southeastern parts of Oromia Region. Vast lowlands, including pastoral and agropastoral areas in the southwest, southeast, east, and north, were inhabited by few people in the year 2000 but are now highly populated. The encroachment on fragile lowland areas increases pressure on the natural resources of these areas.

FIGURE 6.2   Population distribution in Ethiopia, 2000 (persons per square kilometer)

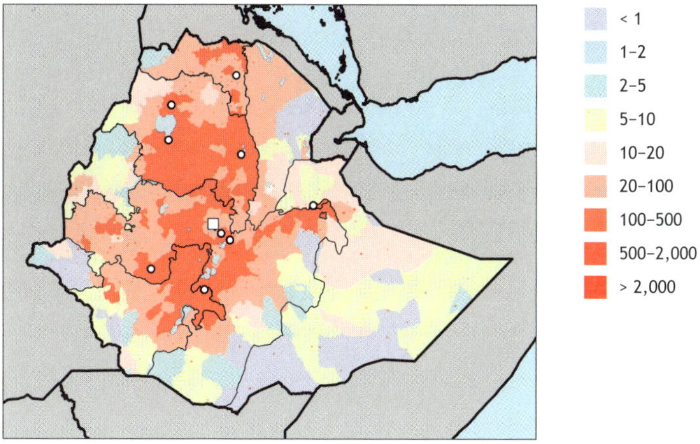

Source: CIESIN et al. (2004).

FIGURE 6.3   Per capita GDP in Ethiopia (constant 2000 US$) and share of GDP from agriculture (percent), 1981-2008

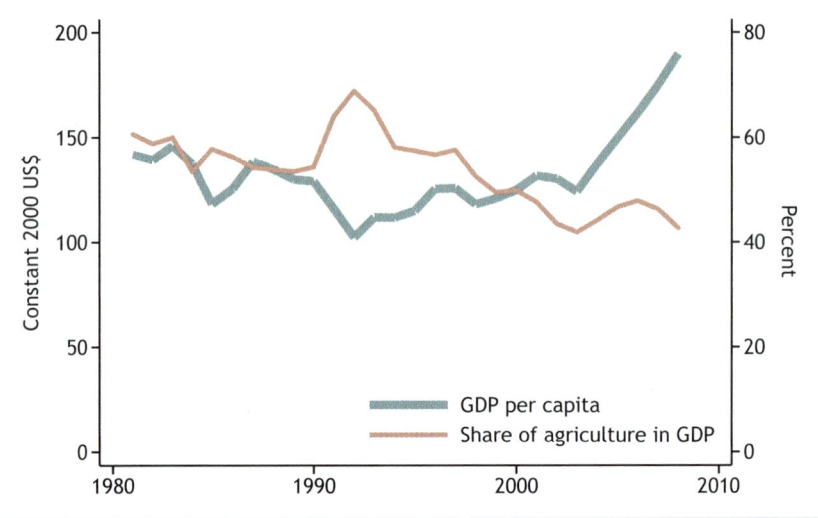

Source: *World Development Indicators* (World Bank 2009).
Note: GDP = gross domestic product; US$ = US dollars.

Income

Figure 6.3 shows great fluctuation in the proportion of GDP in agriculture, with downward trends in recent years. Agriculture as a percentage of total GDP reached its maximum during 1992, following the economic reforms, declined to its historic low in 2003, again rose slightly, and then trended downward in recent years. GDP per capita has fluctuated as well, with its lowest level in the early 1990s and a dramatic increase beginning in 2004. "The agriculture sector's share of GDP declined by three percentage points between 2003/2004 and 2008/2009 and has now been surpassed by services. This impressive growth in services was driven by the rapid expansion in financial intermediation, public administration, and retail business activities" (ADB 2010, 3).

In spite of the decreasing contribution of agriculture to GDP, the figure is still more than 40 percent; the contribution of agriculture to the economy will thus continue to be significant as Ethiopia remains predominantly an agrarian society.

Vulnerability to Climate Change

Table 6.2 provides statistics on education and nutrition for Ethiopia. Statistics on education provide data on several indicators of a population's vulnerability

TABLE 6.2   Education and nutrition statistics for Ethiopia, 2000s

Indicator	Year	Percent
Primary school enrollment: Percent gross (three-year average)	2007	90.8
Secondary school enrollment: Percent gross (three-year average)	2007	30.5
Adult literacy rate	2004	35.9
Under-five malnutrition (weight for age)	2005	34.6

Source: *World Development Indicators* (World Bank 2009).

and resiliency to economic shocks: level of education, literacy, and concentration of labor in poorer or less dynamic sectors. Primary school enrollment is much higher than secondary school enrollment in Ethiopia; however, the adult literacy rate is relatively high.

The percentage of people with full-time paid employment in agriculture is very low; however, the percentage with vulnerable employment (including work on their own farms or as day labors) is still very high. Agriculture (including own-farm labor) employs over 85 percent of the population (Wolde-Giorgis 1997). The percentage of children under age five who are suffering from malnutrition is also high. Huge tasks lie ahead to reduce the level of vulnerability of Ethiopians or to engage them in less vulnerable sectors.

Life Expectancy and Under-Five Mortality

Figure 6.4 shows two noneconomic correlates of poverty: life expectancy and under-five mortality. Generally, life expectancy has been rising steadily in Ethiopia, from less than 40 years in 1960 to about 55 years in 2008. The rapid drop in under-five mortality in two time periods—the early 1960s to the late 1970s and the early 1990s to the present—is associated with a rising trend in life expectancy.

Child mortality decreased gradually from 1960 to 1980, from about 270 to 210 per 1,000 children; the level remained constant into the 1990s and declined to just over 100 children per 1,000 by 2008. The combined effect of declining child mortality and increasing life expectancy has resulted in an overall population increase, putting additional pressure on the land to meet the food requirements of the growing population.

Poverty

Figure 6.5 shows the percentage of the population living below the poverty line (on less than $2 per day). There is great regional variation: more than 90 percent of the people in Tigray, Afar, and Beneshangul earn less than $2 per

FIGURE 6.4   Well-being indicators in Ethiopia, 1960–2008

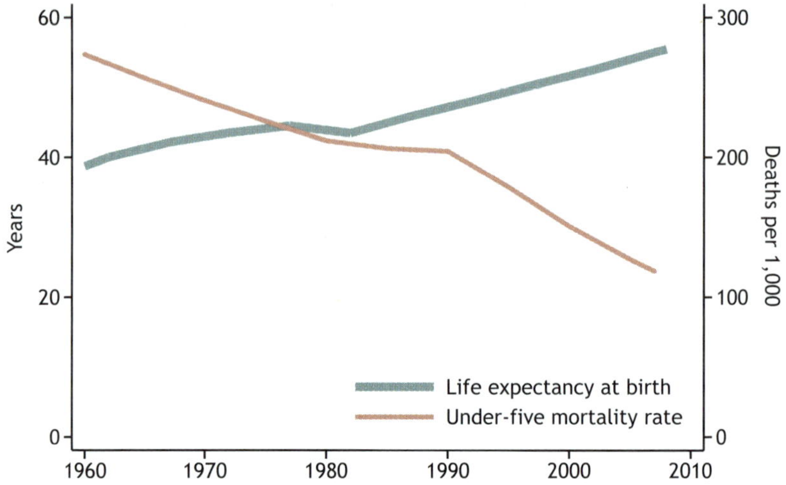

Source: *World Development Indicators* (World Bank 2009).

day, as do 60–70 percent of the people in Oromia and Somali, 70–80 percent in Amhara, and 80–90 percent in SNNP and Gambella states. More than 80 percent of Ethiopian people overall live below the poverty line and hence are vulnerable to small shocks and less able to recover from their impact.

Food Security and Malnourishment

A recent assessment indicates that today much of Ethiopia is facing serious food insecurity. In general, food consumption is lower among the poor, those with smaller agricultural plots, and those living in large families. Food security is attained when all people have physical and economic access to sufficient safe and nutritious food to meet their dietary needs and food preferences for an active and healthy life at all times, without undue risk of losing such access (FAO 1996). Ethiopia has had a structural food deficit since 1980, and about 30 million people live in absolute poverty (Adenew 2003). Approximately 7.8 million Ethiopians received food assistance in 2009/2010 through the Productive Safety Nets Programme. Poor rural households and those without land are the most food insecure (Ethiopia, MOFED 2002b). Children are the most vulnerable and suffer during periods of food deficit following drought. Food shortage in Ethiopia is reportedly the worst among the African countries, as is child malnutrition manifested in the form of wasting and stunting (Ethiopia, MOFED 2002a). Even when some areas have surplus production,

FIGURE 6.5   Poverty in Ethiopia, circa 2005 (percentage of population below US$2 per day)

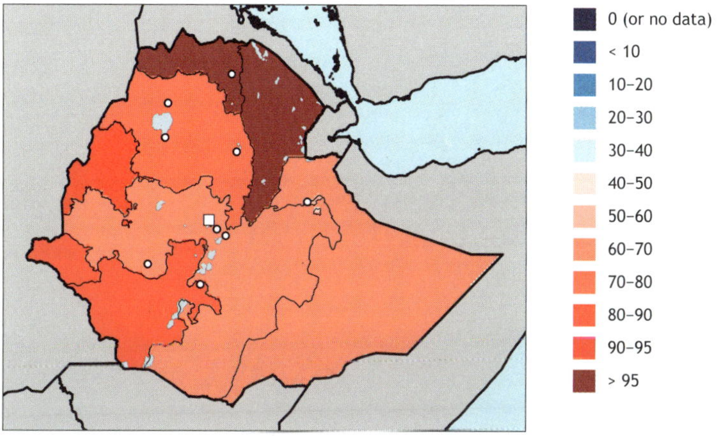

Source: Wood et al. (2010).
Note: Based on 2000 US$ (US dollars) and on purchasing power parity value.

the inadequate infrastructure and limited purchasing power of households constrain access to food, particularly in remote areas.

The standard requirement for subsistence calls for consumption of an average 2,100 kilocalories per day; the average daily per capita calorie requirement needed to maintain the health of the population in Ethiopia is approximately 2,200. Currently, 60 percent of the population consumes fewer calories than this daily physiological requirement. The average daily per capita calorie consumption in Ethiopia is about 1,980—a shortfall of about 220 calories per day, on average, compared with internationally recommended norms (Moreland and Smith 2012). Any lessening of food insecurity in a changing climate will heighten the vulnerability of the people.

Review of Land Use, Potential, and Limitations

Land Use Overview
Ethiopian forest cover has decreased steadily over time, from 30 percent in the 1900s to about 10 percent. The annual loss of highland mountain forest cover has been estimated at about 141,000 hectares, resulting in soil erosion, environmental deterioration, and loss of biodiversity.

Figure 6.6 shows land cover and land use in the country in the year 2000. Shrub-covered closed and open areas dominated the northwestern and

FIGURE 6.6   Land cover and land use in Ethiopia, 2000

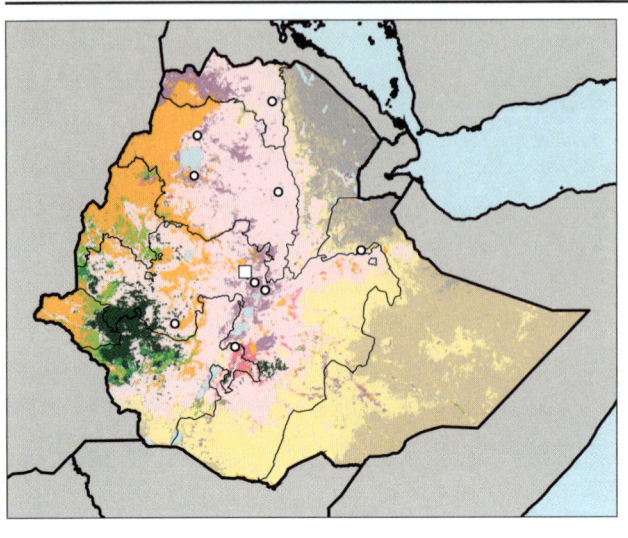

■ Tree cover, broadleaved, evergreen
■ Tree cover, broadleaved, deciduous, closed
■ Tree cover, broadleaved, open
■ Tree cover, broadleaved, needle-leaved, evergreen
■ Tree cover, broadleaved, needle-leaved, deciduous
■ Tree cover, broadleaved, mixed leaf type
■ Tree cover, broadleaved, regularly flooded, fresh water
■ Tree cover, broadleaved, regularly flooded, saline water
■ Mosaic of tree cover/other natural vegetation
■ Tree cover, burnt
■ Shrub cover, closed–open, evergreen
■ Shrub cover, closed–open, deciduous
■ Herbaceous cover, closed–open
■ Sparse herbaceous or sparse shrub cover
■ Regularly flooded shrub or herbaceous cover
■ Cultivated and managed areas
■ Mosaic of cropland/tree cover/other natural vegetation
■ Mosaic of cropland/shrub/grass cover
■ Bare areas
■ Water bodies
■ Snow and ice
■ Artificial surfaces and associated areas
□ No data

Source: GLC2000 (Bartholome and Belward 2005).

FIGURE 6.7   Protected areas in Ethiopia, 2009

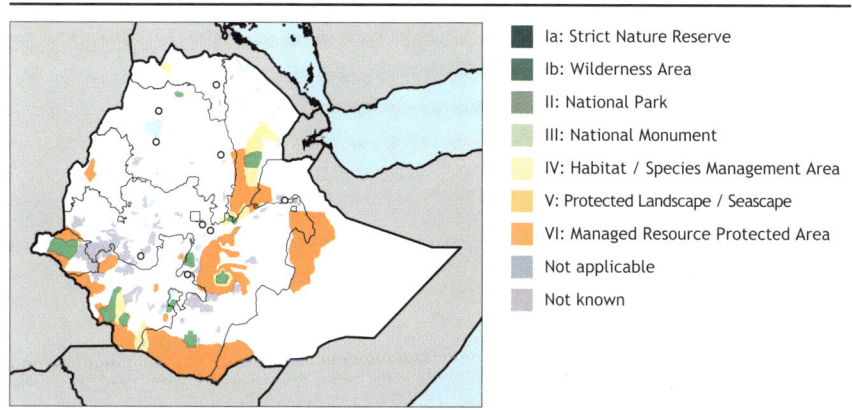

Legend:
- Ia: Strict Nature Reserve
- Ib: Wilderness Area
- II: National Park
- III: National Monument
- IV: Habitat / Species Management Area
- V: Protected Landscape / Seascape
- VI: Managed Resource Protected Area
- Not applicable
- Not known

Sources: Protected areas are from the World Database on Protected Areas (UNEP and IUCN 2009). Water bodies are from the World Wildlife Federation's Global Lakes and Wetlands Database (Lehner and Döll 2004)

parts of the western, southwestern, and east central parts of the country. Sparse herbaceous shrubs now dominate the eastern part, with the remaining dense forest limited to the southern and southwestern sections of the country. The northern, north central, and central areas and parts of the southern areas are now under crop production with little or no tree cover. The northern parts of the highlands are almost devoid of trees. Ethiopia faces a challenging future, because the various forms of land cover react differently to climate change and require unique adaptive strategies.

Figure 6.7 shows protected areas of Ethiopia. Crucial areas in the northern parts of the country are still unprotected, unlike those in the south, southwest, and east and in stretches of the Ethiopian Rift Valley. These fragile sites need to be protected.

Ethiopia has 9 national parks, 3 sanctuaries, 8 reserves, and 18 controlled hunting areas. Due to the declining area under forest, wildlife has been under pressure since the early 1970s. Of the 277 terrestrial mammals found in Ethiopia, 31 are endemic to the country, 20 of which are highland forms. There are 862 bird species recorded in Ethiopia, of which 261 (30.2 percent) are species of international concern; 16 bird species are endemic to Ethiopia, the highest number in Africa south of the Sahara. There are 214 palearctic migrant bird species in Ethiopia, of which a total of 47 are found to summer in Ethiopia (James 2012).

There are about 63 globally recognized endemic bird sites in Ethiopia, mostly in the central highlands, the southern highlands, and the Juba-Sheballe Valley. The Abijata-Shalla Lakes National Park (Southern Rift Valley) has also been proposed as a park due to the high diversity of water birds there. It is estimated that at least 6 reptiles and 34 amphibians are also endemic. Seven mammal and two bird species have been listed by the International Union for the Conservation of Nature as critically endangered. Threats to biodiversity include undervaluation of environmental resources, deforestation due to agricultural expansion and settlement, lack of adequate knowledge of biological resources, and overexploitation.

Wetlands are valuable for rural communities, because they contribute directly to food security in the dry seasons. However, wetlands are being degraded due to human-related activities such as draining for agriculture, cattle grazing, industrial pollution, and unsustainable use. Although there are several sectoral policies, a national wetlands policy is lacking.

An estimated 16.4 percent of the total land area of Ethiopia is under some form of protection. Federal and regional governmental offices as well as environmental nongovernmental organizations are helping local communities reverse the current degradation trends in protected areas. Current interventions aim to promote and strengthen wildlife-based tourism, ecotourism, and nature tourism within the parks; to establish transboundary protected areas; and to develop plans and infrastructure for managing the existing protected areas.

Figure 6.8 shows travel times to cities and towns of various sizes. The upper left-hand map shows that most of the population is remote from large cities. The majority of rural people have to travel considerable distances to towns of more than 100,000 people, limiting their access to good markets. Many rural people, however, live near small towns (with populations in the range of 10,000–25,000). Some of the remote areas (WabeShebele and the Genale Plains) with fertile lands and access to irrigation lack the road infrastructure and markets that could expand their potential for agriculture. The time and transportation costs to move produce to large cities limits access to input and output markets, perpetuating the rural subsistence mode of low-input/low-output production practices. In order to reduce the vulnerability of the rural poor, expansion of road networks is crucial.

Agriculture
Agriculture remains the main activity in the Ethiopian economy and contributes, on average, 47 percent of GDP (down from 66 percent in the 1960s). Ethiopian agriculture employs over 85 percent of the population. Agriculture accounts for about 90 percent of the country's total export earnings. About two-thirds of the total foreign exchange earnings are generated from

FIGURE 6.8   Travel time to urban areas of various sizes in Eritrea, circa 2000

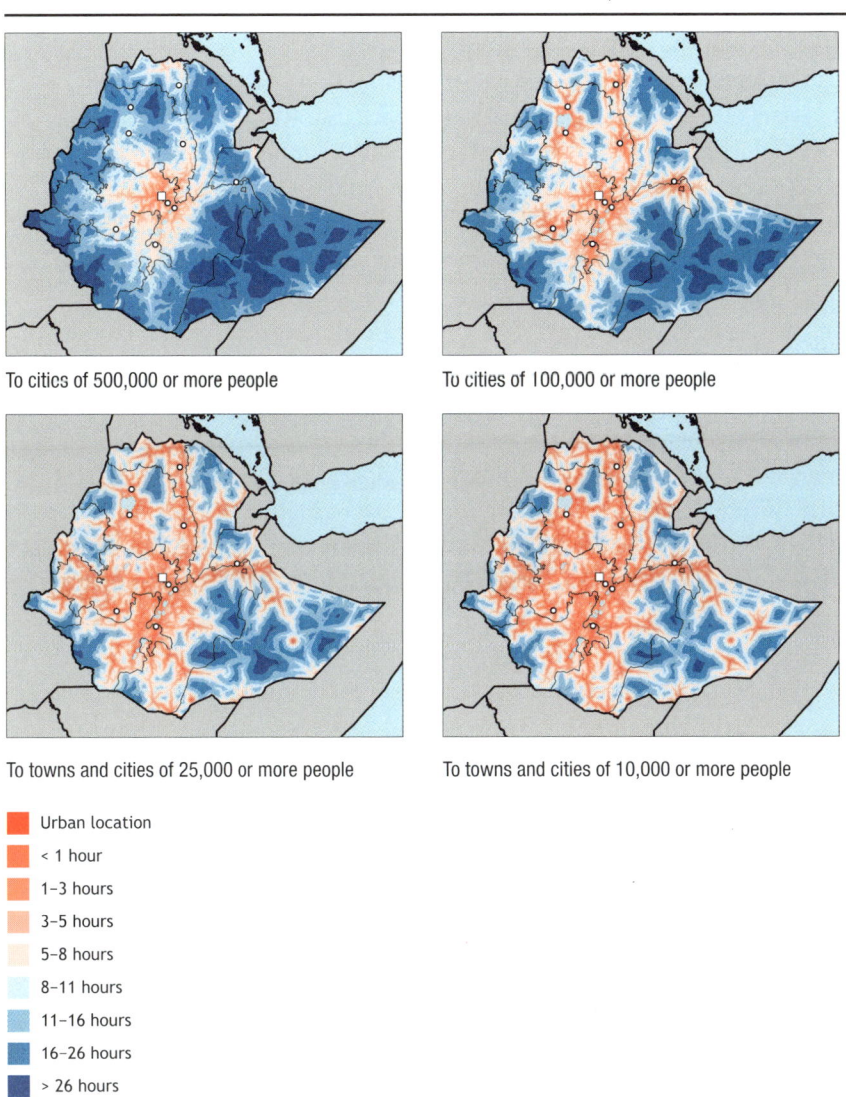

To cities of 500,000 or more people

To cities of 100,000 or more people

To towns and cities of 25,000 or more people

To towns and cities of 10,000 or more people

Urban location
< 1 hour
1–3 hours
3–5 hours
5–8 hours
8–11 hours
11–16 hours
16–26 hours
> 26 hours

Source: Authors' calculations.

coffee exports (Wolde-Giorgis 1997). Smallholder households produce more than 90 percent of the agricultural output and cultivate more than 90 percent of the total cropped land. Crop production is the dominant subsector, accounting for more than 60 percent of agricultural GDP, followed by livestock with 20 percent. Widespread modes of production are subsistence mixed farming in the highlands, in which raising several crops is combined with live-stock rearing, and nomadic pastoralism in the lowlands. It is estimated that 16.5 million hectares (14.8 percent) of 73.6 million hectares (66 percent) of the country's land area is potentially suitable for agricultural production. The potential irrigable land in the country is about 3.7 million hectares. Ethiopia has the largest livestock population in Africa and the tenth-largest in the world, with about 70 million head of livestock.

Tables 6.3–6.5 show key agricultural commodities in terms of area har-vested, the value of the harvest, and food for human consumption (ranked by weight). Table 6.3 shows the harvest area of leading agricultural commodities in Ethiopia during 2006–2008. Teff and other cereals were produced over a larger area than other crops, reflecting teff's wider range of climatic adaptabil-ity, relatively higher prices, and easier storability, in addition to cultural food preferences. Maize also performs well in a wider range of agroecological set-tings. Wheat and maize were the commodities of highest monetary value (see Table 6.4). Although teff was the most important crop, it ranks only fourth in consumption (see Table 6.5) because most rural households sell teff for cash and consume more maize and wheat.

Figure 6.9 shows the estimated yield and growing areas for rainfed maize in 2000. Maize was produced over large areas of the country. However, yields were low, averaging around 2.2 tons per hectare but extending to close to

TABLE 6.3   Harvest area of leading agricultural commodities in Ethiopia, 2006–2008 (thousands of hectares)

Rank	Crop	Percent of total	Harvest area
	Total	100.0	13,035
1	Teff and other cereals	19.5	2,548
2	Maize	12.8	1,663
3	Sorghum	11.4	1,489
4	Wheat	11.1	1,453
5	Barley	7.7	1,001
6	Other roots and tubers	4.8	620
7	Broad beans and horse beans	3.6	469
8	Millet	3.0	388
9	Coffee	2.8	371
10	Chilies and peppers	2.3	293

Source: FAOSTAT (FAO 2010).
Note: All values are based on the three-year average for 2006–2008.

TABLE 6.4   Value of production of leading agricultural commodities in Ethiopia, 2005–2007 (millions of US$)

Rank	Crop	Percent of total	Value of production
	Total	100.0	4,548.0
1	Wheat	12.2	555.9
2	Maize	12.0	545.3
3	Teff and other cereals	11.7	532.1
4	Coffee	9.1	412.8
5	Sorghum	9.1	411.8
6	Other roots and tubers	8.8	400.3
7	Barley	6.2	280.2
8	Broad beans and horse beans	3.0	135.7
9	Sugarcane	2.8	129.3
10	Chilis and peppers	2.0	92.6

Source: FAOSTAT (FAO 2010).
Notes: All values are based on the three-year average for 2005–2007. US$ = US dollars.

TABLE 6.5   Consumption of leading food commodities in Ethiopia, 2003–2005 (thousands of metric tons)

Rank	Crop	Percent of total	Food consumption
	Total	100.0	20,228
1	Other roots and tubers	20.0	4,047
2	Maize	13.7	2,778
3	Wheat	12.4	2,515
4	Teff and other cereals	9.0	1,825
5	Sorghum	8.4	1,692
6	Barley	5.2	1,046
7	Other vegetables	3.5	718
8	Other pulses	3.5	699
9	Other fruits	3.2	647
10	Beer	2.6	521

Source: FAOSTAT (FAO 2010).
Note: All values are based on the three-year average for 2003–2005.

4 tons per hectare in the higher-rainfall areas. The degraded and drier areas of the country produce less than 0.5 tons per hectare. In spite of the considerable potential, yields are low partly due to low levels of erratic rainfall, degraded soil of poor fertility, poor access to inputs, and high input prices resulting from high transport costs.

Figure 6.10 shows the wheat yield and harvest area in Ethiopia. Wheat is produced over a large area; however, its yield is quite low, ranging between 1 and 2 tons per hectare, with slightly higher yields in parts of the higher-rainfall areas in north, northwest, southwest, and central areas of the country. Most areas in the north, south-central, and eastern parts, however, produce 1–2 tons per hectare. The degraded and drier areas produce less than a half-ton per hectare.

Figure 6.11 shows the results for rainfed sorghum. Sorghum is produced over large areas, but its yields are very low, ranging between 0.5 and 2 tons per hectare in most areas in the northern, south-central, western, and eastern parts of the country. More than 67 percent of the arable land is in drought-prone areas, and sorghum appears to be the crop best adapted to the adverse effects of climate. Removal of these drought-related problems as well as poor soil fertility, weeds, and disease is therefore crucial to ensure better yields and thereby food availability and income security for vulnerable zones of the country.

FIGURE 6.9   Yield (metric tons per hectare) and harvest area density (hectares) for rainfed maize in Ethiopia, 2000

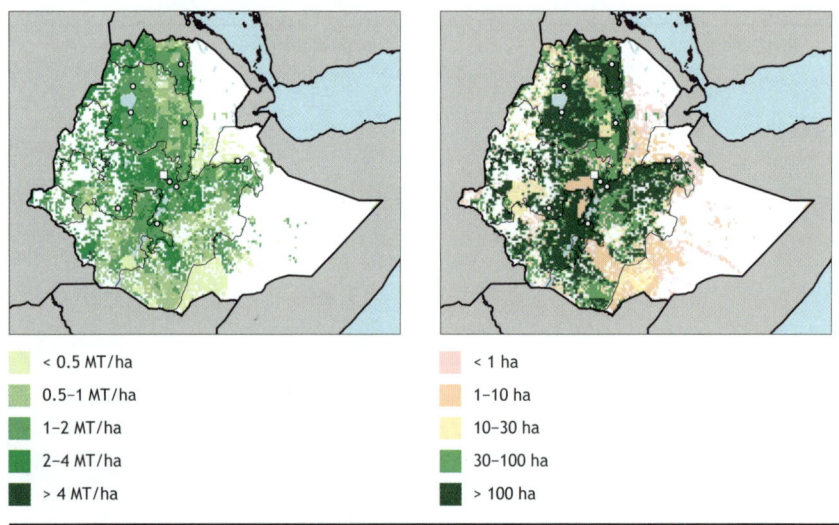

< 0.5 MT/ha	< 1 ha
0.5–1 MT/ha	1–10 ha
1–2 MT/ha	10–30 ha
2–4 MT/ha	30–100 ha
> 4 MT/ha	> 100 ha

Source: SPAM (Spatial Production Allocation Model) (You and Wood 2006; You, Wood, and Wood-Sichra 2006, 2009).
Note: ha = hectare; MT/ha = metric tons per hectare.

FIGURE 6.10   Yield (metric tons per hectare) and harvest area density (hectares) for rainfed wheat in Ethiopia, 2000

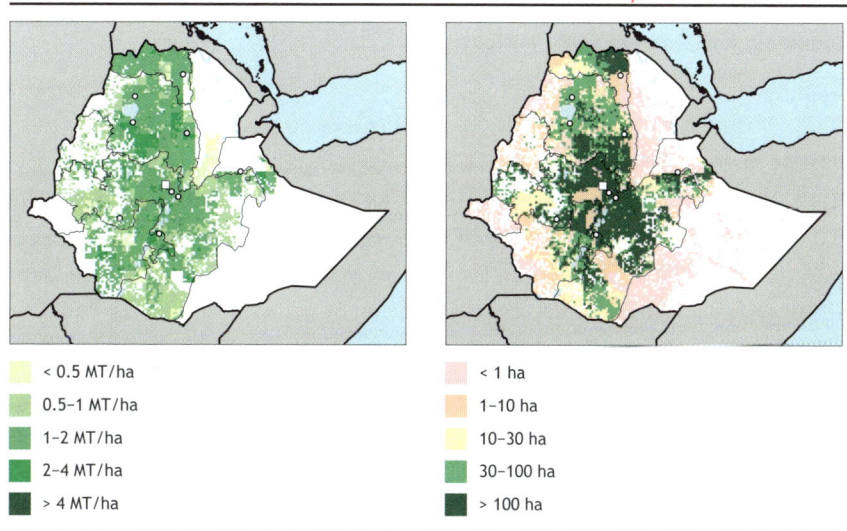

< 0.5 MT/ha	< 1 ha
0.5–1 MT/ha	1–10 ha
1–2 MT/ha	10–30 ha
2–4 MT/ha	30–100 ha
> 4 MT/ha	> 100 ha

Source: SPAM (Spatial Production Allocation Model) (You and Wood 2006; You, Wood, and Wood-Sichra 2006, 2009).
Note: ha = hectare; MT/ha = metric tons per hectare.

FIGURE 6.11   Yield (metric tons per hectare) and harvest area density (hectares) for rainfed sorghum in Ethiopia, 2000

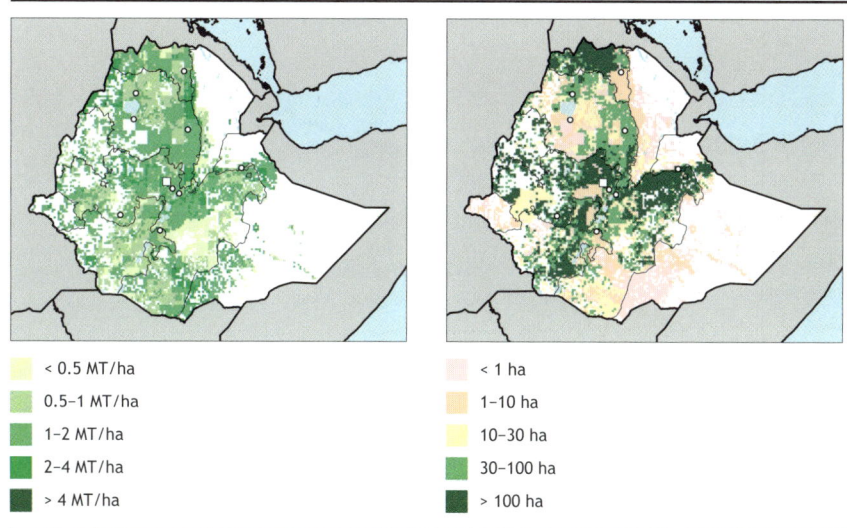

< 0.5 MT/ha	< 1 ha
0.5–1 MT/ha	1–10 ha
1–2 MT/ha	10–30 ha
2–4 MT/ha	30–100 ha
> 4 MT/ha	> 100 ha

Source: SPAM (Spatial Production Allocation Model) (You and Wood 2006; You, Wood, and Wood-Sichra 2006, 2009).
Note: ha = hectare; MT/ha = metric tons per hectare.

Scenarios for the Future

Economic and Demographic Indicators

Population

Figure 6.12 shows population projections by the UN Population Division (UNPOP 2009) through 2050. The Ethiopian population may double by 2050. The population is expected to reach 180 million in the baseline scenario and 205 million in the pessimistic scenario. All the projections show a more or less linear increase in population. Population growth in Ethiopia since the past century has been at the expense of degradation of the natural resource base. Measures to address the rapid pace of population growth and to boost GDP are required to reduce economic vulnerability and increase the adaptive capacity of the people.

Income

Figure 6.13 shows three scenarios for GDP per capita developed by combining three GDP projections with the three population projections of Figure 6.12. The optimistic scenario combines high GDP with low population, the baseline scenario combines the medium GDP projection with the medium

FIGURE 6.12   Population projections for Ethiopia, 2010–2050

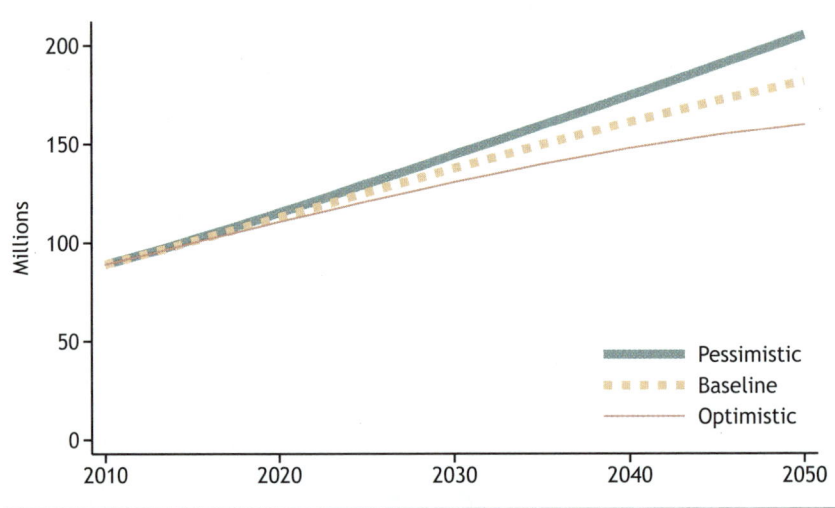

Source: UNPOP (2009).

FIGURE 6.13   Gross domestic product (GDP) per capita in Ethiopia, future scenarios, 2010–2050

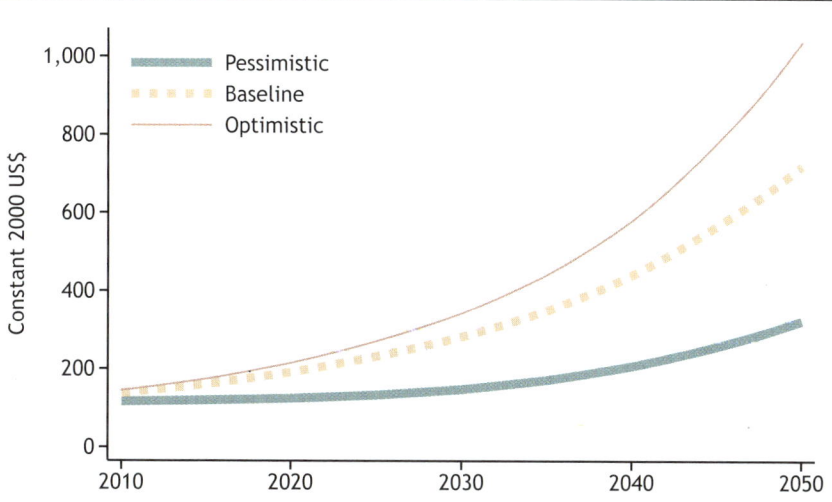

Sources: Computed from GDP data from the World Bank Economic Adaptation to Climate Change project (World Bank 2010), from the Millennium Ecosystem Assessment (2005) reports, and from population data from the United Nations (UNPOP 2009).

population scenario, and the pessimistic scenario combines the low GDP projection with the high population scenario.

Over the past decade, Ethiopia's economic growth has been fluctuating. The modeled results show a low expectation of GDP per capita under each of the three scenarios. The optimistic scenario, assuming the highest rate of GDP growth coupled with the lowest rate of population growth, shows GDP per capita exceeding $1,000 by 2050. The baseline scenario shows GDP per capita at about $700. The pessimistic scenario, assuming the lowest GDP and the highest rate of population growth, shows stunted growth through 2050 but nevertheless an almost doubling of per capita income from 2010. Such stunted growth would represent a vicious circle in which higher vulnerability would imply low adaptive capacity, hampering the growth of the agricultural sector.

Biophysical Analysis

Climate Models

Figure 6.14 shows projected precipitation changes in Ethiopia for 2000–2050 under the four downscaled climate models we use in this book with the A1B scenario.[1] CSIRO Mark 3 shows a decrease in annual precipitation in most highland areas of the country, while all the other models show normal to above-normal rainfall (though CNRM-CM3 shows decreased rainfall in the extreme western area).[2] Indeed, MIROC 3.2 shows a dramatic increase in precipitation over much of Ethiopia. In general, the maps show normal to above-normal levels for almost the entire country.

Figure 6.15 shows projected changes in average daily maximum temperature for the warmest month by 2050. All four GCMs agree that temperatures will rise, but they differ in how much. Two models, CNRM-CM3 and ECHAM 5, show a substantial increase in mean temperature, up to 2.5°C. CNRM-CM3 shows an increase in temperature over most of the west and the north—parts of the country projected to receive low to normal precipitation, indicating that water stresses are likely to increase. ECHAM 5 shows the greatest increase in the northeastern, northwestern, and central portions of the country. MIROC 3.2 and CSIRO Mark 3 show temperature increases for some parts of the country in the range of 1.5°–2°C, but both show that much of the temperature change will be lower, in the 1°–1.5°C range. Increases in temperature are likely to increase rates of evapotranspiration, likely leading to a deficit in the water balance and posing a major challenge to rainfed agriculture.

Extreme events such as droughts, floods, waterlogging, and erosion hazards may further degrade agricultural lands, reducing the quality of arable areas. Policymakers need to consider a mix of possible adaptation strategies to reduce the vulnerability of the agricultural sector and allow it to adapt to the challenges, including contingency plans designed to capitalize on any opportunities that may arise.

Crop Models

We used the Decision Support Software for Agrotechnology Transfer (DSSAT) software to compute crop yields under current temperature and precipitation regimes, and we repeated the simulation exercise for each of the future scenarios for the year 2050. The output for key crops is mapped in Figures 6.16–6.18, comparing crop yields for 2050 under climate change with the projected yields under an unchanged (2000) climate.

---

1   The A1B scenario is a greenhouse gas emissions scenario that assumes fast economic growth, a population that peaks midcentury, and the development of new and efficient technologies, along with a balanced use of energy sources.

2   CNRM-CM3 is National Meteorological Research Center–Climate Model 3. MIROC 3.2 is the Model for Interdisciplinary Research on Climate, developed at the University of Tokyo Center for Climate System Research. CSIRO Mark 3 is a climate model developed at the Australia Commonwealth Scientific and Industrial Research Organisation. ECHAM 5 is a fifth-generation climate model developed at the Max Planck Institute for Meteorology in Hamburg.

FIGURE 6.14   Changes in mean annual precipitation in Ethiopia, 2000–2050, A1B scenario (millimeters)

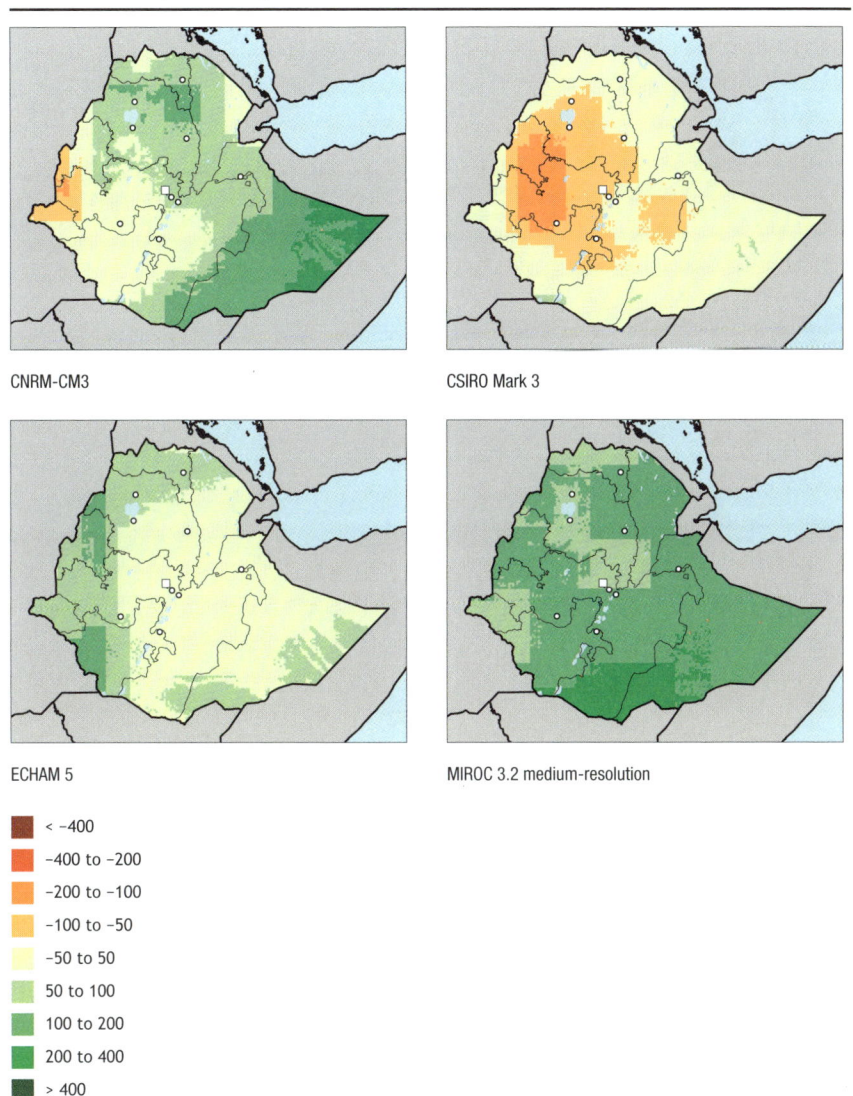

CNRM-CM3

CSIRO Mark 3

ECHAM 5

MIROC 3.2 medium-resolution

- ■ < –400
- ■ –400 to –200
- ■ –200 to –100
- ■ –100 to –50
- ■ –50 to 50
- ■ 50 to 100
- ■ 100 to 200
- ■ 200 to 400
- ■ > 400

Source: Authors' calculations based on Jones, Thornton, and Heinke (2009).

Notes: A1B = greenhouse gas emissions scenario that assumes fast economic growth, a population that peaks midcentury, and the development of new and efficient technologies, along with a balanced use of energy sources; CNRM-CM3 = National Meteorological Research Center–Climate Model 3; CSIRO = climate model developed at the Australia Commonwealth Scientific and Industrial Research Organisation; ECHAM 5 = fifth-generation climate model developed at the Max Planck Institute for Meteorology (Hamburg); GCM = general circulation model; MIROC = Model for Interdisciplinary Research on Climate, developed by the University of Tokyo Center for Climate System Research.

FIGURE 6.15   Change in monthly mean maximum daily temperature in Ethiopia for the warmest month, 2000–2050, A1B scenario (°C)

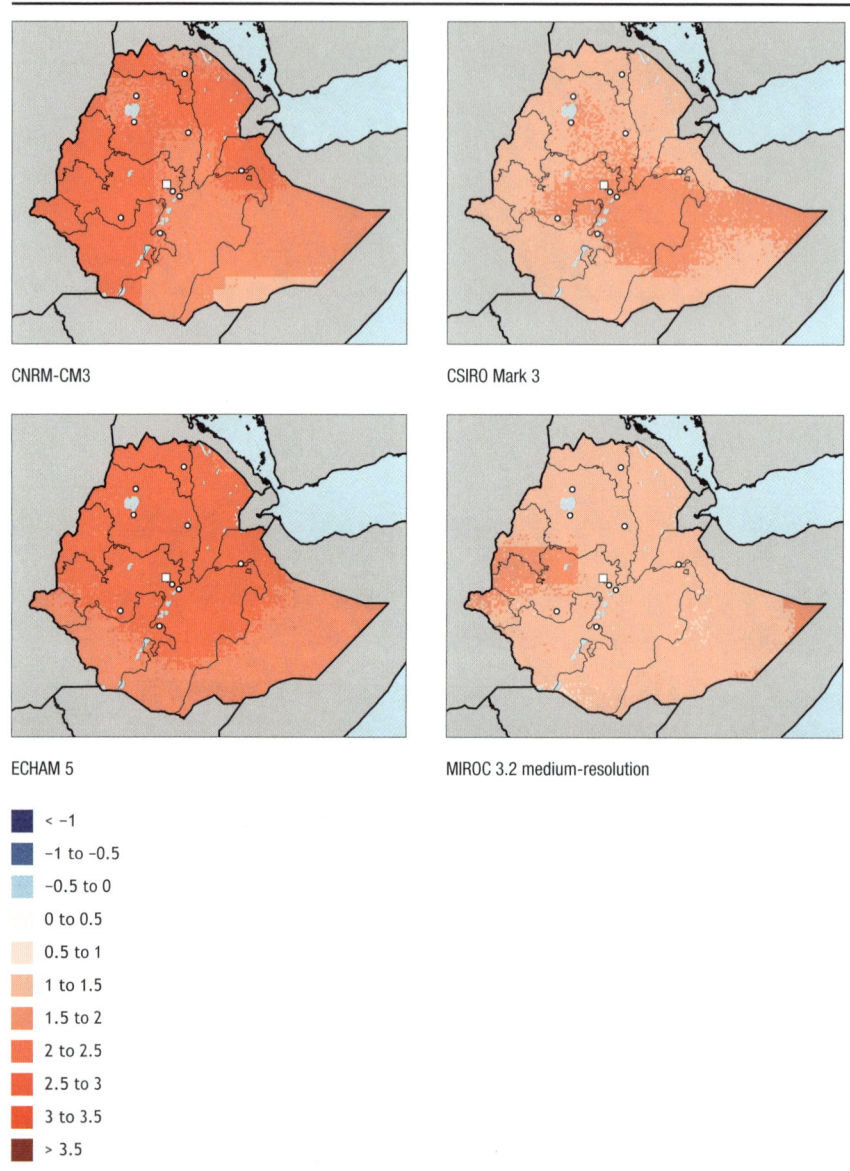

CNRM-CM3

CSIRO Mark 3

ECHAM 5

MIROC 3.2 medium-resolution

■	< –1
■	–1 to –0.5
■	–0.5 to 0
	0 to 0.5
■	0.5 to 1
■	1 to 1.5
■	1.5 to 2
■	2 to 2.5
■	2.5 to 3
■	3 to 3.5
■	> 3.5

Source: Authors' calculations based on Jones, Thornton, and Heinke (2009).

Notes: A1B = greenhouse gas emissions scenario that assumes fast economic growth, a population that peaks midcentury, and the development of new and efficient technologies, along with a balanced use of energy sources; CNRM-CM3 = National Meteorological Research Center–Climate Model 3; CSIRO = climate model developed at the Australia Commonwealth Scientific and Industrial Research Organisation; ECHAM 5 = fifth-generation climate model developed at the Max Planck Institute for Meteorology (Hamburg); GCM = general circulation model; MIROC = Model for Interdisciplinary Research on Climate, developed by the University of Tokyo Center for Climate System Research.

FIGURE 6.16   Yield change under climate change: Rainfed maize in Ethiopia, 2000–2050, A1B scenario

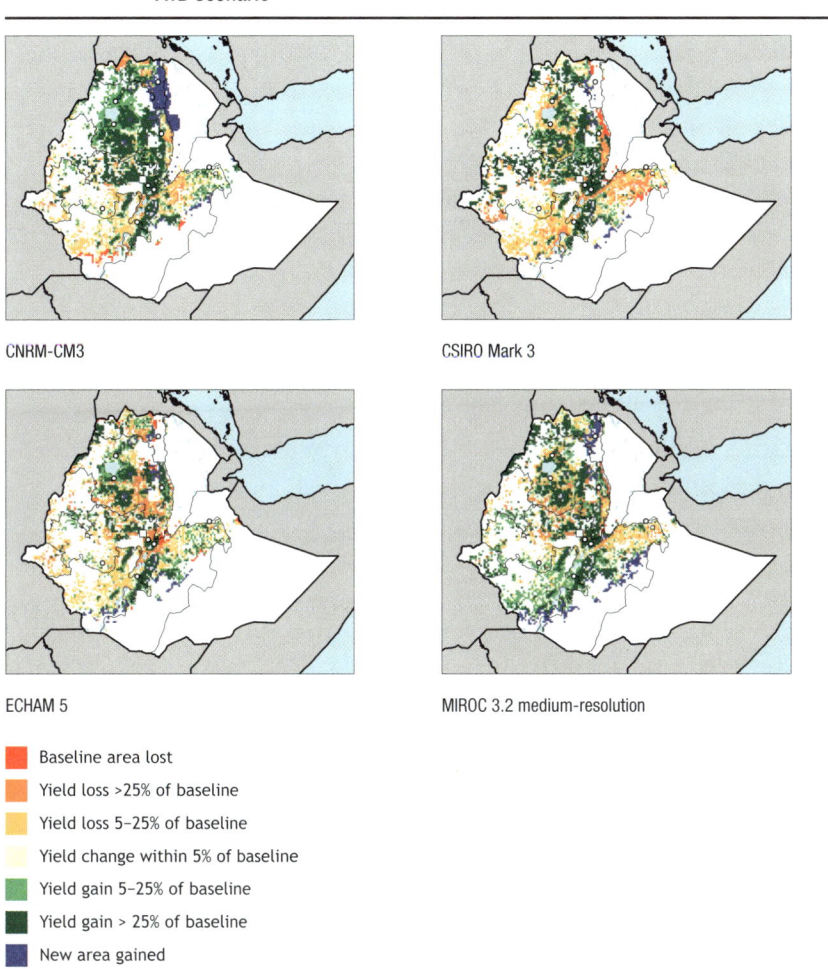

CNRM-CM3

CSIRO Mark 3

ECHAM 5

MIROC 3.2 medium-resolution

- ■ Baseline area lost
- ■ Yield loss >25% of baseline
- ■ Yield loss 5–25% of baseline
- ■ Yield change within 5% of baseline
- ■ Yield gain 5–25% of baseline
- ■ Yield gain > 25% of baseline
- ■ New area gained

Source: Authors' calculations.

Notes: A1B = greenhouse gas emissions scenario that assumes fast economic growth, a population that peaks midcentury, and the development of new and efficient technologies, along with a balanced use of energy sources; CNRM-CM3 = National Meteorological Research Center–Climate Model 3; CSIRO = climate model developed at the Australia Commonwealth Scientific and Industrial Research Organisation; ECHAM 5 = fifth-generation climate model developed at the Max Planck Institute for Meteorology (Hamburg); GCM = general circulation model; MIROC = Model for Interdisciplinary Research on Climate, developed by the University of Tokyo Center for Climate System Research.

Figure 6.16 shows the modeled results for rainfed maize in Ethiopia. All models suggest that there will be a gain in maize yields of more than 25 percent in the eastern highlands at the edge of Great Rift Valley as well as in the north central highlands; to varying degrees they also show some patches of new area gained in the eastern parts of Amhara and Tigray. Three of the four models, however, show an equal amount of existing maize land marginalized or no longer suitable for maize, as well as a marked decrease in the maize yield over the southwestern and eastern parts of central Ethiopia. Both CNRM-CM3 and MIROC 3.2 suggest a considerable gain in maize production area, mainly in eastern Tigray. In contrast, CNRM-CM3 indicates that a strip along the southern periphery of Oromia may lose maize areas completely.

Maize is one of the most important foodcrops for the majority of the rural people in Ethiopia, and its potential vulnerability (indicated by the models) may cause a substantial food deficit. Hence, alternative strategies are needed for adapting maize to potential changes and to ensure the production of sufficient food. However, we can also clearly see that there may be significant gains as a result of climate change. It will be important to monitor areas that in the past have been poor for maize production but might become quite suitable in the future. If farmers are already located in those places, they can take advantage of the changes. If farmers are not present, and particularly if the area is under forest, policymakers have a much more difficult task of deciding whether they are willing to trade forest for agricultural production. It is not all a zero-sum game: policymakers could decide to help return to forest those areas lost to crop production.

Figure 6.17 shows the results of the crop simulation model for rainfed wheat. It suggests substantial reduction in wheat yields and some loss of area, even where rainfall is expected to increase—presumably due to heat stress. Wheat ranks as the most important crop in Ethiopia in terms of monetary value (see Table 6.4), as well as the third most important foodcrop by amount of consumption (see Table 6.5). It is especially disturbing, then, that all four models show future loss of wheat yield: all zones may lose the opportunity for spring wheat production, to levels far below the baseline; some wheat areas may still be available, but they would be very fragile and not so favorable. Also, there may still be possibilities for continued production of wheat in other areas, because Ethiopia offers a wide range of diverse agroecologies.

Figure 6.18 shows an increase in yield and an expansion in area in major sorghum-growing areas. All four models show a decreasing trend in the yield of rainfed sorghum over a very large area in the western and northwestern parts of the country but more than a 25 percent increase in central Ethiopia and isolated chains toward the north, including substantial crop area gained. However, the newly emerging production areas are all in fragile agroenvironments; tremendous work would be required to rehabilitate the degraded areas. Note that CNRM-CM3 and MIROC 3.2 give contrasting results for the southern

FIGURE 6.17   Yield change under climate change: Rainfed wheat in Ethiopia, 2000–2050, A1B scenario

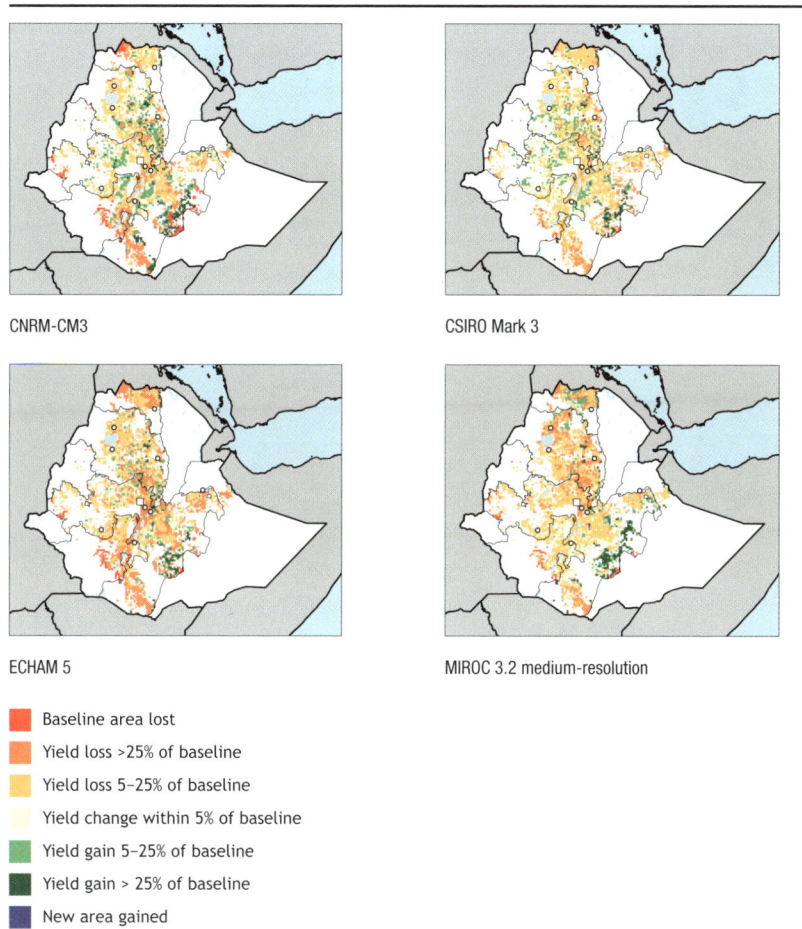

CNRM-CM3

CSIRO Mark 3

ECHAM 5

MIROC 3.2 medium-resolution

- Baseline area lost
- Yield loss >25% of baseline
- Yield loss 5–25% of baseline
- Yield change within 5% of baseline
- Yield gain 5–25% of baseline
- Yield gain > 25% of baseline
- New area gained

Source: Authors' calculations.
Notes: A1B = greenhouse gas emissions scenario that assumes fast economic growth, a population that peaks midcentury, and the development of new and efficient technologies, along with a balanced use of energy sources; CNRM-CM3 = National Meteorological Research Center–Climate Model 3; CSIRO = climate model developed at the Australia Commonwealth Scientific and Industrial Research Organisation; ECHAM 5 = fifth-generation climate model developed at the Max Planck Institute for Meteorology (Hamburg); GCM = general circulation model; MIROC = Model for Interdisciplinary Research on Climate, developed by the University of Tokyo Center for Climate System Research.

and southeastern regions: the former shows a loss of baseline sorghum production area, while the latter shows new opportunities for sorghum over the same area. The four models all show sorghum production declining by 5–25 percent in the western parts of Tigray, Amhara, Oromia, and SNNP, as well as the whole of Benishangul-Gumuz and Gambella states. In addition, two of the models show

FIGURE 6.18 Yield change under climate change: Rainfed sorghum in Ethiopia, 2000–2050, A1B scenario

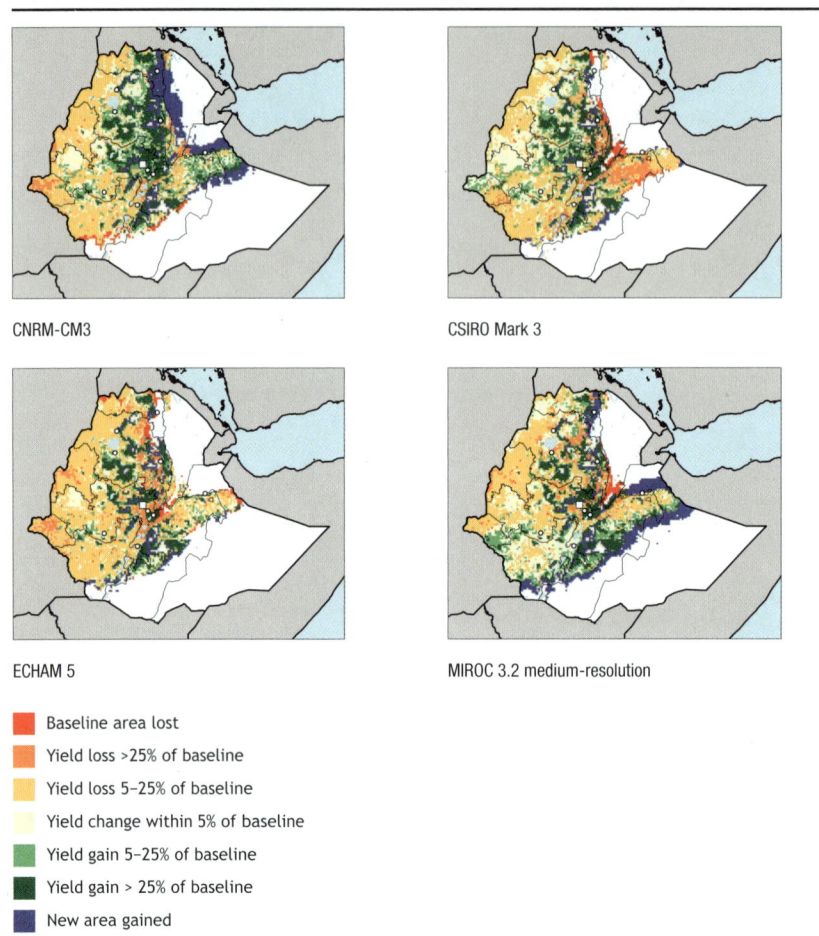

CNRM-CM3

CSIRO Mark 3

ECHAM 5

MIROC 3.2 medium-resolution

■ Baseline area lost
■ Yield loss >25% of baseline
■ Yield loss 5–25% of baseline
□ Yield change within 5% of baseline
■ Yield gain 5–25% of baseline
■ Yield gain > 25% of baseline
■ New area gained

Source: Authors' calculations.

Notes: A1B = greenhouse gas emissions scenario that assumes fast economic growth, a population that peaks midcentury, and the development of new and efficient technologies, along with a balanced use of energy sources; CNRM-CM3 = National Meteorological Research Center–Climate Model 3; CSIRO = climate model developed at the Australia Commonwealth Scientific and Industrial Research Organisation; ECHAM 5 = fifth-generation climate model developed at the Max Planck Institute for Meteorology (Hamburg); GCM = general circulation model; MIROC = Model for Interdisciplinary Research on Climate, developed by the University of Tokyo Center for Climate System Research.

yield declines exceeding 25 percent over large areas of Gambella, and CSIRO Mark 3 shows a considerable decline in most parts of northeastern Oromia.

The analytical results suggest significant impacts of climate change on the production and yields of major crops in terms of both altered production areas and total production. The new areas gained may be less favorable, moreover.

Increasing temperature, even when accompanied by an increase in annual rainfall, may create a higher risk of water deficits due to greater loss through evaporation, though the results in these three figures already reflect the impact of any potential water deficit.

It is unfortunate that DSSAT is not able to model coffee, because coffee is of particular importance to agricultural export earnings in Ethiopia. The effect of climate change on coffee in Ethiopia is ultimately unclear. Higher temperatures would likely adversely effect the yield of coffee in its current locations, yet by 2050, coffee could easily be planted in slightly different locations with cooler temperatures. In the short run, coffee may be adversely affected not only by higher temperatures but also by the coffee berry borer, which likes warmer temperatures. In the longer run, we would expect coffee to be relocated and yields to rise close to where they were prior to climate change.

Vulnerability

Figure 6.19 shows the impact of future GDP and population scenarios on the number of malnourished children under age five in Ethiopia. Figure 6.20 shows the share of children who are malnourished. The number of malnourished children decreases in the optimistic scenario, reflecting a lower

FIGURE 6.19   Number of malnourished children under five years of age in Ethiopia in multiple income and climate scenarios, 2010–2050

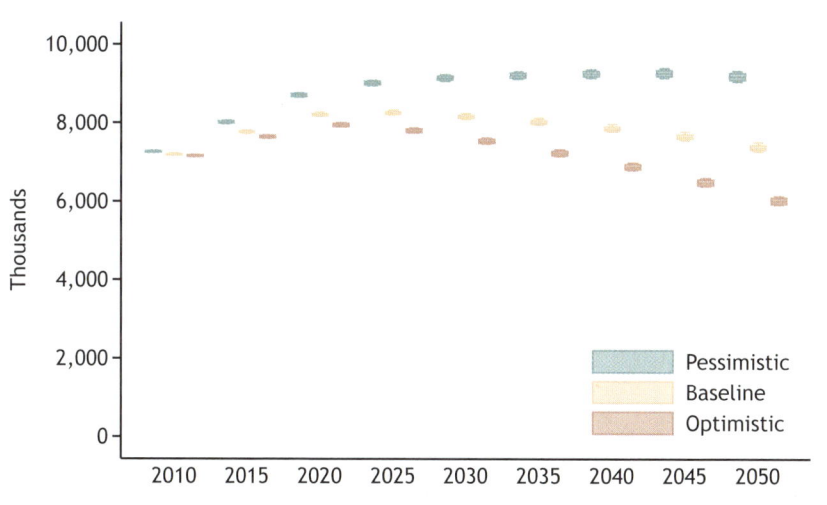

Source: Based on analysis conducted for Nelson et al. (2010).

Note: The box and whiskers plot for each socioeconomic scenario shows the range of effects from the four future climate scenarios.

FIGURE 6.20 Share of malnourished children under five years of age in Ethiopia in multiple income and climate scenarios, 2010–2050

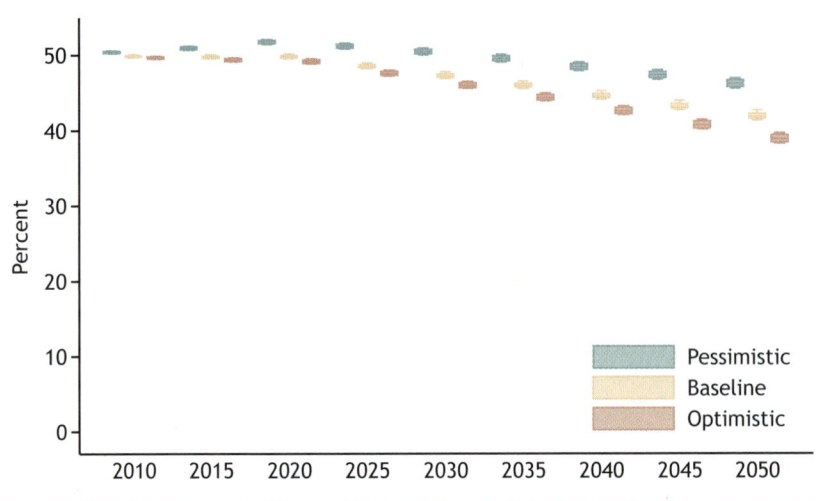

Source: Based on analysis conducted for Nelson et al. (2010).

Note: The box and whiskers plot for each socioeconomic scenario shows the range of effects from the four future climate scenarios.

rate of population growth and improvement in GDP; the baseline scenario shows a more or less constant number of malnourished children. Even though the number of malnourished children increases in the pessimistic scenario between 2010 and 2050, the share declines due to population growth.

Figure 6.21 shows projections of available kilocalories per capita under multiple income and climate scenarios. Clear increases are shown in the optimistic scenario and modest increases in the baseline scenarios but no increase in the pessimistic scenario. The optimistic scenario would achieve the minimal per capita requirement (2,100 kilocalories per day) after 2040, but it will not be possible to attain this target if the pessimistic scenario proves correct.

Agricultural Outcomes

Figure 6.22–6.24 show simulation results from the International Model for Policy Analysis of Agricultural Commodities and Trade (IMPACT) associated with key agricultural crops in Ethiopia. Each featured crop has five graphs showing production, yield, area, net exports, and world price trends for 2010–2050.

The simulation results presented in Figure 6.22 show the maize yield rising somewhat by 2020 and then leveling off. At the same time, the cultivated

FIGURE 6.21   Kilocalories per capita in Ethiopia in multiple income and climate scenarios, 2010–2050

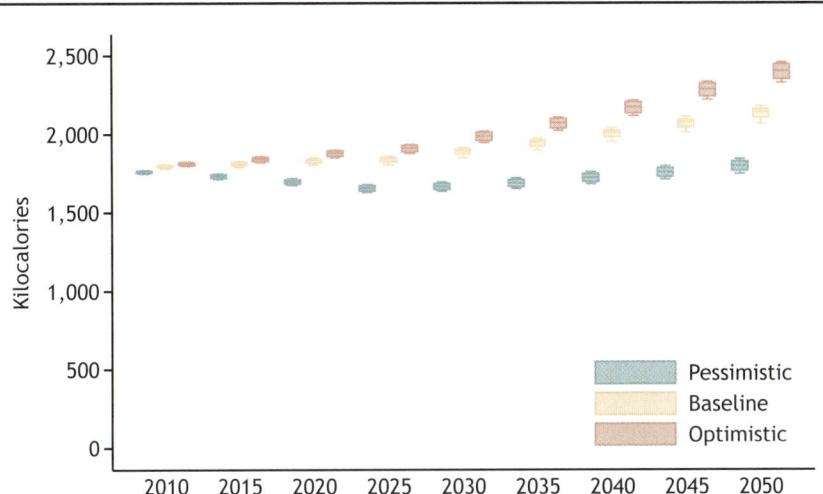

Source: Based on analysis conducted for Nelson et al. (2010).
Note: The box and whiskers plot for each socioeconomic scenario shows the range of effects from the four future climate scenarios.

area of maize declines slightly after 2020. Together these lead production to rise by around 25 percent by 2020, then fall slightly thereafter, with levels in 2050 about the same as those in 2010. Note that some loss of maize area is indicated in the maps projected from DSSAT (see Figure 6.16), but there is some potential gain in maize area in other parts of the country.

The imports of maize into Ethiopia are projected to increase after 2020, despite the attractive world price suggested for maize. The box-and-whisker plots suggest that the interquartile range increases into the future, reflecting divergence of the yields from the climate models. This makes planning for the future more complex and challenging.

For wheat, Figure 6.23 shows that the yield is projected to more than double by 2050. With a slight increase in area, production by 2050 increases to around 2.5 times the 2010 level. Because the range for the production prediction is relatively small, the divergence of the net exports must reflect divergence in consumer demand for wheat. One scenario has Ethiopia as a net exporter, another as most likely a net importer, and the third is undecided as to which way Ethiopia will go.

There appears to be an increasing trend in the export of wheat in the pessimistic scenario. This is attributed to the low GDP in this scenario, which is

FIGURE 6.22   Impact of changes in GDP and population on maize in Ethiopia, 2010–2050

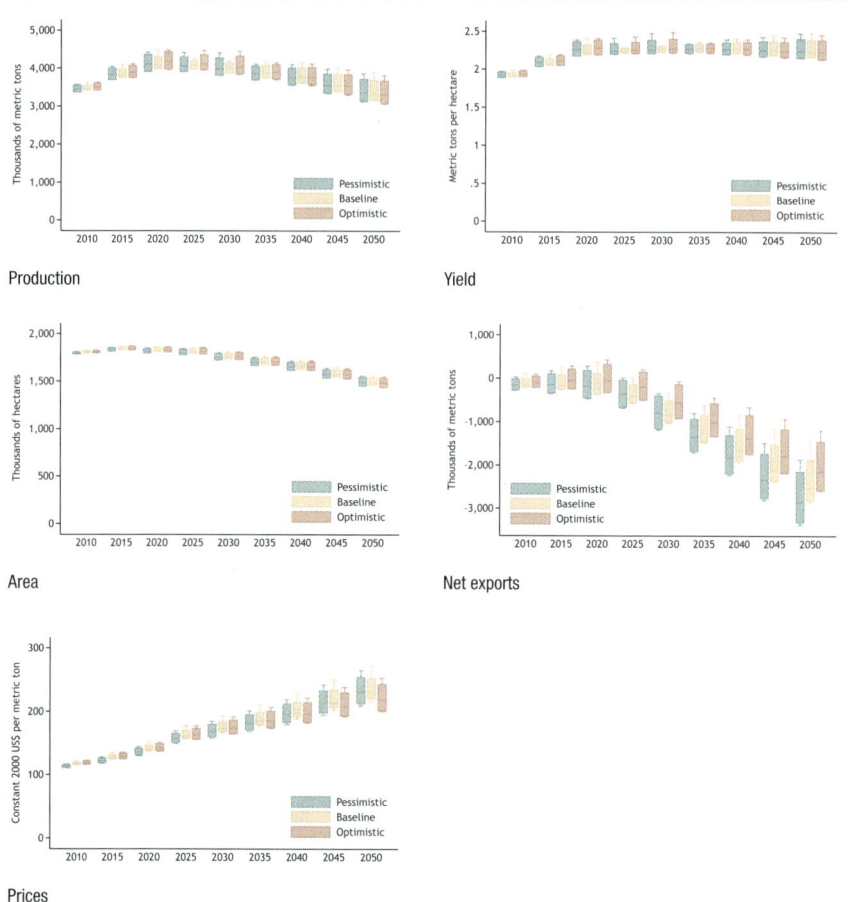

Source: Based on analysis conducted for Nelson et al. (2010).

Notes: The box and whiskers plot for each socioeconomic scenario shows the range of effects from the four future climate scenarios. GDP = gross domestic product; US$ = US dollars.

expected to make it more attractive to export wheat than to consume it. It is also likely that the consumption of maize-based food products will increase and that wheat will be sold. However, in the optimistic scenario, because the GDP is high and the population growth rate is low, consumers may be able to afford to purchase wheat for food, and the county may reduce its wheat exports toward 2050.

The productivity of all crops shows an increasing trend, probably attributed to an assumption of technological progress.

FIGURE 6.23   Impact of changes in GDP and population on wheat in Ethiopia, 2010–2050

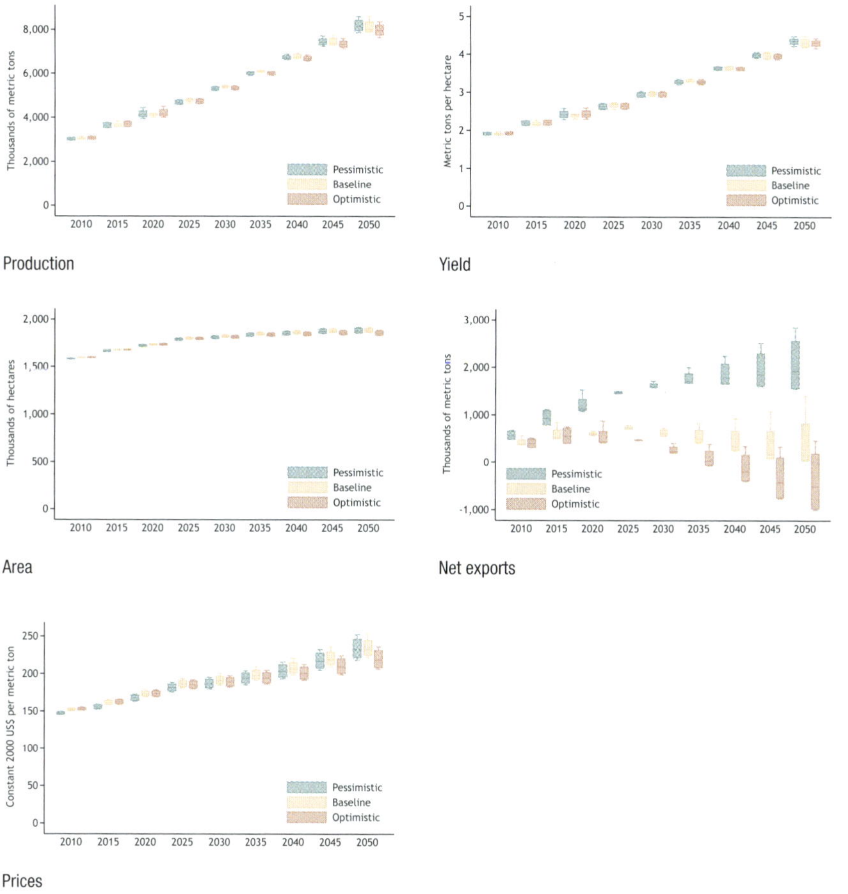

Production

Yield

Area

Net exports

Prices

Source: Based on analysis conducted for Nelson et al. (2010).

Notes: The box and whiskers plot for each socioeconomic scenario shows the range of effects from the four future climate scenarios. GDP = gross domestic product; US$ = US dollars.

Sorghum shows an increasing trend in production, yield, area, net exports, and world prices (Figure 6.24). This is somewhat in agreement with the result presented in Figure 6.11, showing an increase in yield and an expansion in area in major sorghum-growing areas. Sorghum is usually grown in drier areas to which most other high-value cereals and other crops are not adapted. This signals some advantage in sorghum production from expansion into the drier areas of Ethiopia, indicating that there is tremendous work ahead in balancing environmental management and food production needs.

FIGURE 6.24   Impact of changes in GDP and population on sorghum in Ethiopia, 2010–2050

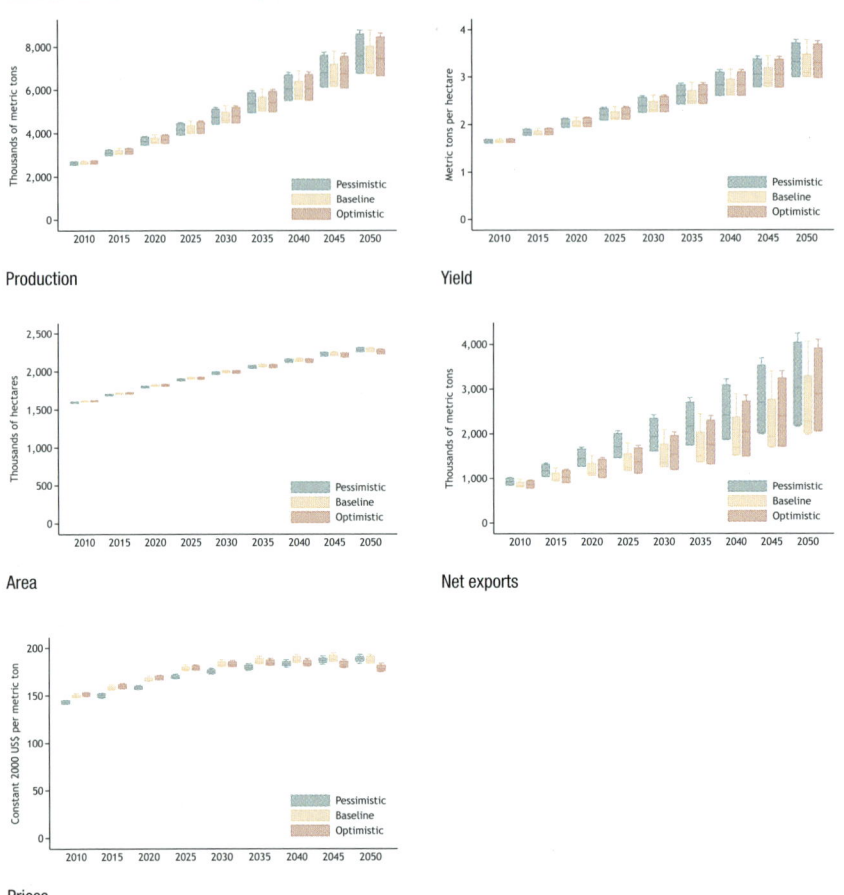

Source: Based on analysis conducted for Nelson et al. (2010).

Notes: The box and whiskers plot for each socioeconomic scenario shows the range of effects from the four future climate scenarios. GDP = gross domestic product; US$ = US dollars.

Conclusions and Policy Recommendations

Projections show Ethiopia's population increasing to as many as 205 million by 2050—more than 2.5 times the present level. Given that over 80 percent of the current population is living below the poverty line, serious problems must be expected from increased demand for food, water, and other basic needs under climate change. But the baseline and optimistic scenarios paint a brighter future for the country, in terms of both smaller population projections and larger GDP per capita projections.

Climate change affects biodiversity as well. Ethiopia, home to a rich and diverse flora and fauna, has already been critically affected by the loss of plant biodiversity. The potential loss of endemic bird species recorded in Ethiopia is of international concern. Protected areas play a role in mitigating climate change by providing buffers against extreme climate events as well as a network of natural habitats to provide pathways for rapid migration and space for evolution and adaptation. Protected areas are likely to be affected by climate change, potentially losing species and even ecosystems.

An essential priority must be conserving natural terrestrial and aquatic ecosystems (both freshwater and saline water bodies), along with restoring degraded ecosystems. These ecosystems play key roles in the global carbon cycle and in adapting to climate change while serving a wide range of functions essential for human well-being.

From the crop model projections it can be inferred that climate change might adversely affect production and cause certain growing areas to be no longer viable. On the positive side, there are areas of projected increases in yield and new areas that could be brought under cultivation in a warmer and sometimes wetter climate. The crop models show that wheat will be one of the hardest-hit crops, with significant losses and very few gains. The variation in outcomes of various models suggests the complexity of the problem of projecting future impacts on agriculture. There is an urgent need to develop alternative adaptation options that will fit into the various plausible scenarios and to design location-specific adaption programs to reach all vulnerable populations.

Many developmental institutions are currently trying to promote technologies for adaptation but in an uncoordinated manner. More collaboration among partners and harmonization of approaches will be required for effective scaling up of proven strategies in order to effectively address smallholders' technological adaptation needs. It is also necessary to adopt cross-sectoral programs that have strong ownership by stakeholders to effectively address climate change adaptation at different levels. There are some efforts underway to climate-proof development efforts and to institutionalize climate change adaptation into research for development as well as into poverty reduction and food security improvement efforts. Nevertheless, implementation of these policies and plans remains to be done.

To meet the challenges of climate change in Ethiopia, we recommend the following:

- Harmonize policies and institutional frameworks affecting climate change adaptation across different approaches and strategies.

- Develop alternative adaptation options for the various plausible scenarios and design multiple adaption programs for the diverse climates of Ethiopia so that all vulnerable populations can be reached.

- Provide rural financing to promote the adoption and use of proven technologies for climate-change adaptation.

- Encourage risk-insuring institutions to insure rainfall risks, especially for smallholder farmers as they adopt improved agricultural production technologies to benefit from potentially increased rainfall.

- Improve the road infrastructure in remote areas to increase farmers' opportunities and access to markets and market information.

- Manage rainwater to prevent potential flooding, waterlogging, erosion, and nutrient leaching under increased rainfall.

- Make reliable climate forecasts available to smallholders to reduce climate-induced risks.

- Integrate efficient agricultural water management practices with productivity-enhancing interventions.

- Integrate indigenous strategies and complex local technical knowledge with science-based knowledge to support adaptation to climate change.

- Promote new crop varieties adapted to drought, such as nutritionally enhanced maize varieties as well as drought-tolerant sorghum, teff, cassava, and market-preferred common bean varieties. Specific nutritionally enhanced crops—such as quality protein maize varieties with high lysine and tryptophan content—could alleviate protein-deficiency problems and under-five malnutrition, widely encountered in rural communities that depend on maize as their staple food.

- Promote dairy goat and poultry farming and silkworm rearing in appropriate agroclimatic conditions.

References

ADB (African Development Bank). 2010. "Ethiopia's Economic Growth Performance: Current Situation and Challenges." *Economic Brief* 1 (5). http://afdb.org/fileadmin/uploads/afdb/ Documents/Publications/ECON %20Brief_Ethiopias %20Economic %20growth.pdf.

Adenew, B. 2003. *The Food Security Role of Agriculture in Ethiopia.* FAO–ROA (Food and Agriculture Organization–Roles of Agriculture) Research Project at EEA/EEPRI (Ethiopian Economic Association /Ethiopian Economic and Policy Research Institute). Addis Ababa.

Bartholome, E., and A. S. Belward. 2005. "GLC2000: A New Approach to Global Land Cover Mapping from Earth Observation Data." *International Journal of Remote Sensing* 26 (9–10): 1959–1977.

CIESIN (Center for International Earth Science Information Network), Columbia University, IFPRI (International Food Policy Research Institute), World Bank, and CIAT (Centro Internacional de Agricultura Tropical). 2004. *Global Rural–Urban Mapping Project (GRUMP), Alpha Version: Population Density Grids.* Palisades, NY, US: Socioeconomic Data and Applications Center (SEDAC), Columbia University. http://sedac.ciesin.columbia.edu/gpw.

Ethiopia, MOFED (Ministry of Finance and Economic Development). 2002a. *Ethiopia: Sustainable Development and Poverty Reduction Program.* Addis Ababa.

———. 2002b. *Food Security Strategy.* Addis Ababa.

FAO (Food and Agriculture Organization of the United Nations). 1996. *World Food Summit Plan of Action.* Rome. http://fao.org/docrep/003/w3613e00.htm#PoA.

———. 2010. FAOSTAT. Rome. http://faostat.fao.org.

Jones, P. G., P. K. Thornton, and J. Heinke. 2009. *Generating Characteristic Daily Weather Data Using Downscaled Climate Model Data from the IPCC's Fourth Assessment.* Project report for the International Livestock Research Institute. Geneva: International Panel on Climate Change.

Lehner, B., and P. Döll. 2004. "Development and Validation of a Global Database of Lakes, Reservoirs, and Wetlands." *Journal of Hydrology* 296 (1–4): 1–22.

Millennium Ecosystem Assessment. 2005. *Ecosystems and Human Well-being: Synthesis.* Washington, DC: Island Press. http://www.maweb.org/en/Global.aspx.

Moreland, S., and E. Smith. 2012. *Modeling Climate Change, Food Security and Population.* Chapel Hill, NC, US: MEASURE Evaluation PRH. Study Summary: "Improving Access to Family Planning Can Promote Food Security in the Face of Ethiopia's Changing Climate." Accessed June 6, 2012. http://www.cpc.unc.edu/measure/publications/sr-12-69/at_download/document.

Nelson, G. C., M. W. Rosegrant, A. Palazzo, I. Gray, C. Ingersoll, R. Robertson, S. Tokgoz, et al. 2010. *Food Security, Farming, and Climate Change to 2050: Scenarios, Results, Policy Options.* Washington, DC: International Food Policy Research Institute.

UNEP (United Nations Environment Programme) and IUCN (International Union for Conservation of Nature). 2009. World Database on Protected Areas (WDPA) Annual Release 2009. No longer available online.

UNPOP (United Nations Secretariat, Department of Economic and Social Affairs, Population Division). 2009. *World Population Prospects: The 2008 Revision.* Accessed April 6, 2010. http://esa.un.org/unpp.

Wolde-Georgis, T. 1997. "El Niño and Drought Early Warning in Ethiopia." *Internet Journal of African Studies* 2 (1–7).

Wood, S., G. Hyman, U. Deichmann, E. Barona, R. Tenorio, Z. Guo, S. Castano, O. Rivera, E. Diaz, and J. Marin. 2010. "Sub-national Poverty Maps for the Developing World Using International Poverty Lines: Preliminary Data Release." Accessed May 6. http://povertymap.info.

World Bank. 2009. *World Development Indicators.* Accessed May 2011.

———. 2010. *Economics of Adaptation to Climate Change: Synthesis Report.* Washington, DC. http://climatechange.worldbank.org/content/economics-adaptation-climate-change-study-homepage.

You, L., and S. Wood. 2006. "An Entropy Approach to Spatial Disaggregation of Agricultural Production." *Agricultural Systems* 90 (1–3): 329–347.

You, L., S. Wood, and U. Wood-Sichra. 2006. "Generating Global Crop Distribution Maps: From Census to Grid." Paper presented at the International Association of Agricultural Economists Conference, Brisbane, Australia, August 11–18.

———. 2009. "Generating Plausible Crop Distribution and Performance Maps for Sub-Saharan Africa Using a Spatially Disaggregated Data Fusion and Optimization Approach." *Agricultural Systems* 99 (2–3): 126–140.

Chapter 7

KENYA

Michael Makokha Odera, Timothy S. Thomas, Michael Waithaka,
and Miriam Kyotalimye

Kenya is an ecologically diverse nation located on the Indian Ocean in East Africa. Most of the land is classified as arid or semiarid, yet at higher elevations lush montane forests are found. Kenya is home to the second-highest peak in Africa, glacier-capped Mount Kenya. Although agriculture is declining in importance in terms of gross domestic product (GDP), approximately 75 percent of the country's labor force is still devoted to agriculture. Because of this and because of Kenya's relatively slow growth in GDP per capita since 1980, adapting to the impact of climate change on agriculture is of critical importance to its population over the next 30 or 40 years.

In this chapter we review some national statistics that will help characterize the capacity of the population to adapt to climate change and describe Kenya's land base and current agricultural context. Then we show results of crop models that incorporate climate projections from recent global models. Finally, we incorporate those results into a global model for food and for agricultural supply and demand that takes into account demographic and economic projections for Kenya. Along the way, we identify areas of potential losses and gains.

The chapter draws heavily from the *National Irrigation Board Strategic Plan* (Kenya, Ministry of Water and Irrigation 2008), the *Kenya Agricultural Research Institute (KARI) Strategic Plan* (KARI 2009), the *Agricultural Sector Development Strategy* (Kenya 2010), the *National Climate Change Response Strategy* (Kenya, Ministry of Environment and Mineral Resources, 2010), the "Draft National Arid and Semi-Arid Lands Policy" (Kenya 2011a), the "Draft National Food and Nutrition Security Policy" (Kenya 2011b), and the "Draft National Irrigation Policy" (Kenya 2011c).

In the past decade or so, Kenya has developed policies, strategies, and programs in the agricultural and environment sectors with direct and indirect bearing on adaptation of the agricultural sector to climate change. The *National Climate Change Response Strategy* (NCCRS) (Kenya 2010) aims at mainstreaming climate change issues throughout all economic sectors and ensuring coordinated implementation of climate change mitigation and adaptation activities. The NCCRS also acknowledges the importance of effective communication, education, and public awareness programs in improving the resilience of communities and productive sectors for adaptation to climate change.

The Agriculture Sector Coordination Unit has developed the *Agricultural Sector Development Strategy* (Kenya, Ministry of Environment and Mineral Resources, 2010). The strategy has articulated subsector strategic focus around key themes that have a bearing on the adaptive response of the agricultural sector and all subsectors therein to the vulnerability occasioned by climate change. The Ministry of Environment and Mineral Resources, Office of the Prime Minister, and Ministry of Development of Northern Kenya and other Arid Lands have developed a five-year Natural Resources Management Program (2010–2014) whose overall objective is to contribute to reduttced poverty in the context of "Kenya Vision 2030" (Kenya 2007b) to safeguard the state of the environment and promote sustainable management of natural resources, including adaptation to climate change.

Review of the Current Situation and Trends

Economic and Demographic Indicators

Population
Figure 7.1 shows total and rural population counts (left axis) and the share of the urban population (right axis). The population trends and growth rates given in Figure 7.1 and Table 7.1 show a general upward trend in the urban, rural, and total populations, with the urban population rising from 14 percent in 1960 to 20 percent in 2008, while the total population increased from about 8 million people in 1960 to approximately 38 million in 2008. The rural population increased from approximately 7 million in 1960 to 30 million in 2008. From the population growth rates given in Table 7.1, we note that in all the four decades and eight years under study, the urban growth rate was higher than the rural growth rate, indicating a general trend of rural to urban migration or high relative urban child delivery rates over the period of analysis. There has also been a declining trend in growth rate over the temporal scale across all three categories of population. The highest urban growth rate was experienced in the first two decades. After that, the growth rates decreased between 1980 and 1999, only to rise marginally between 2000 and 2008. The total population growth rate took an upward trend in the first two decades

FIGURE 7.1   Population trends in Kenya: Total population, rural population, and percent urban, 1960–2008

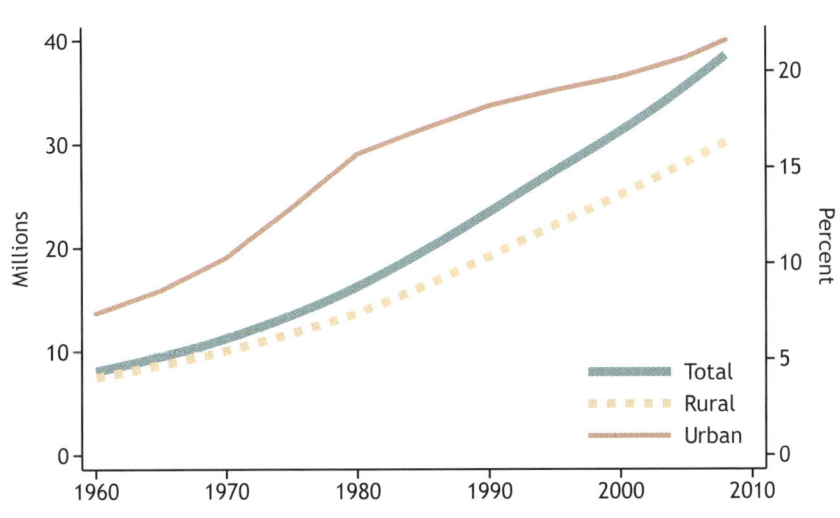

Source: *World Development Indicators* (World Bank 2009).

FIGURE 7.2   Population distribution in Kenya, 2000 (persons per square kilometer)

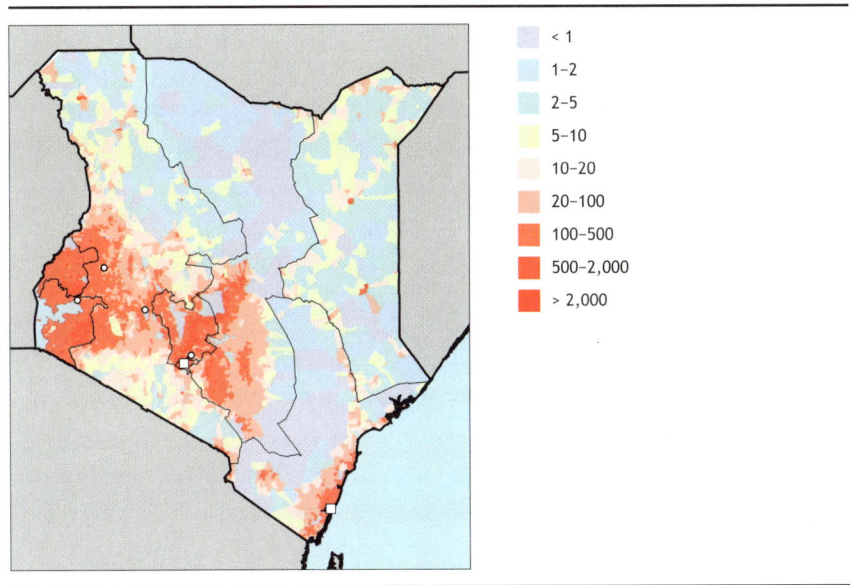

Source: CIESIN et al. (2004).

TABLE 7.1   Population growth rates in Kenya, 1960–2008 (percent)

Decade	Total growth rate	Rural growth rate	Urban growth rate
1960–1969	3.3	3.0	6.6
1970–1979	3.7	3.1	7.9
1980–1989	3.7	3.4	5.2
1990–1999	2.9	2.7	3.7
2000–2008	2.6	2.3	3.8

Source: Authors' calculations based on *World Development Indicators* (World Bank 2009).

under analysis, only to level off between 1980 and 1989, then decreased between 1990 and 2008. The rural population growth rate took on an upward trend in the first three decades, reaching its peak in 1989, and then assumed a downward trend thereafter until 2008.

Figure 7.2 provides additional information concerning population density. A glance at this figure reveals a relatively high population density in agriculturally medium- to high-potential areas of central Kenya, the coastal strip, the Rift Valley, and western Kenya. The population density in the arid and semi-arid upper parts of Eastern Province is between 1 and 2 persons per square kilometer. This can be partly explained by the pastoral lifestyle in the region.

Income

Figure 7.3 shows trends in GDP per capita and the proportion of GDP from agriculture. Agriculture is included as an indicator because it is vulnerable to the impacts of climate change and is a key indicator of the level of development of the country, whose mainstay economic activity is agriculture.

Generally, the graph shows a rapid upward trend in GDP per capita in the period between 1960 and 1990, increasing from approximately $260 in 1960 to around $450 some 30 years later. This was followed by a general decline until 2005 ($400). Since 2005, there has been an upward trend in GDP per capita. During the same period, the proportion of agriculture's contribution to GDP declined from approximately 40 percent in 1960 to near 20 percent around 2008, though the sudden drop from 26 percent in 2007 represents the impact of the 2008–2009 drought. A comparison of the trends of GDP per capita and the proportionate contribution of agriculture to GDP as given in Figure 7.3 clearly brings out the inverse relationship between growth in GDP and the proportionate contribution of agriculture to GDP. The only exception to the general trend was a short period in the second half of the 1970s when both GDP per capita and proportionate contribution of

FIGURE 7.3   Per capita GDP in Kenya (constant 2000 US$) and share of GDP from
agriculture (percent), 1960–2008

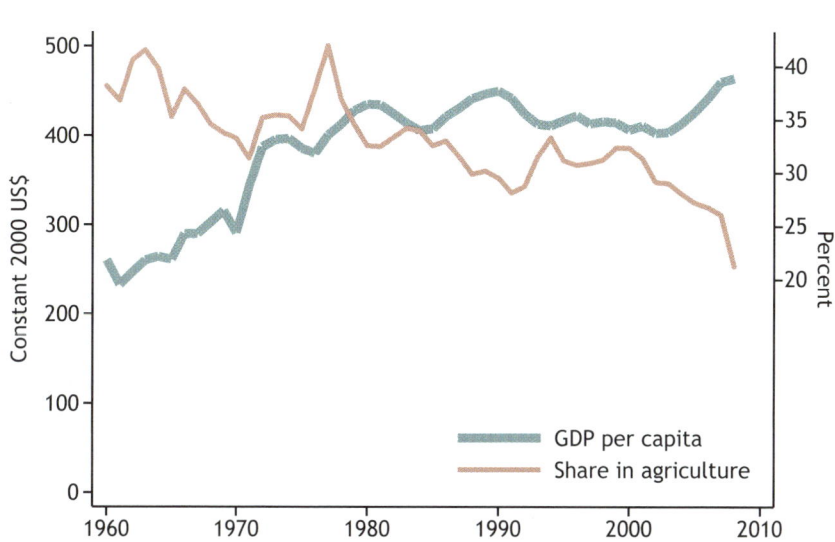

Source: *World Development Indicators* (World Bank 2009).
Note: GDP = gross domestic product; US$ = US dollars.

agriculture to GDP concurrently assumed an upward trend. However, the scenario of the inverse relationship is well demonstrated in the period after the 2002 elections, where we observe a sharp decline in the proportionate contribution of agriculture to GDP as the general GDP per capita rose.

Vulnerability to Climate Change

Vulnerability is defined in many ways. According to one definition, vulnerability is the lack of ability to recover from a stress. In the agricultural sector, poor people are particularly vulnerable to the stresses of an uncertain climate. In this case, vulnerability means susceptibility to a negative outcome. The vulnerability of agriculture to climate change is considered its susceptibility to some form of natural hazard and is determined primarily by its capacity to respond to that hazard. Vulnerability is considered at many levels, ranging from the level of the individual household to national and regional levels.

The extent to which climate change may damage or harm the agricultural sector depends on the sector's sensitivity and ability to adapt to new conditions. The adaptive capacity of the agricultural sector to climate change in Kenya is low mainly due to limited economic resources for investment in more

resilient production systems, low levels of technological development or adoption of developed technology, heavy reliance on rainfed agriculture, frequent
droughts and floods, endemic crop and livestock diseases, frequent incidences
of pest infestation, relatively high levels of postharvest losses, and the general poverty among the majority of smallholder producers. Adaptation of the
agricultural sector to climate change will entail planned or automatic changes
in processes, practices, or structures that minimize potential damage or take
advantage of opportunities associated with changes in climate.

Table 7.2 provides data on indicators of vulnerability and resiliency to
economic shocks that revolve around the level of education of the population and their literacy. Kenya's high enrollment rate in primary schools
(112.6 percent) offers a critical means by which knowledge and sensitization
on climate change issues and adaptation strategies could easily be communi-
cated to a wider population. The high level of secondary schools enrollment
(52.8 percent) and the high adult literacy rate (73.6 percent) offer other critical means for imparting knowledge and creating awareness on adaptation to
climate change.

Figure 7.4 shows two noneconomic correlates of poverty, life expectancy
and under-five mortality. A look at this figure reveals that life expectancy
increased gradually in Kenya from 1960 up to the early 1980s, rising from the
low level of about 48 years (1960) and leveling off at approximately 60 years
(1980). This status was maintained until the early 1990s, when it started
decreasing gradually to a minimum of about 50 years in 2002. Thereafter,
it assumed an upward trend to approximately 55 years by 2008. The under-
five mortality rate improved from the high of about 200 deaths per 1,000 in
1960 to a low of 100 deaths per 1,000 in 1990; thereafter it increased to
approximately 130 deaths per 1,000 by 2008. The trends in Figure 7.4 show an
inverse relationship between life expectancy at birth and under-five mortality
rates over a temporal scale. The high rate of under-five mortality is associated

TABLE 7.2   Education and nutrition statistics for Kenya, 2000s

Indicator	Year	Percent
Primary school enrollment: Percent gross (three-year average)	2007	112.6
Secondary school enrollment: Percent gross (three-year average)	2007	52.8
Adult literacy rate	2000	73.6
Under-five malnutrition (weight for age)	2003	16.5

Source: *World Development Indicators* (World Bank 2009).

FIGURE 7.4   Well-being indicators in Kenya, 1960–2008

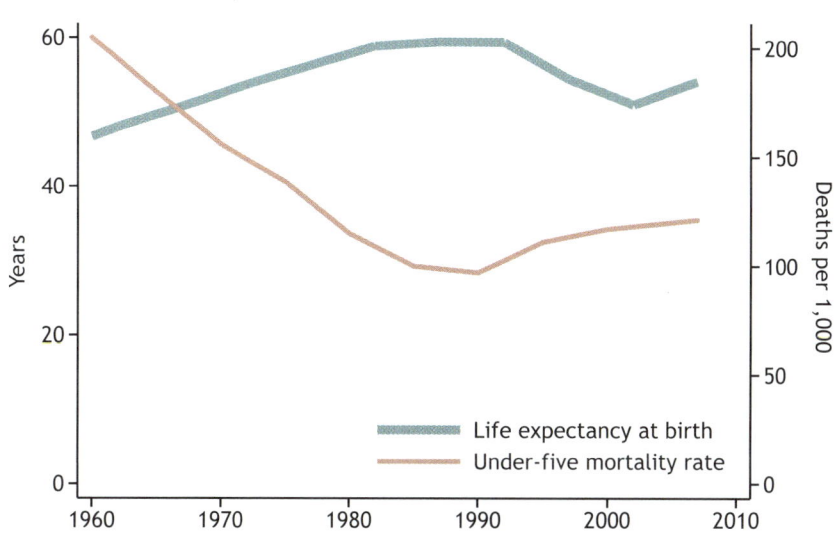

Source: *World Development Indicators* (World Bank 2009).

with malnutrition, among other factors. This implies that enhancing the resiliency of the agricultural sector to climate change could help mitigate malnutrition and thus reduce the high under-five mortality rates.

Figure 7.5 shows the proportion of the population living on less than
$2 per day. The scenario in the figure reveals a situation in which between
40 and 70 percent of the population in Rift Valley, Eastern, and Nyanza
Provinces and in the lower parts of Coast and Western Provinces live on less
than $2 per day. In the semiarid North Eastern Province, the upper parts
of Coast Province, and some parts of Rift Valley and Central Provinces,
between 20 and 40 percent of the population live on less than $2 per day. In
small sections of Central, Rift Valley, and Coast Provinces, between 10 and
20 percent of the population survive on less than $2 per day. In the metro-
politan Nairobi Province, between 10 and 20 percent live on less than $2 per
day. The spatial distribution of poverty seen in Figure 7.5 shows that propor-
tionately fewer inhabitants of the predominantly pastoralist North Eastern
Province, the upper Coast Province, and some parts of the upper Eastern and
Rift Valley Provinces live on less than $2 compared to the agropastoralists
in the southern parts of Eastern Province and the lower parts of Coast and
Rift Valley Provinces and the agriculturalist and mixed farming inhabitants

FIGURE 7.5   Poverty in Kenya, circa 2005 (percentage of population below US$2 per day)

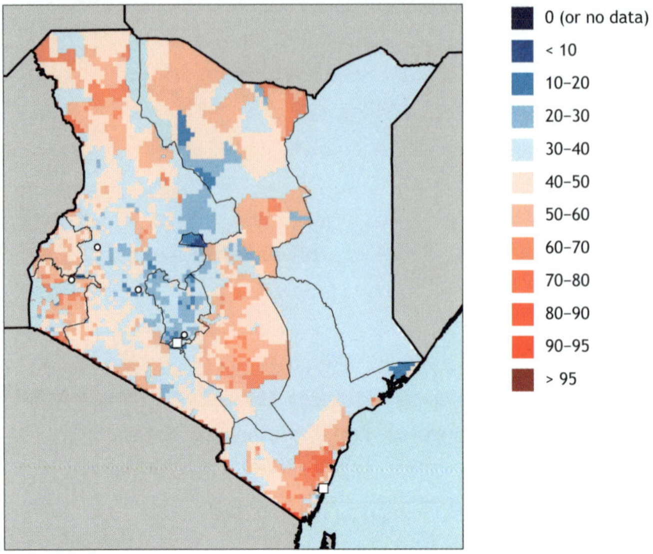

Source: Wood et al. (2010).
Note: Based on 2005 US$ (US dollars) and on purchasing power parity value.

of Nyanza and Western Provinces. Cash crop farming dominates Central Province and parts of Eastern Province neighboring Mount Kenya, which, together with the Metropolitan Nairobi Province, have the smallest proportion of their residents (between 10 and 30 percent) living on less than $2 per day.

Review of Land Use, Potential, and Limitations

Land Use Overview

Land is an important factor of production because it provides the foundation for all other activities, such as agriculture, water distribution, settlement, tourism, wildlife management, forestry, and infrastructure development. Land issues are important to the social, economic, and political development of Kenya. Over the years, the administration and management of land in Kenya have been challenging because of the lack of a comprehensive national land policy, worsened by the existence of many land laws, some of which are conflicting. This has led to fragmentation of the land, breakdown in land administration, and disparities in land ownership. Other challenges include deterioration of land quality, squatting, landlessness, underuse and abandonment of agricultural land, tenure insecurity, and conflict. To address these challenges, the Government of Kenya has developed policies and legal and institutional reforms regarding the security of land tenure, land use and development, and sustainable conservation of the environment.

Figure 7.6 shows land cover and land use in Kenya as of 2000. The largest proportion of the country falls into two categories of land cover: (1) herbaceous cover, closed-open, and (2) sparse herbaceous or shrub cover. The two types of land cover are ideal for pastoralism involving sheep, goats, and camels, which characterizes the livelihoods of the arid and semiarid lands (ASALs) that cover 83 percent of the country's land area. Cultivated and managed areas are to be found in agriculturally high- to medium-potential areas of the country. Areas with tree cover of broadleaved evergreen or deciduous closed types are found in the five major water catchments, which also serve as the sources of major rivers in the country, dubbed "the five water towers," and along the coastal strip. The country's grain basket, Rift Valley and Western Provinces, is characterized by mosaic cover: cropland, trees, or other natural vegetation; cropland or shrubs; or grass.

Natural resources from wildlife and forestry in the protected areas play two basic roles in development: support to subsistence livelihoods, mainly as sources of firewood, and sources of development resources, mainly in the form of earnings received from the supply of goods and services to the tourism sector that is associated with wildlife. A major focus in the development and use of natural resources is to ensure their sustainability and the stability of the resource base. To achieve this objective, the Government of Kenya has developed policies and institutional and legal frameworks that aim at protecting, conserving, and managing forest and wildlife resources; strengthening forestry and wildlife research; promoting extension and training; and designing, supporting, and implementing forestry and wildlife flagship projects in Vision 2030. Notable among the policies and institutional and legal documents are a "Draft Forest Policy" (Kenya 2005) and a "Draft Wildlife Policy" (Kenya 2007a).

Figure 7.7 shows the locations of protected areas in Kenya, including national parks, water masses, game reserves, and forests. These locations provide important protection for fragile ecosystems, which also serve as sources of natural resources of economic importance such as freshwater and fuelwood, pastures for livestock and wildlife, and centers of attraction

FIGURE 7.6   Land cover and land use in Kenya, 2000

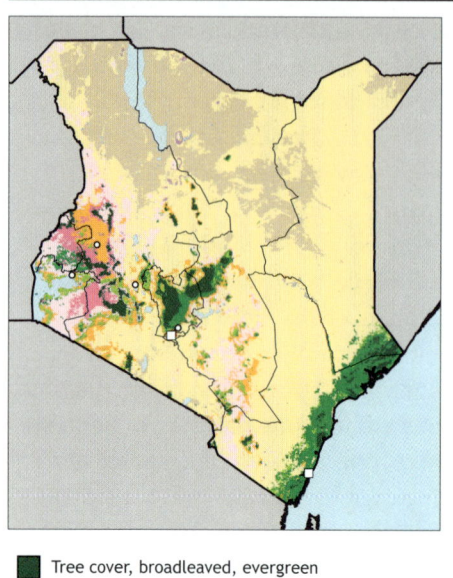

Tree cover, broadleaved, evergreen

Tree cover, broadleaved, deciduous, closed

Tree cover, broadleaved, open

Tree cover, broadleaved, needle-leaved, evergreen

Tree cover, broadleaved, needle-leaved, deciduous

Tree cover, broadleaved, mixed leaf type

Tree cover, broadleaved, regularly flooded, fresh water

Tree cover, broadleaved, regularly flooded, saline water

Mosaic of tree cover/other natural vegetation

Tree cover, burnt

Shrub cover, closed–open, evergreen

Shrub cover, closed–open, deciduous

Herbaceous cover, closed–open

Sparse herbaceous or sparse shrub cover

Regularly flooded shrub or herbaceous cover

Cultivated and managed areas

Mosaic of cropland/tree cover/other natural vegetation

Mosaic of cropland/shrub/grass cover

Bare areas

Water bodies

Snow and ice

Artificial surfaces and associated areas

No data

Source: GLC2000 (Bartholome and Belward 2005).

FIGURE 7.7   Protected areas in Kenya, 2009

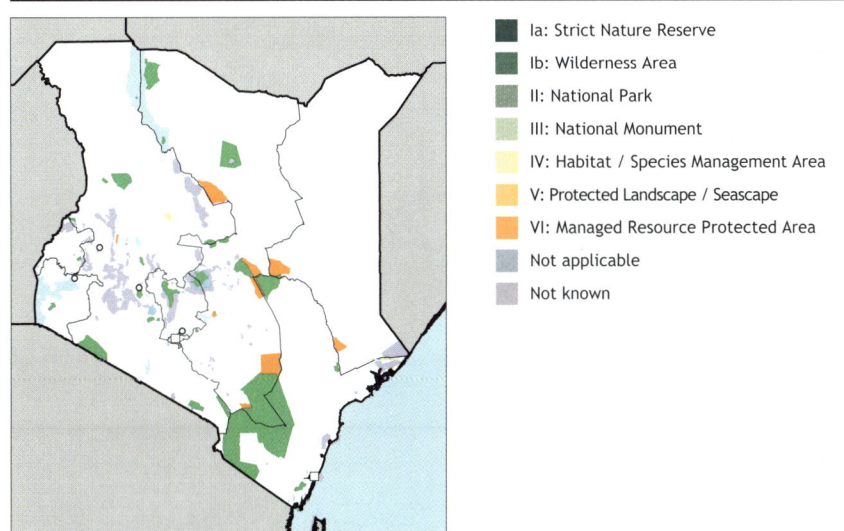

Ia: Strict Nature Reserve

Ib: Wilderness Area

II: National Park

III: National Monument

IV: Habitat / Species Management Area

V: Protected Landscape / Seascape

VI: Managed Resource Protected Area

Not applicable

Not known

Sources: Protected areas are from the World Database on Protected Areas (UNEP and IUCN 2009). Water bodies are from the World Wildlife Federation's Global Lakes and Wetlands Database (Lehner and Döll 2004).

for service-oriented economic activities such as tourism. The main protected areas depicted in the figure include the country's national parks, namely Tsavo National Park, the Maasai Mara, Marsabit Game Reserve, Turkana Game Reserve, Meru National Park, and parts of Amboseli National Park. The International Union for the Conservation of Nature (IUCN) Category VI managed resource–protected areas include parts of Tsavo East, Samburu Game Reserve, and Tana River Game Reserve; other IUCN areas include those around Mount Kenya and Mount Elgon in the central and western parts of the country, respectively, the southern tip of the Kenya–Somalia border, Samburu in the midnorth region, and parts of South Rift. Water and biodiversity from protected areas can contribute to adaptation to climate change through interventions such as the use of irrigation agriculture and genetic materials from the wild in breeding for adaptation to the adverse effects of climate change. However, a majority of the country is not protected, which implies that species of ecological importance, including the endangered ones, that fall in such areas are susceptible to destructive human activities.

Figure 7.8 shows travel times to towns and cities of various sizes, which provide potential markets for agricultural products. Policymakers need to keep in mind the importance of transport costs when considering the

FIGURE 7.8   Travel time to urban areas of various sizes in Kenya, circa 2000

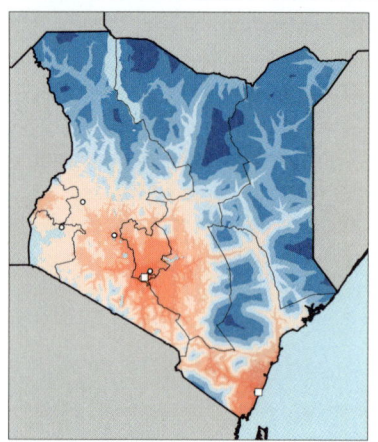

To cities of 500,000 or more people

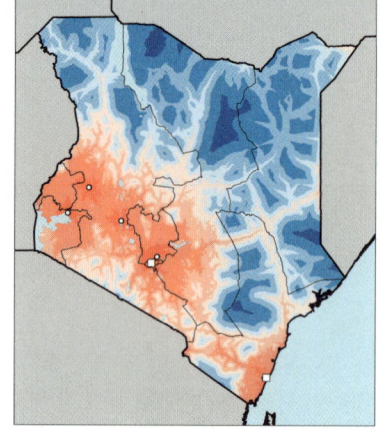

To cities of 100,000 or more people

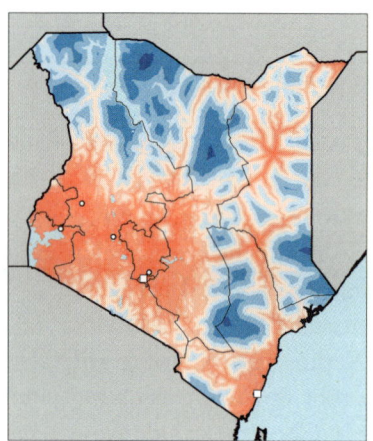

To towns and cities of 25,000 or more people

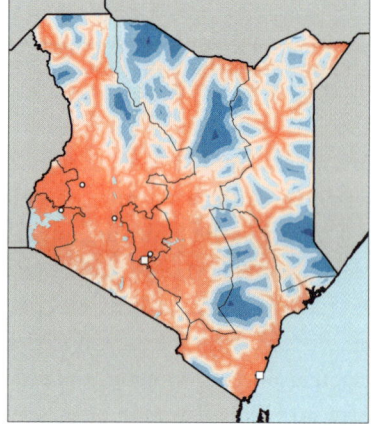

To towns and cities of 10,000 or more people

Urban location
< 1 hour
1–3 hours
3–5 hours
5–8 hours
8–11 hours
11–16 hours
16–26 hours
> 26 hours

Source: Authors' calculations.

potential for agricultural expansion. Fertile but unused land that is far from markets represents potential land for expansion only if transport infrastructure is provided and the land does not conflict with the preservation priorities shown in Figure 7.7. The resilience of the agricultural sector to shocks occasioned by climate change depends to a large extent on the availability of transport infrastructure. A glimpse at Figure 7.8 shows that farmers from agriculturally high- to medium-potential areas of Central, Rift Valley, Western, and Nyanza Provinces and some parts of Coast Province need less time to reach the more lucrative markets in towns with more than 100,000 inhabitants. The predominantly pastoralist communities in ASALs spend upward of 26 hours to reach such markets with their animals, under the assumptions used in the model, and possibly longer if the animals are herded to market. Towns with populations of between 10,000 and 25,000 inhabitants could easily be reached by both agricultural producers and pastoralists, because the time needed to reach them is in the range of 1–8 hours for most parts of the country.

Agriculture

Tables 7.3 and 7.4 show key agricultural commodities in terms of area harvested as a percentage of the total area harvested and value of production. Considering the area harvested and its contribution to overall food security, Table 7.3 reveals that maize and beans are the most important crops in Kenya, contributing 37.5 percent and 17.9 percent, respectively, of the harvested areas.

A look at Table 7.4 reveals that in terms of the value of the crops produced, maize, tea, and potatoes are the top three commodities, contributing 17.9 percent, 16.4 percent, and 8.8 percent, respectively, to the total value of crops produced, and are strategic to income generation from farming.

Figure 7.9 shows the estimated yield and harvest areas for rainfed maize. The picture given is one of low productivity, with areas such as Northern, North Eastern, and upper Coast Provinces producing less than 0.5 tons per hectare, though there are many areas throughout the country producing up to 4 tons per hectare, including areas near Lake Victoria, Mount Elgon, the central Kenya highlands, southwestern Coast Province, and even northernmost North Eastern Province. In terms of the land area sown with the crop, the model predicts a fairly high preference for maize production, with nearly half of the country committing from 30 to more than 100 hectares of land to the crop. The areas predicated to commit more than 100 hectares to maize include the traditional grain basket of northern Rift, Western, Central, and

TABLE 7.3   Harvest area of leading agricultural commodities in Kenya, 2006–2008 (thousands of hectares)

Rank	Crop	Percent of total	Harvest area
	Total	100.0	4,631
1	Maize	37.5	1,734
2	Beans	17.9	828
3	Pigeon peas	3.9	182
4	Coffee	3.5	163
5	Tea	3.3	151
6	Cowpeas	3.2	147
7	Sorghum	3.0	141
8	Wheat	2.7	127
9	Potatoes	2.6	119
10	Millet	2.3	106

Source: FAOSTAT (FAO 2010).
Note: All values are based on the three-year average for 2006–2008.

TABLE 7.4   Value of production for leading agricultural commodities in Kenya, 2005–2007 (millions of US$)

Rank	Crop	Percent of total	Value of production
	Total	100.0	3,650.7
1	Maize	17.9	651.7
2	Tea	16.4	597.6
3	Potatoes	8.8	321.2
4	Beans	6.7	243.3
5	Sugarcane	3.9	143.3
6	Cabbages and other brassicas	3.5	127.7
7	Pineapples	3.4	125.1
8	Coffee	3.1	114.0
9	Avocados	3.1	111.4
10	Tomatoes	3.0	109.1

Source: FAOSTAT (FAO 2010).
Notes: All values are based on the three-year average for 2005–2007. US$ = US dollars.

FIGURE 7.9 Yield (metric tons per hectare) and harvest area density (hectares) for rainfed maize in Kenya, 2000

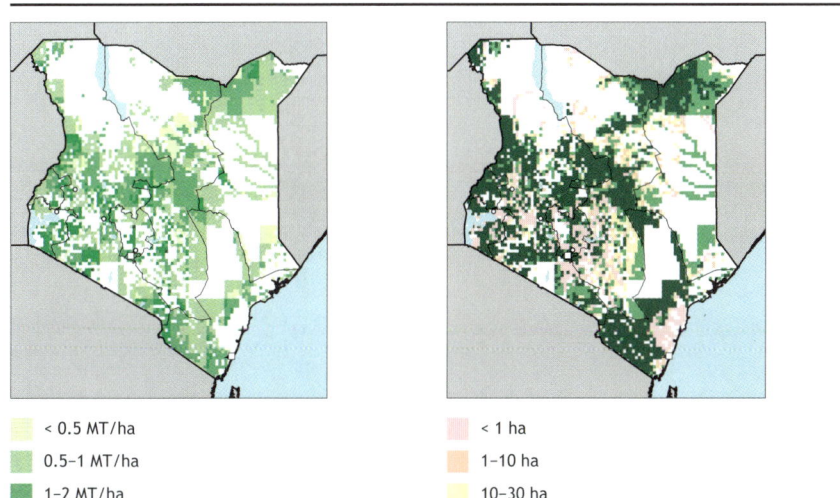

< 0.5 MT/ha	< 1 ha
0.5–1 MT/ha	1–10 ha
1–2 MT/ha	10–30 ha
2–4 MT/ha	30–100 ha
> 4 MT/ha	> 100 ha

Source: SPAM (Spatial Production Allocation Model) (You and Wood 2006; You, Wood, and Wood-Sichra 2006, 2009).
Note: ha = hectare; MT/ha = metric tons per hectare.

southern Rift Provinces and the Central Kenya Highlands. The only exceptions in this category are areas on the border between Kenya and Ethiopia and the southwestern Coast Province. There is a clear positive correlation between areas willing to plant relatively large areas with maize and the productivity of the crop.

Figure 7.10 shows where rainfed beans are grown in Kenya, along with their productivity, which for the country averages around half a metric ton per hectare (FAO 2010).

Scenarios for the Future

Economic and Demographic Indicators

Population
Figure 7.11 shows population projections for Kenya by the UN Population Division through the year 2050. Population projections predict a 150 percent increase between the base year and 2050 when the high-variant scenario

FIGURE 7.10 Yield (metric tons per hectare) and harvest area density (hectares) for rainfed beans in Kenya, 2000

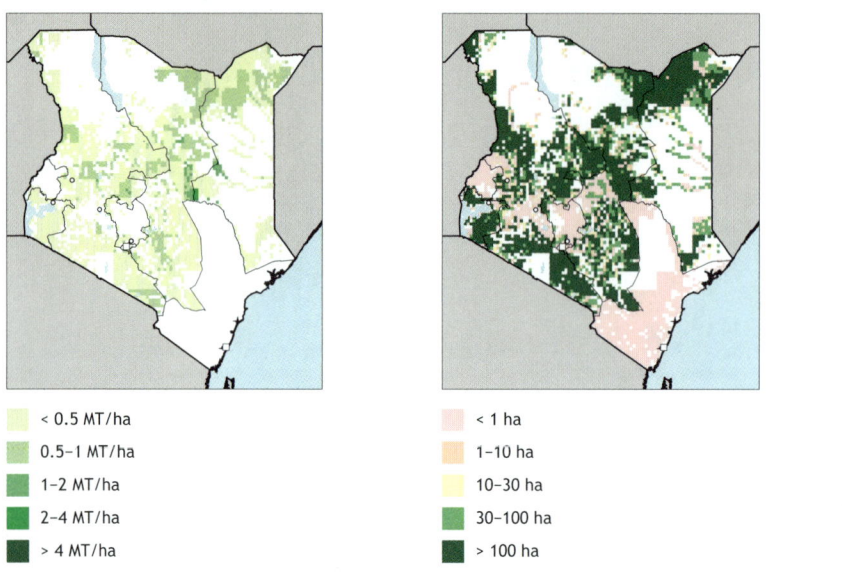

< 0.5 MT/ha		< 1 ha	
0.5–1 MT/ha		1–10 ha	
1–2 MT/ha		10–30 ha	
2–4 MT/ha		30–100 ha	
> 4 MT/ha		> 100 ha	

Source: SPAM (Spatial Production Allocation Model) (You and Wood 2006; You, Wood, and Wood-Sichra 2006, 2009).
Note: ha = hectare; MT/ha = metric tons per hectare.

FIGURE 7.11 Population projections for Kenya, 2010–2050

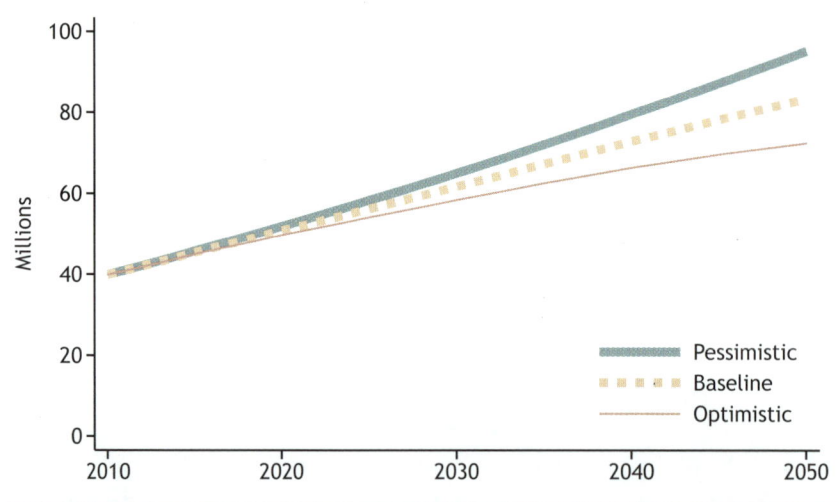

Source: UNPOP (2009).

is considered, doubling of the population within the same period when the medium-variant scenario is considered, and an increase by 75 percent in the same period when the low-variant scenario is considered. The predictions from the three scenarios paint a picture of the likely pressure on productive resources if the country is challenged to remain food secure with increased numbers to feed.

Income

Figure 7.12 shows the three scenarios for GDP per capita used in this analysis. These are the result of combining three GDP projections with the three population projections of Figure 7.11 from the United Nations Population Division. The optimistic scenario combines high GDP with low population. The baseline scenario combines the medium GDP projection with the medium population projection. The pessimistic scenario combines the low GDP projection with the high population projection.

The emerging scenario is that the country will be able to achieve a high level of GDP per capita only if it embraces policies that will stimulate economic growth while concurrently checking high rates of population growth.

FIGURE 7.12 Gross domestic product (GDP) per capita in Kenya, future scenarios, 2010–2050

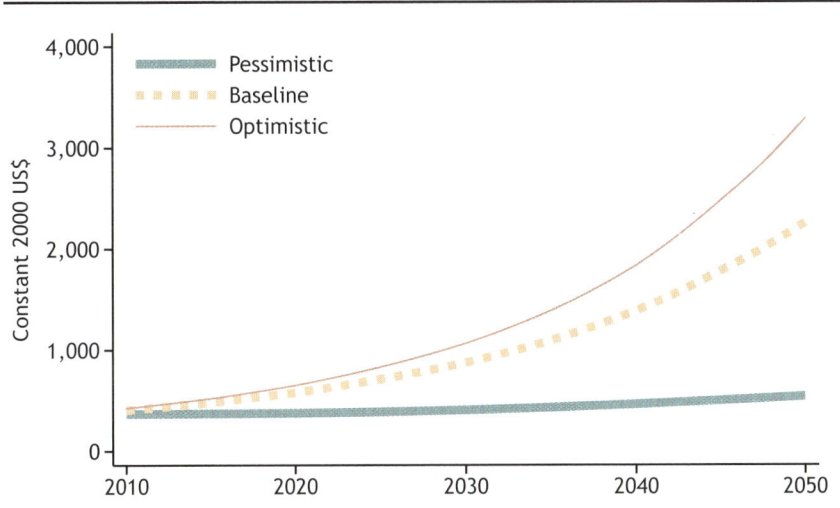

Sources: Computed from GDP data from the World Bank Economic Adaptation to Climate Change project (World Bank 2010), from the Millennium Ecosystem Assessment (2005) reports, and from population data from the United Nations (UNPOP 2009).
Note: US$ = US dollars.

The pessimistic scenario projects very slow growth in per capita GDP, which will increase by just under 50 percent between 2010 and 2050, rising to $543.

Biophysical Analysis

Climate Models

The data used in our analysis draws from four downscaled general circulation models (GCMs). Figure 7.13 shows projected precipitation changes under the four downscaled climate models, which were used with the A1B scenario.[1]

The CNRM-CM3 model predicts a significant reduction in annual rainfall in areas neighboring Lake Victoria, but elsewhere the story is more positive.[2] No change is anticipated for the Rift Valley. The model predicts a promising situation for the current ASALs of Coast, Eastern, and North Eastern Provinces, where there will be an increase of between 100 and 200 millimeters of precipitation. The CSIRO Mark 3 model predicts a change in rainfall for most of the country, with a very moderate increase around Turkana District and in the southwestern part of the country.

The predictions of the ECHAM 5 model are similar to those of the CSIRO Mark 3 model in that much of the country is projected to maintain the same rainfall as it has been receiving. Parts of the relatively dry mid to upper eastern, northwestern, and Lake Victoria region will benefit from gains of between 100 and 200 millimeters of precipitation. The MIROC 3.2 model presents the most optimistic scenario, predicting that nearly the entire country will experience an increase of between 100 and 300 millimeters of precipitation, with the extreme northern and southwestern parts of the country gaining most. The predictions of MIROC 3.2 forecast an era of enhanced resilience to the adverse effects of climate change.

Figure 7.14 shows changes in mean daily maximum temperature for the warmest month in Kenya. The CNRM-CM3 model predicts that the current agriculturally medium- to high-potential areas of the country and the northern

---

1   The A1B scenario is a greenhouse gas emissions scenario that assumes fast economic growth, a population that peaks midcentury, and the development of new and efficient technologies, along with a balanced use of energy sources.

2   CNRM-CM3 is National Meteorological Research Center–Climate Model 3. MIROC 3.2 is the Model for Interdisciplinary Research on Climate, developed at the University of Tokyo Center for Climate System Research. CSIRO Mark 3 is a climate model developed at the Australia Commonwealth Scientific and Industrial Research Organisation. ECHAM 5 is a fifth-generation climate model developed at the Max Planck Institute for Meteorology in Hamburg.

FIGURE 7.13   Changes in mean annual precipitation in Kenya, 2000–2050, A1B scenario
(millimeters)

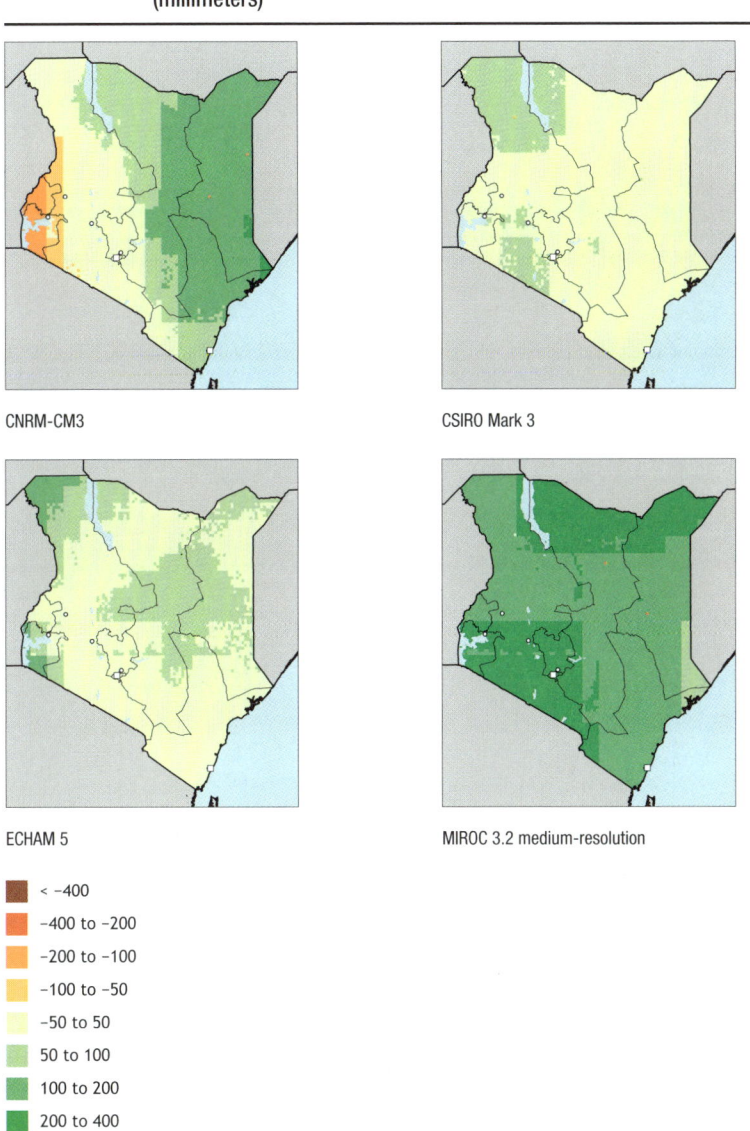

CNRM-CM3

CSIRO Mark 3

ECHAM 5

MIROC 3.2 medium-resolution

■ < −400
■ −400 to −200
■ −200 to −100
■ −100 to −50
■ −50 to 50
■ 50 to 100
■ 100 to 200
■ 200 to 400
■ > 400

Source: Authors' calculations based on Jones, Thornton, and Heinke (2009).

Notes: A1B = greenhouse gas emissions scenario that assumes fast economic growth, a population that peaks midcentury, and the development of new and efficient technologies, along with a balanced use of energy sources; CNRM-CM3 = National Meteorological Research Center–Climate Model 3; CSIRO = climate model developed at the Australia Commonwealth Scientific and Industrial Research Organisation; ECHAM 5 = fifth-generation climate model developed at the Max Planck Institute for Meteorology (Hamburg); GCM = general circulation model; MIROC = Model for Interdisciplinary Research on Climate, developed by the University of Tokyo Center for Climate System Research.

FIGURE 7.14   Changes in monthly mean maximum daily temperature in Kenya for the warmest month, 2000–2050, A1B scenario (°C)

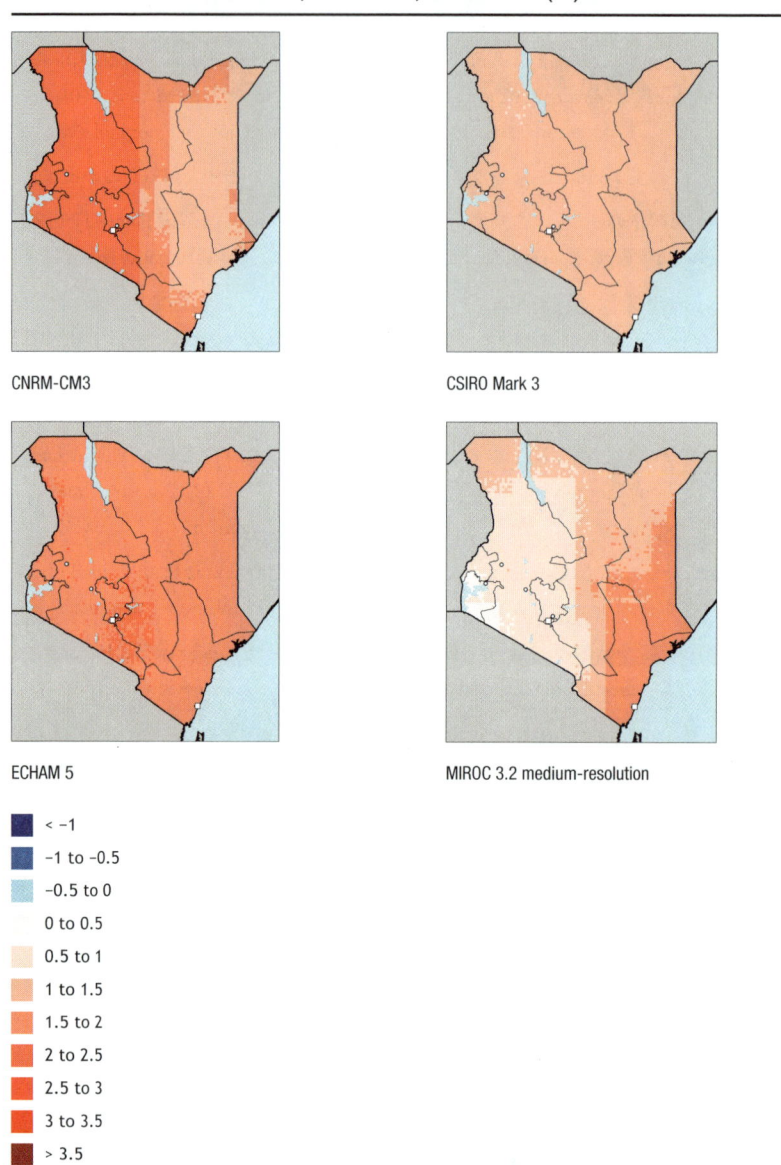

CNRM-CM3

CSIRO Mark 3

ECHAM 5

MIROC 3.2 medium-resolution

< −1
−1 to −0.5
−0.5 to 0
0 to 0.5
0.5 to 1
1 to 1.5
1.5 to 2
2 to 2.5
2.5 to 3
3 to 3.5
> 3.5

Source: Authors' calculations based on Jones, Thornton, and Heinke (2009).

Notes: A1B = greenhouse gas emissions scenario that assumes fast economic growth, a population that peaks midcentury, and the development of new and efficient technologies, along with a balanced use of energy sources; CNRM-CM3 = National Meteorological Research Center–Climate Model 3; CSIRO = climate model developed at the Australia Commonwealth Scientific and Industrial Research Organisation; ECHAM 5 = fifth-generation climate model developed at the Max Planck Institute for Meteorology (Hamburg); GCM = general circulation model; MIROC = Model for Interdisciplinary Research on Climate, developed by the University of Tokyo Center for Climate System Research.

ASALs will experience a temperature increase of between 1.5° and 2.5°C over the prediction period. This is likely to lead to temperature stress, disease, an increase in pests, and high levels of evapotranspiration. The eastern part of the country will be affected less severely, with predicted temperature increases of between 1° and 1.5°C.

The CSIRO Mark 3 model predicts a fairly uniform increase in temperature of 1°–1.5°C across the country, while the ECHAM 5 model predicts a similar spatial spread but with a slightly higher increase in temperature of between 1.5° and 2°C, with small patches of the area experiencing an increase of between 2° and 2.5°C.

The MIROC 3.2 model predicts a situation in which most of the currently agriculturally medium- to high-potential areas will experience a marginal increase in temperature of between 0.5° and 1°C, while the easternmost part of the eastern, northeastern, and southern parts of Coast Province will experience increases of between 1° and 2°C, which is likely to result in heat stress.

Crop Models

The Decision Support System for Agrotechnology Transfer (DSSAT) crop model was employed to compute yields under current temperature and precipitation regimes. The exercise was then repeated for future scenarios for the year 2050. For all locations, crop variety, soil, and management practices were held constant. The future yield results from DSSAT were then compared to the current or baseline yield results from DSSAT.

The output for key crops is mapped in Figures 7.15 and 7.16. The comparison is between the crop yields for 2050 with climate change and the yields with the 2000 climate. In Figure 7.15 we see the model predictions for rainfed maize. There is much variation between models, and in most models there is observable geographic variation within them as well. The MIROC 3.2 model is the most optimistic of the four, projecting yield increases in most areas, including large areas with a yield increase of more than 25 percent. The CNRM-CM3 GCM predicts loss of area in part of Rift Valley Province. These lost areas may be due to temperature rise. The ECHAM 5 GCM predicts a yield reduction in Coast Province of between 5 and 25 percent.

All models predict yield gains in areas that have not previously been able to cultivate maize. These are areas that were too dry for successful maize production. With new areas becoming available for maize cultivation, it seems important that policymakers consider encouraging people to cultivate maize in these areas sometime in the future. Some of these might be areas that are currently

FIGURE 7.15   Yield change under climate change: Rainfed maize in Kenya, 2000–2050, A1B scenario

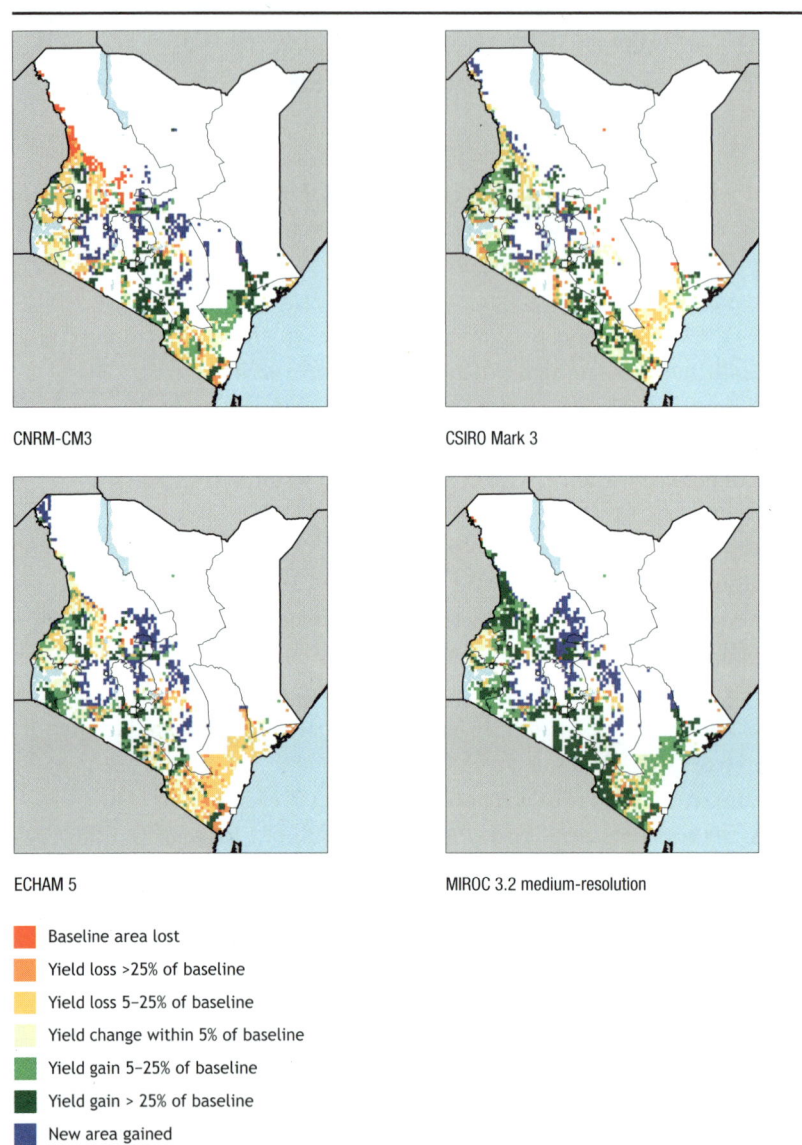

CNRM-CM3

CSIRO Mark 3

ECHAM 5

MIROC 3.2 medium-resolution

■ Baseline area lost
■ Yield loss >25% of baseline
■ Yield loss 5–25% of baseline
  Yield change within 5% of baseline
■ Yield gain 5–25% of baseline
■ Yield gain > 25% of baseline
■ New area gained

Source: Authors' calculations.

Notes: A1B = greenhouse gas emissions scenario that assumes fast economic growth, a population that peaks midcentury, and the development of new and efficient technologies, along with a balanced use of energy sources; CNRM-CM3 = National Meteorological Research Center–Climate Model 3; CSIRO = climate model developed at the Australia Commonwealth Scientific and Industrial Research Organisation; ECHAM 5 = fifth-generation climate model developed at the Max Planck Institute for Meteorology (Hamburg); GCM = general circulation model; MIROC = Model for Interdisciplinary Research on Climate, developed by the University of Tokyo Center for Climate System Research.

FIGURE 7.16   Yield change under climate change: Rainfed wheat in Kenya, 2000–2050, A1B scenario

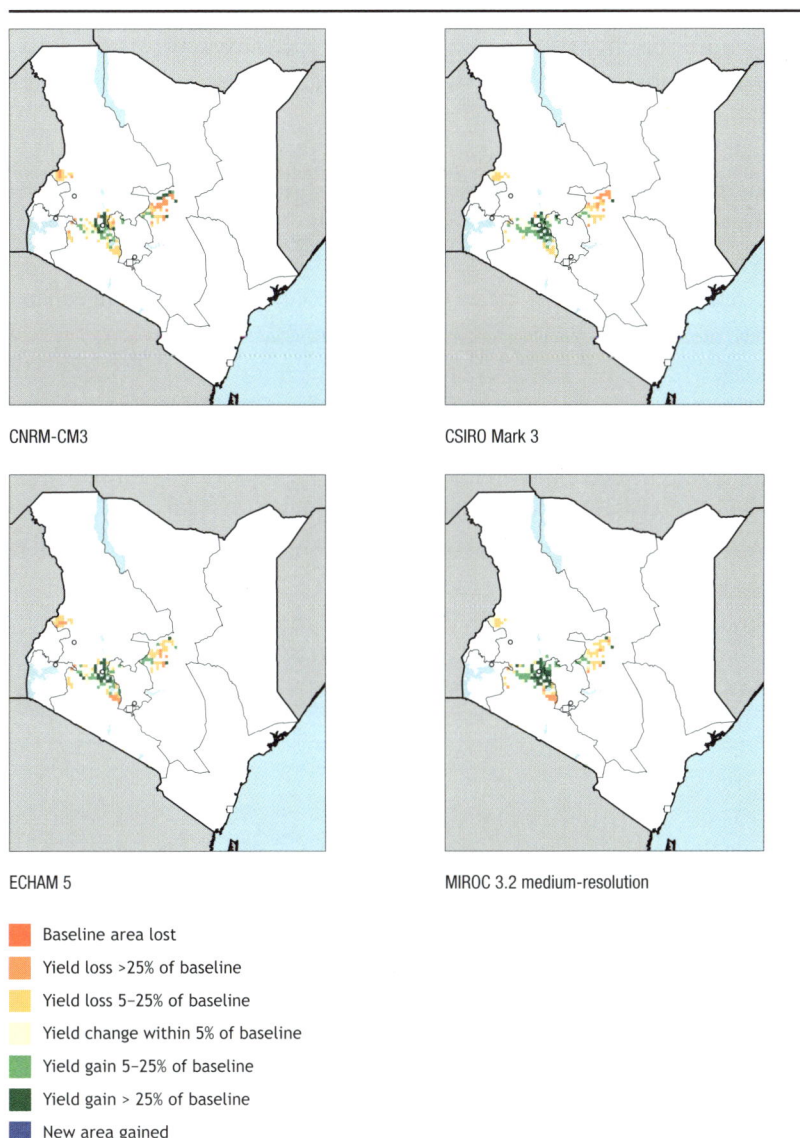

CNRM-CM3

CSIRO Mark 3

ECHAM 5

MIROC 3.2 medium-resolution

■ Baseline area lost
■ Yield loss >25% of baseline
■ Yield loss 5–25% of baseline
□ Yield change within 5% of baseline
■ Yield gain 5–25% of baseline
■ Yield gain > 25% of baseline
■ New area gained

Source: Authors' calculations.

Notes: A1B = greenhouse gas emissions scenario that assumes fast economic growth, a population that peaks midcentury, and the development of new and efficient technologies, along with a balanced use of energy sources; CNRM-CM3 = National Meteorological Research Center–Climate Model 3; CSIRO = climate model developed at the Australia Commonwealth Scientific and Industrial Research Organisation; ECHAM 5 = fifth-generation climate model developed at the Max Planck Institute for Meteorology (Hamburg); GCM = general circulation model; MIROC = Model for Interdisciplinary Research on Climate, developed by the University of Tokyo Center for Climate System Research.

receiving precipitation that is not adequate to support maize production, like most of the ASALs. The point is that climate change may cause farmers to abandon some areas they are currently cultivating and move to new areas with a potential for maize production. Laws and procedures to facilitate such movement should be enacted in advance.

The yield change maps for wheat presented in Figure 7.16 are in general agreement, with losses predicted for the areas north of Mount Kenya and east of Mount Elgon. The four models also predict yield increases for wheat in a small area in Central Rift Valley neighboring Central Province. In general, the predictions show that maize will do better under climate change than will wheat.

Vulnerability

Figure 7.17 shows the impact of future GDP and population scenarios on the number of malnourished children under age five. Figure 7.18 shows the share of children who are malnourished. Under the pessimistic scenario, the number of malnourished children under age five increases to a maximum of 1.7 million by 2015 before gradually reducing to a minimum of 1.6 million, then gradually rising to a peak of 1.8 million by 2050. Under the baseline scenario, the

FIGURE 7.17   Number of malnourished children under five years of age in Kenya in multiple income and climate scenarios, 2010–2050

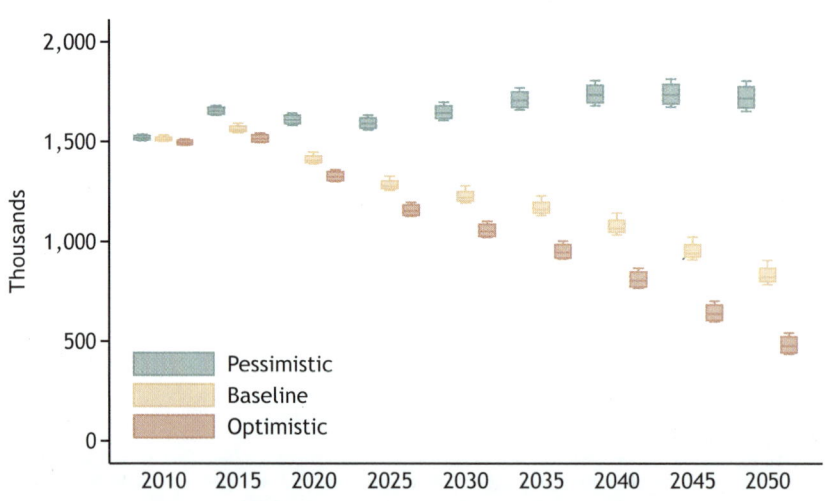

Source: Based on analysis conducted for Nelson et al. (2010).

Note: The box and whiskers plot for each socioeconomic scenario shows the range of effects from the four future climate scenarios.

FIGURE 7.18   Share of malnourished children under five years of age in Kenya in multiple income and climate scenarios, 2010–2050

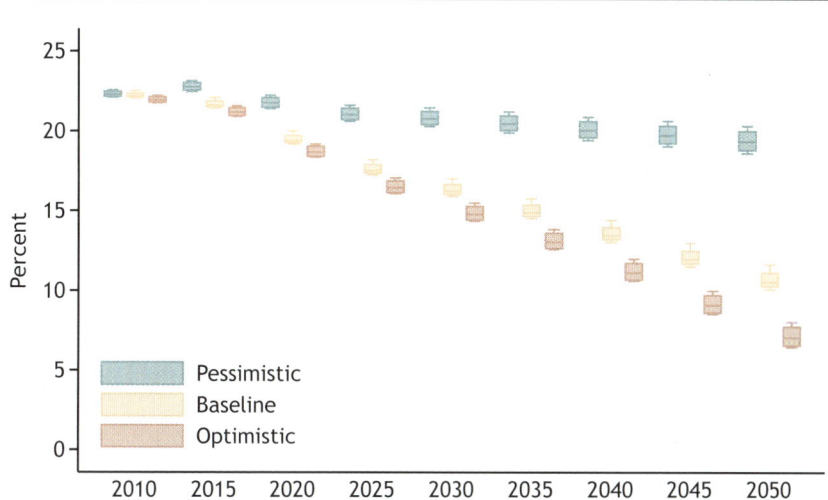

Source: Based on analysis conducted for Nelson et al. (2010).

Note: The box and whiskers plot for each socioeconomic scenario shows the range of effects from the four future climate scenarios.

numbers will increase from 1.5 million to a peak of 1.6 million before declining across a temporal scale to a minimum of 0.8 million by 2050. Under the optimistic scenario, the numbers are predicted to decrease from a maximum of 1.5 million in 2010 to a minimum of 0.5 million by 2050. The share of malnourished children declines steadily because of population growth.

Figure 7.19 shows the kilocalories per capita available to each person in Kenya. A look at trends in available kilocalories per capita shows a slow downward trend under the pessimistic scenario. The available kilocalories decline from a high of 1,950 in 2010 to a low level of 1,650 in 2050. During this period, in the pessimistic scenario, the main staple, maize, almost doubles in price, while income rises less than 50 percent. When looking at both the optimistic and the baseline scenarios, mean calorie consumption rises slowly for the first 20 years, after which it assumes a relatively rapid upward trend up to 2050.

Agricultural Outcomes

Figure 7.20 shows simulation results of the impact of changes in GDP and population on maize in Kenya. A glance at Figure 7.20 reveals the following: (1) there is a gradual upward trend in the production of maize over time under

FIGURE 7.19   Kilocalories per capita in Kenya in multiple income and climate scenarios, 2010–2050

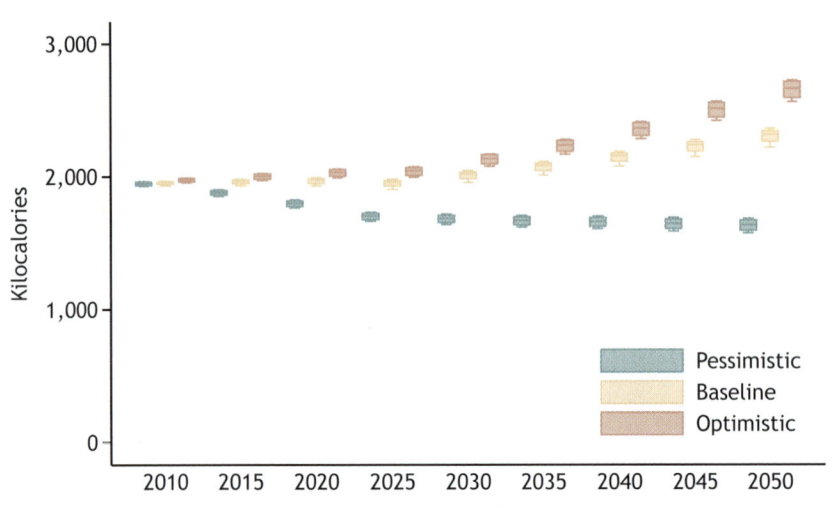

Source: Based on analysis conducted for Nelson et al. (2010).

Note: The box and whiskers plot for each socioeconomic scenario shows the range of effects from the four future climate scenarios.

all three prediction scenarios; (2) the productivity of maize assumes an upward trend across the temporal scale for all three scenarios; (3) the area sown with maize rises slightly until around 2030, then declines to 2050, ending approximately where it began; (4) the variation between climate models is large for net exports, though we can at least say that the median rises to around 2025, then declines, but with 2050 levels slightly higher than those for 2010; and (5) the trend in the world price of maize assumes an upward trend across the temporal scale for all three scenarios.

Conclusions and Policy Recommendations

The trends in the contribution of agriculture to GDP in Kenya show that the overall importance of agriculture declined from 1960 to 2008. Nevertheless, it is still an important sector for the employment of many people and for keeping more people from slipping into poverty. These auxiliary roles of agriculture are likely to remain important for quite some time into the future.

In the four climate models we used in this study, we noted significant differences between them and within them, geographically. The CNRM-CM3 model, for example, projects a much warmer future in the higher-productivity

FIGURE 7.20   Impact of changes in GDP and population on maize in Kenya, 2010–2050

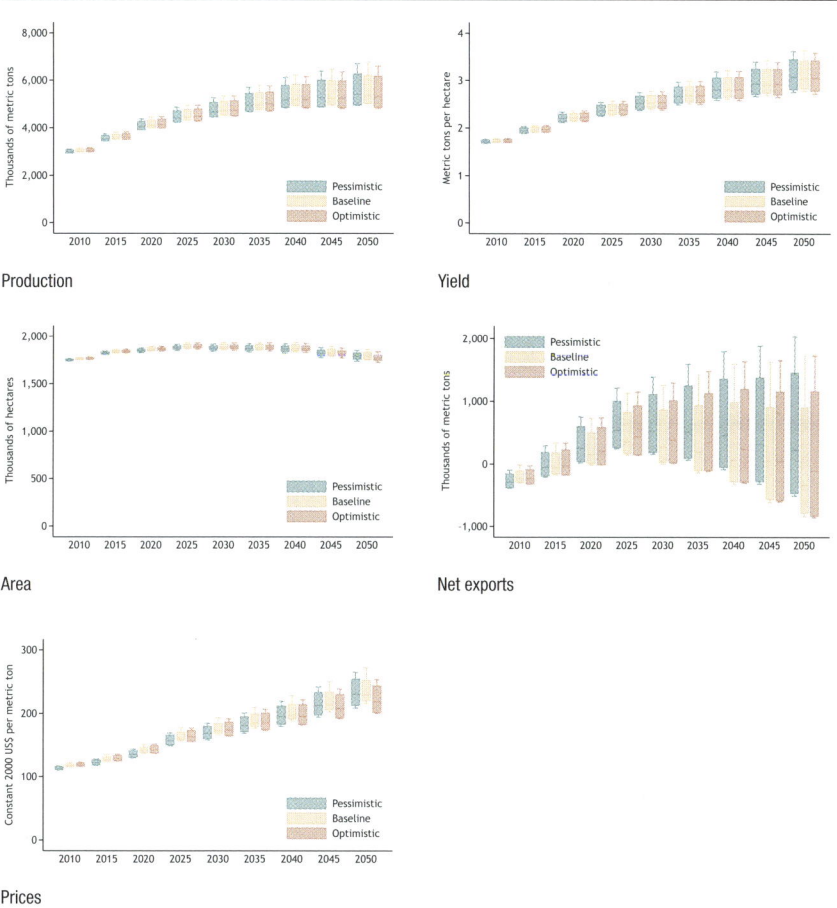

Production

Yield

Area

Net exports

Prices

Source: Based on analysis conducted for Nelson et al. (2010).

Notes: The box and whiskers plot for each socioeconomic scenario shows the range of effects from the four future climate scenarios. GDP = gross domestic product; US$ = US dollars.

areas of Kenya, with either little change in rainfall or a decline; meanwhile, the MIROC 3.2 model projects little temperature rise and a wetter future for Kenya, particularly in some of the higher-productivity areas. It is no surprise, then, that when they are used in crop models for rainfed maize, the results based on the CNRM-CRM3 model predicted yield reductions for large areas of Kenya and even a loss of some areas for the planting rainfed maize, while the results based on the MIROC 3.2 model predicted that there will be areas of yield increases, some quite large. Such differences between climate models suggest that policies aimed at helping farmers adapt to climate change must be flexible, easily reversible, and perhaps capable of offering differing regional solutions based on the different impacts of climate change across regions.

The International Model for Policy Analysis of Agricultural Commodities and Trade results, which takes into consideration technological improvements, projects an increase of around 80 percent in maize yield between 2010 and 2050, which seems to be a large enough increase to meet in-country demand for maize. Many assumptions were made to predict rates of technological improvements, and it is possible that if policymakers fail to sufficiently fund agricultural research institutions or extension agencies, the actualized rate of technological improvement will be slower, leading to slower growth in yield and to farmers' not producing enough food to meet demand.

In a rapidly changing climate, farmers cannot rely on traditional means of learning and transmitting knowledge. That is why research and extension are so important, not only to create new varieties or perhaps simply test internationally developed varieties but also to propagate the seeds and keep farmers informed of developments so that they can take advantage of them. The high enrollment rates in primary and secondary schools offer a critical means for sensitization of the population to adaptation to climate change in the agricultural sector. Furthermore, they suggest that farmers will be able to understand more complicated and varied messages from extension agents that will likely need to be presented to them as changes in climate become more apparent.

Although irrigation agriculture has the potential to serve as an important adaptive intervention against known and anticipated adverse effects of climate change, the situation in Kenya is such that irrigation accounts for only 1.7 percent of the total land area under agriculture, which calls for an upscaling of irrigation.

Stresses on the rural and agricultural sectors will be less severe if the population growth rate is slowed and the other sectors of the economy continue growing. This suggests a need for policies that encourage family planning, including educating adults and ensuring that girls receive opportunities for secondary education. It also suggests that there is a need for continued work on improving the business environment in Kenya so that manufacturing and services can thrive.

Yield improvements are contingent on farmers' having access to inputs at the lowest market prices possible and being able to sell their outputs at the highest possible prices. One of the important ways to help ensure such efficiencies is continued investment in infrastructure.

As this study has shown, farmers in Kenya can potentially thrive even with climate change. Their probability of success would be greatly enhanced with a supportive policy environment. Adopting the proposals just presented, or ones that are similar, will go a long way toward creating such an environment.

References

Bartholome, E., and A. S. Belward. 2005. "GLC2000: A New Approach to Global Land Cover Mapping from Earth Observation Data." *International Journal of Remote Sensing* 26 (9–10): 1959–1977.

CIESIN (Center for International Earth Science Information Network), Columbia University, IFPRI (International Food Policy Research Institute), World Bank, and CIAT (Centro Internacional de Agricultura Tropical). 2004. *Global Rural–Urban Mapping Project (GRUMP), Alpha Version: Population Density Grids.* Palisades, NY, US: Socioeconomic Data and Applications Center (SEDAC), Columbia University. http://sedac.ciesin.columbia.edu/gpw.

FAO (Food and Agriculture Organization of the United Nations). 2010. FAOSTAT. Rome. http://faostat.fao.org.

Jones, P. G., P. K. Thornton, and J. Heinke. 2009. *Generating Characteristic Daily Weather Data Using Downscaled Climate Model Data from the IPCC's Fourth Assessment.* Project report for the International Livestock Research Institute. Geneva: International Panel on Climate Change.

KARI (Kenya Agricultural Research Institute). 2009. *KARI Strategic Plan, 2009–2014.* Nairobi.

Kenya. 2005. "Draft Forest Policy." Nairobi.

———. 2007a. "Draft Wildlife Policy." Nairobi.

———. 2007b. "Kenya Vision 2030: A Globally Competitive and Prosperous Kenya." Accessed November 7, 2011. www.kilimo.go.ke/kilimo_docs/pdf/Kenya_VISION_2030-final.pdf.

———. 2010. *Agricultural Sector Development Strategy, 2010–2020.* Nairobi.

———. 2011a. "Draft National Arid and Semi-Arid Lands Policy." Nairobi.

———. 2011b. "Draft National Food and Nutrition Security Policy." Nairobi.

———. 2011c. "Draft National Irrigation Policy." Nairobi.

Kenya, Ministry of Environment and Mineral Resources. 2010. *National Climate Change Response Strategy.* Nairobi.

Kenya, Ministry of Water and Irrigation. 2008. *National Irrigation Board Strategic Plan, 2008–2012.* Nairobi.

Lehner, B., and P. Döll. 2004. "Development and Validation of a Global Database of Lakes, Reservoirs, and Wetlands." *Journal of Hydrology* 296 (1–4): 1–22.

Millennium Ecosystem Assessment. 2005. *Ecosystems and Human Well-being: Synthesis.* Washington, DC: Island Press. www.maweb.org/en/Global.aspx.

Nelson, G. C., M. W. Rosegrant, A. Palazzo, I. Gray, C. Ingersoll, R. Robertson, S. Tokgoz, et al. 2010. *Food Security, Farming, and Climate Change to 2050: Scenarios, Results, Policy Options.* Washington, DC: International Food Policy Research Institute.

StataCorp. 2009. Stata: Release 11. Statistical Software. College Station, TX, US.

UNEP (United Nations Environment Programme) and IUCN (International Union for Conservation of Nature). 2009. World Database on Protected Areas (WDPA) Annual Release 2009. No longer available online.

UNPOP (United Nations Secretariat, Department of Economic and Social Affairs, Population Division). 2009. *World Population Prospects: The 2008 Revision.* Accessed April 06, 2010. http://esa.un.org/unpp.

Wood, S., G. Hyman, U. Deichmann, E. Barona, R. Tenorio, Z. Guo, S. Castano, O. Rivera, E. Diaz, and J. Marin. 2010. "Sub-national Poverty Maps for the Developing World Using International Poverty Lines: Preliminary Data Release." Accessed May 6, 2010. http://povertymap.info.

World Bank. 2009. *World Development Indicators.* Accessed May 2011. http://data.worldbank.org/data-catalog/world-development-indicators.

———. 2010. *Economics of Adaptation to Climate Change: Synthesis Report.* Washington, DC. http://climatechange.worldbank.org/content/economics-adaptation-climate-change-study-homepage.

You, L., and S. Wood. 2006. "An Entropy Approach to Spatial Disaggregation of Agricultural Production." *Agricultural Systems* 90 (1–3): 329–347.

You, L., S. Wood, and U. Wood-Sichra. 2006. "Generating Global Crop Distribution Maps: From Census to Grid." Paper presented at the International Association of Agricultural Economists Conference, Brisbane, Australia, August 11–18.

———. 2009. "Generating Plausible Crop Distribution and Performance Maps for Sub-Saharan Africa Using a Spatially Disaggregated Data Fusion and Optimization Approach." *Agricultural Systems* 99 (2–3): 126–140.

Chapter 8

MADAGASCAR

Mireille Rahaingo Vololona, Miriam Kyotalimye, Timothy S. Thomas, and Michael Waithaka

M adagascar is located in the southwestern Indian Ocean. The island has a total area of 587,041 square kilometers, with a coastline stretching 5,603 kilometers. The relief of the country is quite varied and often uneven, although no peak is more than 3,000 meters above sea level. Some of its regions are dominated by plains and plateaus extending into the vast delta areas (Madagascar, Ministère de l'Environnement et des Forêts 2010).

Overall, the climate of Madagascar is tropical with regional variations. Average annual temperatures range between 23° and 27°C depending on altitude. Precipitation is determined by the monsoon and trade winds blowing across various parts of the island. The eastern and northwestern coasts are dominated by the southeasterly trade winds that blow constantly during the winter, carrying heavy rains and producing annual precipitation ranging from 2,000 millimeters to 3,700 millimeters. The central plateau and the western coast receive rain from the monsoon winds prevalent during the summer, with precipitation ranging from 1,000 millimeters to 1,500 millimeters per year. The southern part of the island, however, does not get wind movement and consequently receives as little as 350 millimeters per year; in places, it is a semidesert. The central plateau enjoys a tropical mountain climate with well-differentiated seasons (Madagascar, Ministère de l'Environnement et des Forêts 2010).

Although Madagascar's economy is agrarian, much of the land is unsuitable for cultivation because of mountainous terrain, extensive lateralization, and inadequate or irregular rainfall. Only about 5 percent of the land area is cultivated at any one time, of which 16 percent is irrigated. In addition to providing livelihoods for two-thirds of the population, agriculture contributes 29 percent of the nation's GDP. The economy also benefits from trade and a small and uncompetitive industrial sector (USDoS 2011).

Most farmers practice small-scale subsistence on small family plots in rural areas. A variety of foodcrops—including rice, cassava, bananas, maize, and sweet potatoes—is grown. Nonetheless, the yields are insufficient to meet domestic demand. For instance, per capita rice production declined from 1.2 tons in 1975 to only 0.9 ton in 2006 (Rapport National d'Investissement Madagascar 2008). Sisal and sugarcane dominate large-scale production. "Slash-and-burn" (shifting) agriculture is a common practice resulting in environmental degradation and forest loss. The technique has been perpetuated by the lack of adequate infrastructure in many rural regions, limiting access to information, agricultural inputs, credit, and markets (Erdmann 2003). This situation makes adaptation to climate change more challenging, because it restricts options for agricultural diversification and keeps farmers reliant on rain and thus more vulnerable to changing seasonal and precipitation patterns.

Agriculture in Madagascar is challenged by extreme weather events, including droughts and cyclones, as well as other climate-related disasters, including landslides and locust plagues (FAO 2000, 2010b; Preventionweb 2011). For example, in early 2000, a series of three particularly savage cyclones affected more than 1 million people and caused damage to agricultural infrastructure of nearly $85 million. The cyclones devastated the production of livestock and major crops, including rice, maize, cloves, vanilla, and coffee; they also affected other sectors such as environment and health (FAO 2000). Assessing the nation's vulnerability to these events is critical, given that such climate phenomena are likely to continue to occur, with damaging consequences for the economic development of the country and the livelihoods of its people.

Although Madagascar is a signatory to both the United Nations Framework Convention on Climate Change and the Kyoto Protocol, and although Madagascar's National Adaptation Programme of Action was adopted in 2006, institutional progress on climate change has been slow due to limited finances and political uncertainty.

Review of the Current Situation and Trends

Economic and Demographic Indicators

Population

Figure 8.1 shows Madagascar's total and rural population (left axis) and the share of the urban population (right axis). Figure 8.2 shows the population distribution in 2000 (in terms of persons per square kilometer). Table 8.1 provides additional information concerning rates of population growth.

The Malagasy population was estimated at 19.1 million people in 2008. The rural, urban, and total populations were growing at an increasing rate in the four

FIGURE 8.1   Population trends in Madagascar: Total population, rural population, and percent urban, 1960–2008

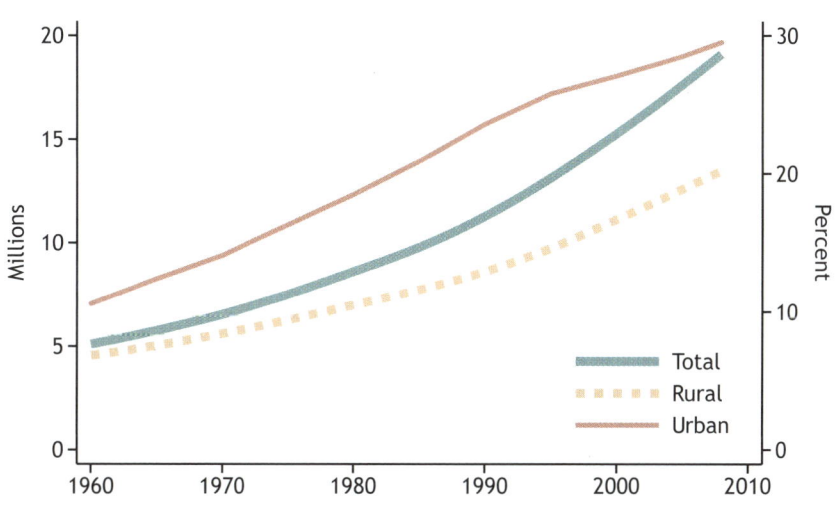

Source: *World Development Indicators* (World Bank 2009).

FIGURE 8.2   Population distribution in Madagascar, 2000 (persons per square kilometer)

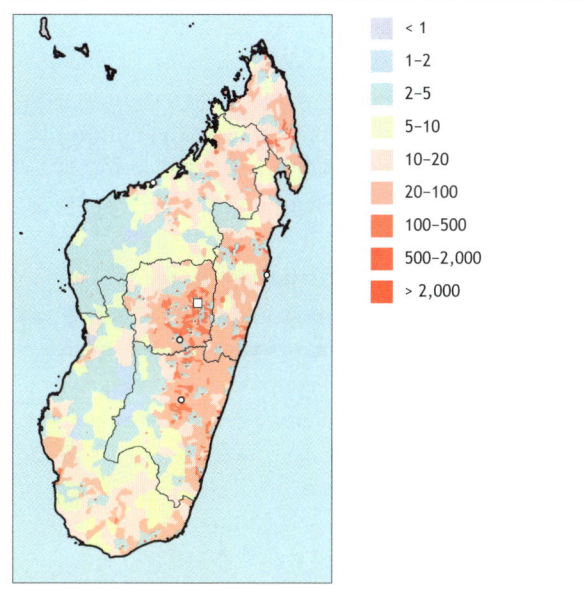

< 1
1–2
2–5
5–10
10–20
20–100
100–500
500–2,000
> 2,000

Source: CIESIN et al. (2004).

TABLE 8.1   Population growth rates in Madagascar, 1960–2008 (percent)

Decade	Total growth rate	Rural growth rate	Urban growth rate
1960–1969	2.5	2.1	5.4
1970–1979	2.8	2.2	5.5
1980–1989	2.7	2.0	5.1
1990–1999	3.1	2.6	4.5
2000–2008	2.8	2.4	3.9

Source: Authors' calculations based on *World Development Indicators* (World Bank 2009).

decades from 1960 through 1999 (see Figure 8.1). This growth is attributed to high fertility rates along with improved health services that led to better maternal health and a drop in mortality for children under five years of age. The past decade has seen growth in all population categories at decreasing rates: overall growth was 2.69 percent in 2008 compared to 2.86 percent in 1988. This was due to the reduced fertility rates brought about by increased use of family planning methods (Sharp and Kruse 2011). By 2005, at least 18 percent of women of childbearing age were using contraceptives, and the average fertility rate was down to 5.4 children per woman (although the rate was as high as 7–10 children in some rural areas) (IMF 2007). Although most of the people in Madagascar continue to live in the rural areas, rural population growth rates are much slower than urban rates. The urban population constituted 29.5 percent of the total population in 2008—up from 10.6 percent in 1960. Urbanization coupled with an increasing intensity of extreme weather events in a country with low adaptive capacity poses challenges relating to urban water scarcity and a higher disease burden due to the consequent poor sanitation. Currently only 35 percent of the population has access to safe drinking water.

Figure 8.2 shows the geographic distribution of the population in Madagascar; estimations are based on census data and other sources. Madagascar has a relatively low population density, estimated at 32.8 inhabitants per square kilometer as of 2008 (PNUD 2010). Generally the population is unequally distributed, with the eastern and central highlands more densely populated than the western parts of the island. The reason is that the eastern and central terrain (plains in Andapa and Alaotra and the forest zones along the coast) are suitable for agriculture, particularly irrigated rice farming and exporting. The western part is suitable mainly for livestock rearing with the exception of the rich alluvial plains of Betsiboka, Bas Mangoky, the delta of Dabara, and Betsiriry Valley.

FIGURE 8.3   Per capita GDP in Madagascar (constant 2000 US$) and share of GDP from agriculture (percent), 1960–2008

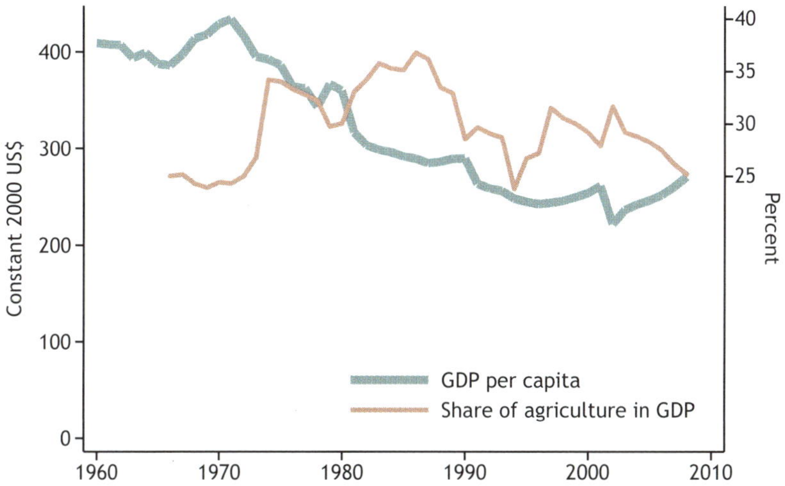

Source: *World Development Indicators* (World Bank 2009).
Note: GDP = gross domestic product; US$ = US dollars.

Income

Figure 8.3 shows trends in gross domestic product (GDP) per capita and the proportion of GDP from agriculture. Generally, Madagascar has been characterized by declining GDP per capita for the past half century due to an underdeveloped economy. Sources of GDP growth include light manufacturing, tourism and ecotourism (based on the country's rich biodiversity, natural habitats, and indigenous lemurs), and textiles. The country is also the global leader in vanilla production and exports. Its per capita GDP was $400 in the 1960s but decreased to less than $300 in the 1980s. The political crisis of 2002 sent the per capita GDP to an all-time low, below $250. Economic and structural reforms adopted after that period have contributed to gains in per capita GDP, shown in Figure 8.3 as a sharp rise after 2002. In 2009 Madagascar's GDP per capita was estimated at $438 (USDoS 2011).[1]

Agriculture remains important to Madagascar's economic growth. Crop production, livestock, and fisheries represent, on average, 30 percent of the wealth produced in the country. Trends show a nearly constant proportion of

---

1 This differs from the amount shown in Figure 8.3 because the graph is in constant 2000 US$.

GDP from agriculture over the past five decades, suggesting that there was a low level of economic development over that period, though agriculture's share in GDP declined from 33 to 25 percent between 1988 and 2008. Generally, the country is overreliant on rice, reflecting a lack of research and training in potential opportunities for agricultural diversification. The continued dependence on agriculture, an undiversified economy, and low per capita GDP would make it difficult for the Malagasy population and the country to counter the adverse effects of climate change.

Vulnerability to Climate Change

Table 8.2 provides some data on additional indicators of the population's vulnerability and resiliency to economic shocks: the level of education, literacy, and concentration of labor in poorer or less dynamic sectors.

Madagascar has a relatively high level of adult literacy; however, its primary school enrollment is significantly higher than its secondary school enrollment. This is partly attributed to adult literacy programs such as ASAMA and Ambohitsoratra, as well as to the government policy requiring children from age 6 to 14 to attend school and the launch of the "free education for all" program in 2003. The high proportion of basic education suggests that the country has the capacity to take advantage of climate change mitigation and adaptation strategies through schooling. This is critical given that most of the population is employed in agriculture and other vulnerable work. However, completion of school is only 19 percent at the primary level and 7 percent at the secondary level (IMF 2007); this has been blamed on a curriculum that is not sufficiently diversified to meet the needs of the various players in the economy.

Madagascar has one of the highest rates of malnutrition in the region, affecting 37 percent of children under age five as of 2004. According to Sharp and Kruse (2011), Malagasy children are prone to the condition early in life.

TABLE 8.2 Education and labor statistics for Madagascar, 2000s

Indicator	Year	Percent
Primary school enrollment: Percent gross (three-year average)	2007	141.4
Secondary school enrollment: Percent gross (three-year average)	2007	26.4
Adult literacy rate	2000	70.7
Percent employed in agriculture	2005	82.0
Percent with vulnerable employment (in agriculture on own farm or as a day laborer)	2005	86.4
Under-five malnutrition (weight for age)	2004	36.8

Source: *World Development Indicators* (World Bank 2009).

By the age of 24 months, more than half the children are nutritionally at risk. This condition is worse among the poorer households. The level of adult malnutrition is also high: an estimated 65 percent of the population lacks access to the required 2,300 kilocalories per day. One consequence of malnutrition is reduced agricultural productivity, which promotes food insecurity.

Figure 8.4 presents two noneconomic correlates of poverty, life expectancy and under-five mortality, in which Madagascar has shown steady improvement over the past five decades. Child mortality has fallen, and life expectancy at birth has steadily increased—from only 51.4 years in 1988 to 55.5 years in 2005 and 60.1 years in 2008. The country's progress on these indicators is attributed to advances in health policies and facilities; both the vaccination rate and access to nutritional supplements exceed 80 percent. Climate change, however, presents direct risks to health. Extreme events such as cyclones can destroy health infrastructure and lives, with possible secondary effects of cholera, malaria, and dysentery epidemics, which have associated economic consequences. The cyclone in the year 2000, for example, showed clearly how climate change could affect mortality and life expectancy and increase vulnerability (Reliefweb 2000).

Figure 8.5 shows the proportion of the population living on less than $2 per day. In Madagascar poverty is widespread, with two-thirds of the population living under the poverty line (Madagascar, Ministère de l'Economie et de

FIGURE 8.4   Well-being indicators in Madagascar, 1960–2008

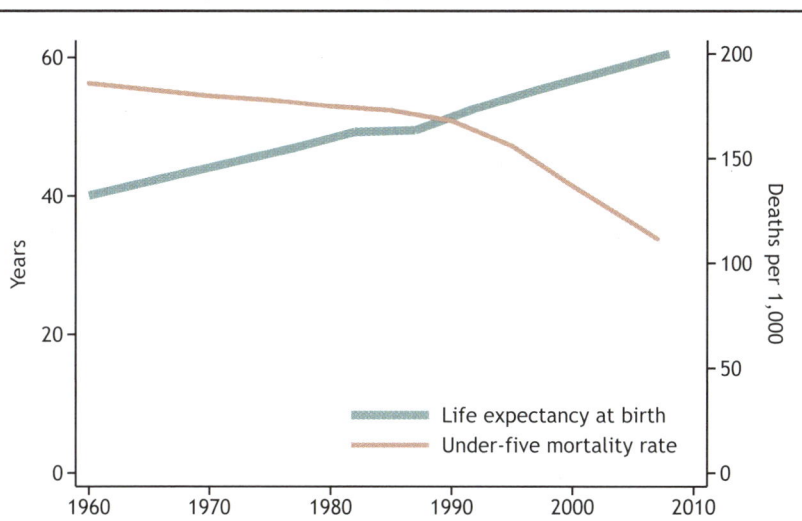

Source: *World Development Indicators* (World Bank 2009).

FIGURE 8.5   Poverty in Madagascar, circa 2005 (percentage of population below US$2 per day)

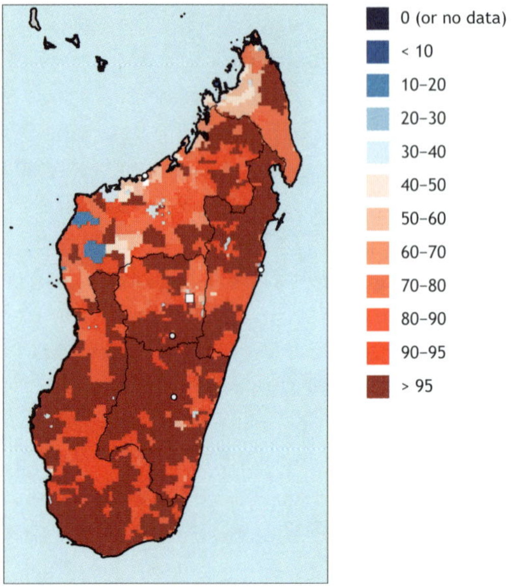

Legend:
- 0 (or no data)
- < 10
- 10–20
- 20–30
- 30–40
- 40–50
- 50–60
- 60–70
- 70–80
- 80–90
- 90–95
- > 95

Source: Wood et al. (2010).
Note: Based on 2005 US$ (US dollars) and on purchasing power parity value.

l'Industrie 2011). The situation has improved, however, since 2003, when at least 85.1 percent of the population was below the poverty line (IMF 2007). Vulnerability is greater in the rural areas, where 74 percent live in poverty compared to 54 percent in urban areas (Sharp and Kruse 2011).

According to Vision 2025 Madagascar (2007), regional disparities also exist: the eastern, southern, and southeastern parts of the island still show poverty rates of more than 80 percent. Indeed, in most regions of the country, more than 95 percent of the people live on less than $2 per day (Figure 8.5). The lowest proportions of poverty are in the western parts of the country, which record about 20 percent below the poverty line. High poverty levels suggest a generally low level of resilience to climate change.

Review of Land Use, Potential, and Limitations

Land Use Overview

Figure 8.6 shows land cover and land use as of 2000. Madagascar, with 587,041 square kilometers of land, has just 5 percent cultivable land and

FIGURE 8.6   Land cover and land use in Madagascar, 2000

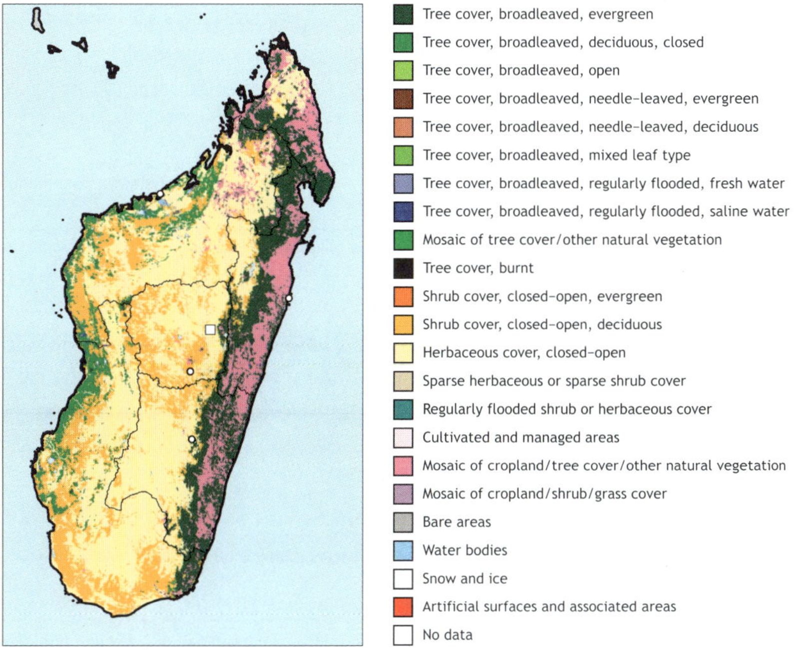

Tree cover, broadleaved, evergreen
Tree cover, broadleaved, deciduous, closed
Tree cover, broadleaved, open
Tree cover, broadleaved, needle–leaved, evergreen
Tree cover, broadleaved, needle–leaved, deciduous
Tree cover, broadleaved, mixed leaf type
Tree cover, broadleaved, regularly flooded, fresh water
Tree cover, broadleaved, regularly flooded, saline water
Mosaic of tree cover/other natural vegetation
Tree cover, burnt
Shrub cover, closed–open, evergreen
Shrub cover, closed–open, deciduous
Herbaceous cover, closed–open
Sparse herbaceous or sparse shrub cover
Regularly flooded shrub or herbaceous cover
Cultivated and managed areas
Mosaic of cropland/tree cover/other natural vegetation
Mosaic of cropland/shrub/grass cover
Bare areas
Water bodies
Snow and ice
Artificial surfaces and associated areas
No data

Source: GLC2000 (Bartholome and Belward 2005).

1 percent in perennial crops. The remaining 94 percent is pastures, forests, and uncultivated lands. Much of the country is covered by savanna or pseudo-steppe. Forest zones and export crops are located in the east; foodcrops are grown on the highlands.

Forest resources occupy 12.3 million hectares, or 18.5 percent of the land area (République de Madagascar, Ministère de l'Environnement et des Forêts 2010), containing a wealth of invaluable biodiversity in flora and fauna. Of this area, 98.2 percent comprises natural to degraded forest; 1.8 percent is planted forest. Only 1.8 million hectares are subject to conservation efforts supported by the system of protected areas and managed by the National Association of Protected Areas. The western coast is bordered with mangrove swamps. Dry vegetation with thorn bushes covers the extreme southwestern and southern parts of the country (Madagascar, Ministère de l'Environnement et des Forêts 2010). The decline in forest cover is estimated at 200,000 hectares per year, mostly due to conversion to agricultural land use through slash-and-burn culti-vation practices. It is important to slow forest loss, to sustain biodiversity as a source of germplasm for generating technologies to respond to climate change.

Figure 8.7 shows the locations of protected areas, including parks and reserves. These locations serve as biodiversity conservation sites and support the tourism industry. Madagascar has one of the highest global rankings in flora and fauna biodiversity and endemism. In the forest ecosystem, 85 percent of flora, 39 percent of birds, 91 percent of reptiles, 99 percent of amphibians, and 100 percent of lemurs are endemic. Encroachment thus has severe consequences for biodiversity conservation.

Madagascar has several types of protected areas, categorized according to the guidelines of the International Union for the Nature Conservation. These categories include Ia, Strict Nature Reserve; Ib, Wilderness Area; II, National Park; III, National Monument; IV, Habitat or Species Management Area; V, Protected Landscape or Seascape; and VI, protected area with sustainable use of natural resources. Some protected areas are controlled by the government, some are coadministered, others are private protected areas, and still others are managed by communities.

Protected areas are subject to human encroachment and especially deforestation. This is partly due to the high level of dependence on wood for energy

FIGURE 8.7   Protected areas in Madagascar, 2009

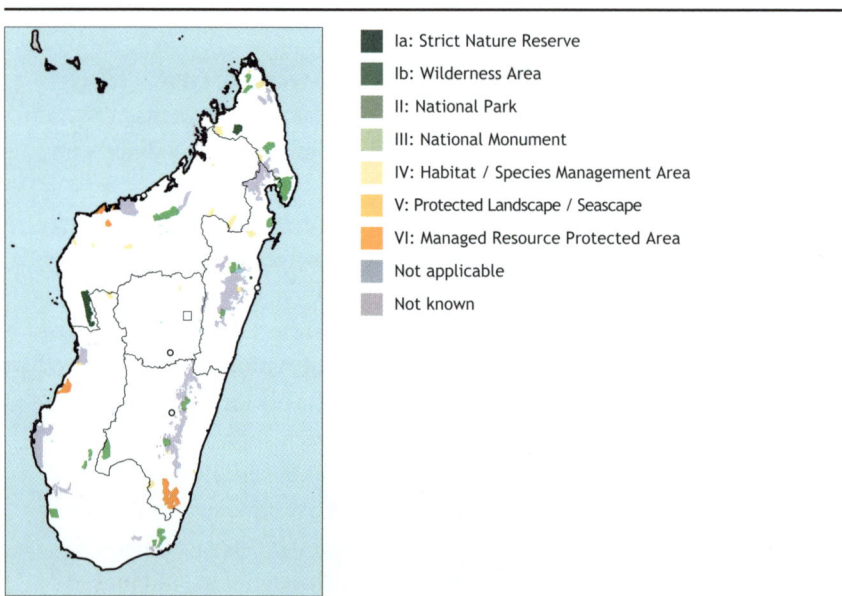

Sources: Protected areas are from the World Database on Protected Areas (UNEP and IUCN 2009). Water bodies are from the World Wildlife Federation's Global Lakes and Wetlands Database (Lehner and Döll 2004).

sources. Renewable energy sources such as biofuel and solar are nonexistent, and electricity coverage is only 4 percent. Another factor is limited land ownership; only 10 percent of the population held a land title or certificate of occupancy as of 2005. Efforts are being carried out to better manage forest resources. The National Plan of Environmental Action advocates for the creation and management of 2.65 million hectares of protected areas. It also supports reducing deforestation by 75 percent, along with systematic application of environmental safeguard measures for private- and public-sector investments, especially in the sensitive areas. Environmental instrument panels were created to measure the state of degradation and progress related to protection. However, a network of protected areas is expensive: the annual management cost of the protected areas is estimated at approximately $14 million, starting from 2012 (World Bank 2010b).

Figure 8.8 shows travel times to cities and towns of various sizes, which provide us with a tool to consider the issue of accessibility in regard to potential markets for agricultural products. Overall trade in Madagascar would increase by 20 percent with proper maintenance of infrastructure (IMF 2007). By 2005 the country had 6,300 kilometers of asphalt roads and 2,277 kilometers of maintained rural or gravel roads. By 2005, at least 35 percent of communities had access to a year-round road, and the number of isolated areas had declined from 59 percent in 2003 to 33 percent. The road network is not uniformly distributed. In places close to the urban centers, particularly in Antananarivo, Fianarantsoa, and Taomasina Provinces, it takes less than three hours to go from one area to another. Farming populations located in these provinces find it easier to access agricultural markets, with direct impacts on poverty levels. Provinces like Antsirananaare are more remote: it takes 16–26 hours to access urban markets from these locations. Extreme climate events, such as floods that damage infrastructure, lead to increased travel costs and time for farmers in this area.

Agriculture

Table 8.3 shows key agricultural commodities in terms of area harvested, and Table 8.4 shows food consumption ranked by weight. Rice occupies first place in terms of agricultural production, followed by cassava and maize. Rice growing constitutes the primary economic activity for the majority of rural farmers, and rice is produced widely in the country. Adequate temperatures, good agricultural practices, and sufficient water volume are necessary to ensure a good harvest. In Madagascar the crop grows well on compact soils under irrigated systems. Yields range from 1.80 to 2.57 tons per hectare, reflecting low use of such inputs as fertilizers, improved seed, and appropriate machinery.

FIGURE 8.8   Travel time to urban areas of various sizes in Madagascar, circa 2000

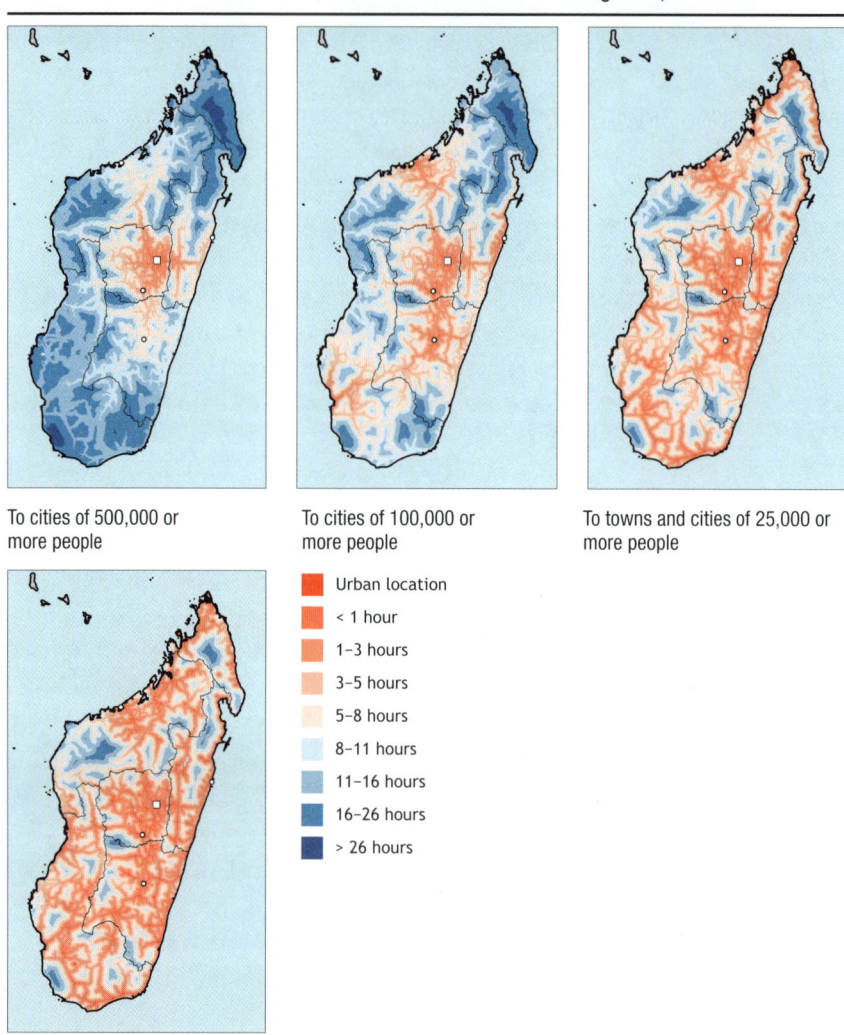

To cities of 500,000 or
more people

To cities of 100,000 or
more people

To towns and cities of 25,000 or
more people

To towns and cities of 10,000 or
more people

Legend:
- Urban location
- < 1 hour
- 1–3 hours
- 3–5 hours
- 5–8 hours
- 8–11 hours
- 11–16 hours
- 16–26 hours
- > 26 hours

Source: Authors' calculations.

TABLE 8.3 Harvest area of leading agricultural commodities in Madagascar, 2006–2008 (thousands of hectares)

Rank	Crop	Percent of total	Harvest area
	Total	100.0	2,885
1	Rice	43.1	1,244
2	Cassava	11.0	317
3	Maize	9.6	277
4	Sweet potatoes	4.3	125
5	Coffee	4.2	122
6	Beans	2.9	84
7	Sugarcane	2.8	82
8	Vanilla	2.3	67
9	Bananas	2.0	58
10	Groundnuts	1.9	55

Source: FAOSTAT (FAO 2010a).
Note: All values are based on the three-year average for 2006–2008.

TABLE 8.4 Consumption of leading food commodities in Madagascar, 2003–2005 (thousands of metric tons)

Rank	Crop	Percent of total	Food consumption
	Total	100.0	6,919
1	Cassava	30.9	2,138
2	Rice	25.5	1,764
3	Other fruits	5.5	383
4	Sweet potatoes	5.3	370
5	Maize	4.3	300
6	Other vegetables	4.1	284
7	Bananas	3.9	268
8	Potatoes	2.1	144
9	Beef	1.8	121
10	Sugar	1.6	114

Source: FAOSTAT (FAO 2010a).
Note: All values are based on the three-year average for 2003–2005.

Rice is the preferred staple food for the large majority of Malagasy people. However, in terms of consumption, rice falls behind cassava (see Table 8.4). Although rice is produced in large quantities, the country is a net importer (FINTRAC 2008). Cassava is the second-leading source of calories, representing nearly 14 percent of consumption; for poor households, especially in the south, it constitutes more than 25 percent of the calories consumed. In times of food shortage, families fall back on cassava as a means of coping with hunger and non-affordable rice prices (Dostie, Randriamamonjy, and Rabenasolo 1999).

Figures 8.9 and 8.10 show production of rice in Madagascar. Rainfed rice is produced throughout the country. However, the highest yields occur mainly in the central and northern parts of Antananarivo Province, where the largest harvest areas are achieved. Production of irrigated rice is scattered throughout the country. The effects of climate change on precipitation levels will require more reliance on irrigation for this crop.

FIGURE 8.9   Yield (metric tons per hectare) and harvest area density (hectares) for rainfed rice in Madagascar, 2000

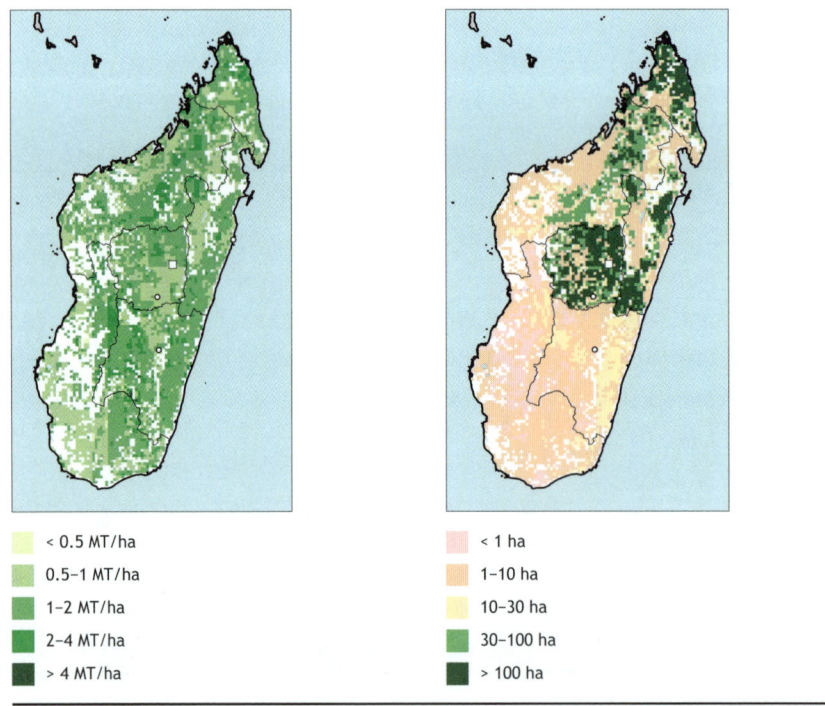

< 0.5 MT/ha	< 1 ha
0.5–1 MT/ha	1–10 ha
1–2 MT/ha	10–30 ha
2–4 MT/ha	30–100 ha
> 4 MT/ha	> 100 ha

Source: SPAM (Spatial Production Allocation Model) (You and Wood 2006; You, Wood, and Wood-Sichra 2006, 2009).
Note: ha = hectare; MT/ha = metric tons per hectare.

FIGURE 8.10   Yield (metric tons per hectare) and harvest area density (hectares) for
irrigated rice in Madagascar, 2000

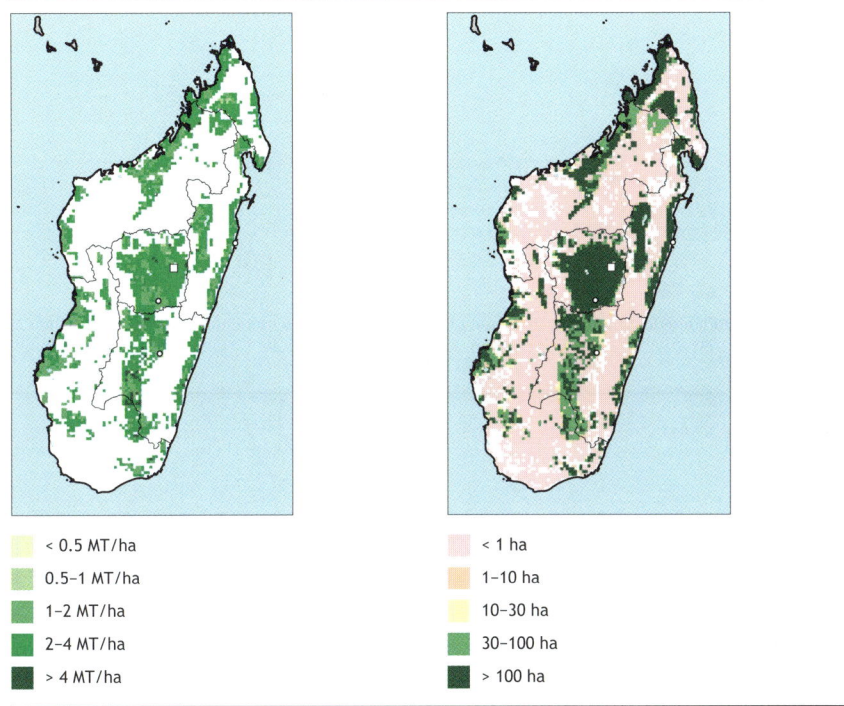

< 0.5 MT/ha	< 1 ha
0.5–1 MT/ha	1–10 ha
1–2 MT/ha	10–30 ha
2–4 MT/ha	30–100 ha
> 4 MT/ha	> 100 ha

Source: SPAM (Spatial Production Allocation Model) (You and Wood 2006; You, Wood, and Wood-Sichra 2006, 2009).
Note: ha = hectare; MT/ha = metric tons per hectare.

Cassava grows in various types of soils, in areas that are not prone to flooding. The plant is favored by temperatures between 25° and 30°C. Figure 8.11 shows that cassava is produced in most of the country; the highest level of production is in Fianarantsoa, which provides 35 percent of national output (Dostie, Randriamamonjy, and Rabenasolo 1999). Cassava is prone to cassava mosaic disease, which is severe during periods of high temperature and little rainfall (Ranomenjanahary, Ramelison, and Seruwagi 2005). The traditional export crops, such as vanilla, coffee, cloves, and pepper, are cultivated on the eastern coast and in the northeast, as are litchis, one of the newer export crops. Other commercial crops, including sugarcane, groundnuts, tobacco, and cotton, were once well developed in the northwest and the south of the island but have declined.

Maize is concentrated in a few areas of the country, but despite its limited distribution geographically, in terms of harvested area, is ranked third (Figure 8.12); yields range from under 0.5 to 2.0 tons per hectare. Production is highest in Antsiranana Province in the north.

FIGURE 8.11   Yield (metric tons per hectare) and harvest area density (hectares) for rainfed cassava in Madagascar, 2000

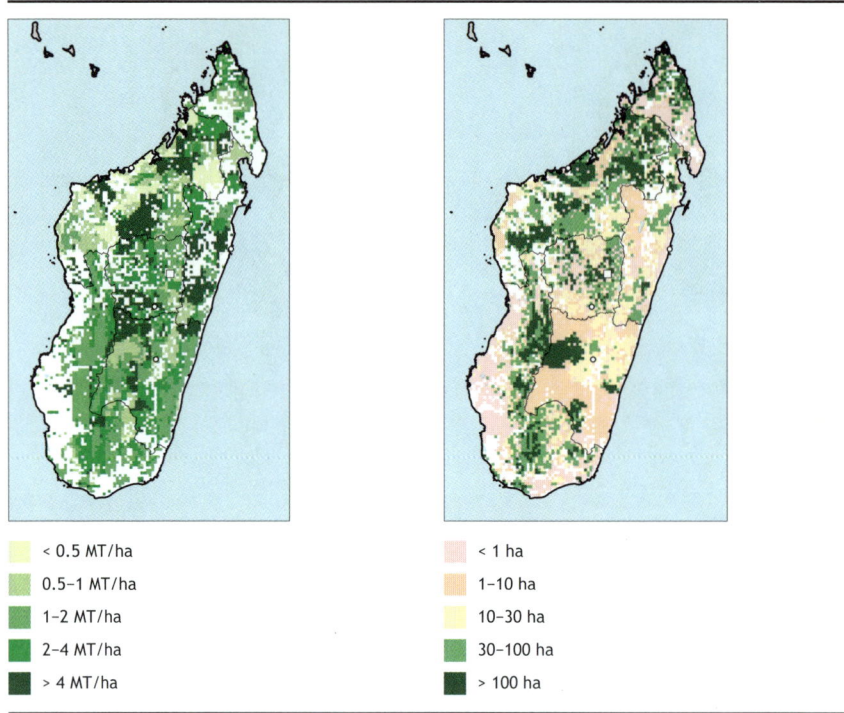

< 0.5 MT/ha		< 1 ha	
0.5–1 MT/ha		1–10 ha	
1–2 MT/ha		10–30 ha	
2–4 MT/ha		30–100 ha	
> 4 MT/ha		> 100 ha	

Source: SPAM (Spatial Production Allocation Model) (You and Wood 2006; You, Wood, and Wood-Sichra 2006, 2009).
Note: ha = hectare; MT/ha = metric tons per hectare.

Scenarios for the Future

Economic and Demographic Indicators

Population

Figure 8.13 shows population projections by the UN Population Division through 2050. The projections for the Malagasy population for 2050 range from just under 40 million people to almost 52 million people. According to Sharp and Kruse (2011), factors determining health and survival have improved greatly, to the extent that political and economic crises have had negligible effects on population growth. Moreover, a substantial proportion of the population is young people, with high fertility rates: at least 75 percent of women have had a child by age 16. Although awareness of family planning services is relatively high among teenagers (45 percent), access is still low. Hence

FIGURE 8.12   Yield (metric tons per hectare) and harvest area density (hectares) for rainfed maize in Madagascar, 2000

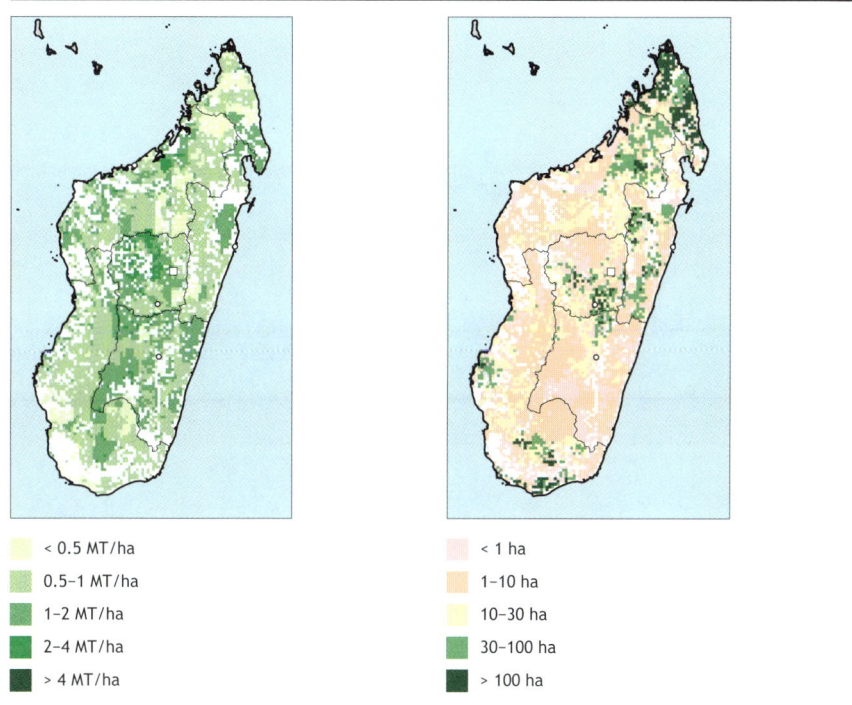

< 0.5 MT/ha	< 1 ha
0.5–1 MT/ha	1–10 ha
1–2 MT/ha	10–30 ha
2–4 MT/ha	30–100 ha
> 4 MT/ha	> 100 ha

Source: SPAM (Spatial Production Allocation Model) (You and Wood 2006; You, Wood, and Wood-Sichra 2006, 2009).
Note: ha = hectare; MT/ha = metric tons per hectare.

it is likely that population growth will follow the high-variant projection, with the population more than doubling by 2050.

Income

Figure 8.14 shows the three scenarios of GDP per capita used for this study. The optimistic scenario combines high GDP with low population. The baseline scenario combines the medium GDP projection with the medium population projection. Finally, the pessimistic scenario combines the low GDP projection with the high population projection.

GDP per capita is shown increasing in all scenarios, with the pessimistic scenario showing a slow increase to $650 by 2050 but the optimistic scenario showing a rise all the way to $1,740. Madagascar has an underdeveloped economy, and its per capita GDP has been declining for nearly half a century.

FIGURE 8.13   Population projections for Madagascar, 2010–2050

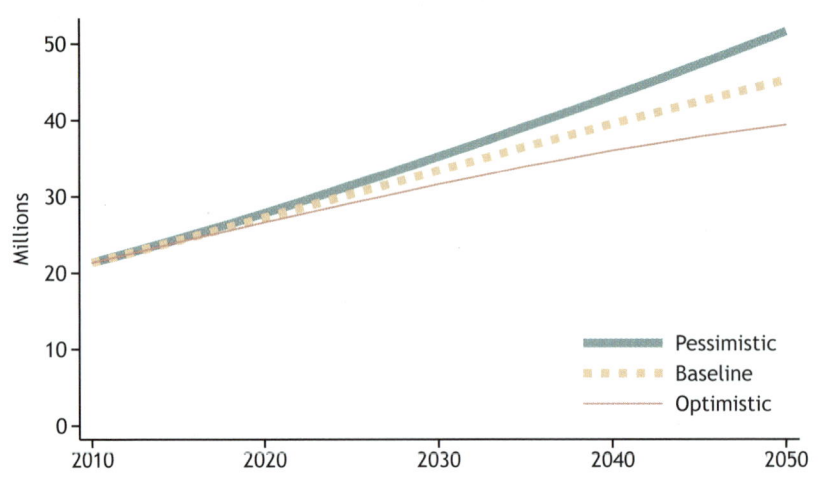

Source: UNPOP (2009).

FIGURE 8.14   Gross domestic product (GDP) per capita future scenarios for Madagascar, 2010–2050

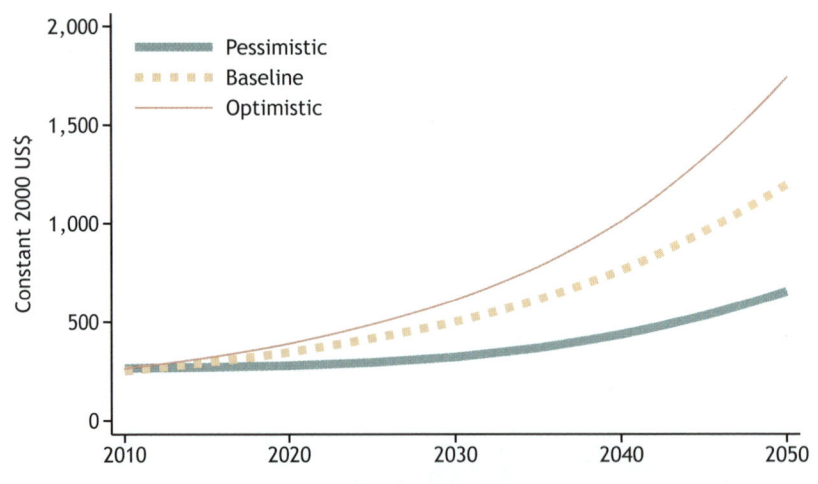

Sources: Computed from GDP data from the World Bank Economic Adaptation to Climate Change project (World Bank 2010a), from the Millennium Ecosystem Assessment (2005) reports, and from population data from the United Nations (UNPOP 2009).
Note: US$ = US dollars.

Structural reforms have brought some improvements, but much of the country remains poor. Recurring political crises have not helped the situation. Most of the population remains employed in agriculture, which is vulnerable to climatic effects. To go beyond the pessimistic scenario would require a stable environment with aggressive policies aimed at improving agricultural productivity and diversification of the economy to raise incomes, especially for the poor. To attain maximum impact, it is important that policies integrate strategies for climate change adaptation.

Biophysical Analysis

Climate Models

This chapter uses four downscaled climate models for the A1B scenario.[2] Figure 8.15 shows projected precipitation changes. For the northern part of the country, the models show either no change or an increase in precipitation. For the southern parts of the country, rainfall either remains relatively unchanged or decreases. Both the increases in the north and the decreases in the south range from −50 to −200 millimeters. Reduced rainfall has consequences for agricultural production, given that most of the key crops are rainfed. It is likely that farmers will need to produce more irrigated rice and less rainfed rice. A lower amount of rainfall will also mean increased cassava mosaic disease, which will reduce crop harvests. In short, without adaptive options, the future is likely to see increases in food insecurity. Because the south already has much lower rainfall than the north, this reduction could be a significant blow to people cultivating many of the annual crops that are currently cultivated there.

Figure 8.16 shows increases in temperature ranging from 0.5° to 3°C, almost throughout the country. These predictions include ranges that are much lower than earlier predictions of temperature increases of 2.5°–3°C (Madagascar, Ministère de l'Environnement, des Eaux et Forêts 2006). There are differences between the models. The CSIRO Mark 3 model has a median temperature change of 1.3°C, cooler than the median for the ECHAM 5 model, 1.7°C, and considerably cooler than the median for both the CNRM-CM3 and the

---

2 The A1B scenario is a greenhouse gas emissions scenario that assumes fast economic growth, a population that peaks midcentury, and the development of new and efficient technologies, along with a balanced use of energy sources.

FIGURE 8.15   Changes in mean annual precipitation in Madagascar, 2000–2050, A1B scenario (millimeters)

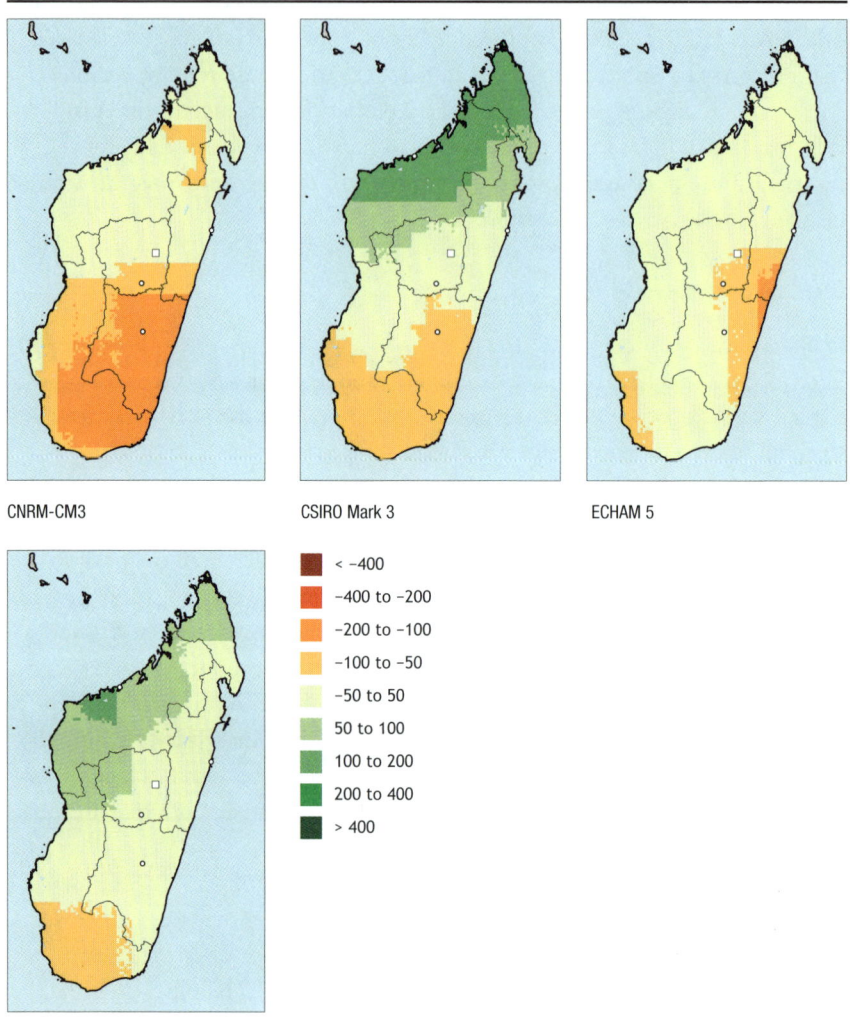

CNRM-CM3

CSIRO Mark 3

ECHAM 5

< −400

−400 to −200

−200 to −100

−100 to −50

−50 to 50

50 to 100

100 to 200

200 to 400

> 400

MIROC 3.2 medium-resolution

Source: Authors' calculations based on Jones, Thornton, and Heinke (2009).

Notes: A1B = greenhouse gas emissions scenario that assumes fast economic growth, a population that peaks midcentury, and the development of new and efficient technologies, along with a balanced use of energy sources; CNRM-CM3 = National Meteorological Research Center–Climate Model 3; CSIRO = climate model developed at the Australia Commonwealth Scientific and Industrial Research Organisation; ECHAM 5 = fifth-generation climate model developed at the Max Planck Institute for Meteorology (Hamburg); GCM = general circulation model; MIROC = Model for Interdisciplinary Research on Climate, developed by the University of Tokyo Center for Climate System Research.

FIGURE 8.16   Changes in monthly mean maximum daily temperature in Madagascar for the warmest month, 2000–2050, A1B scenario (°C)

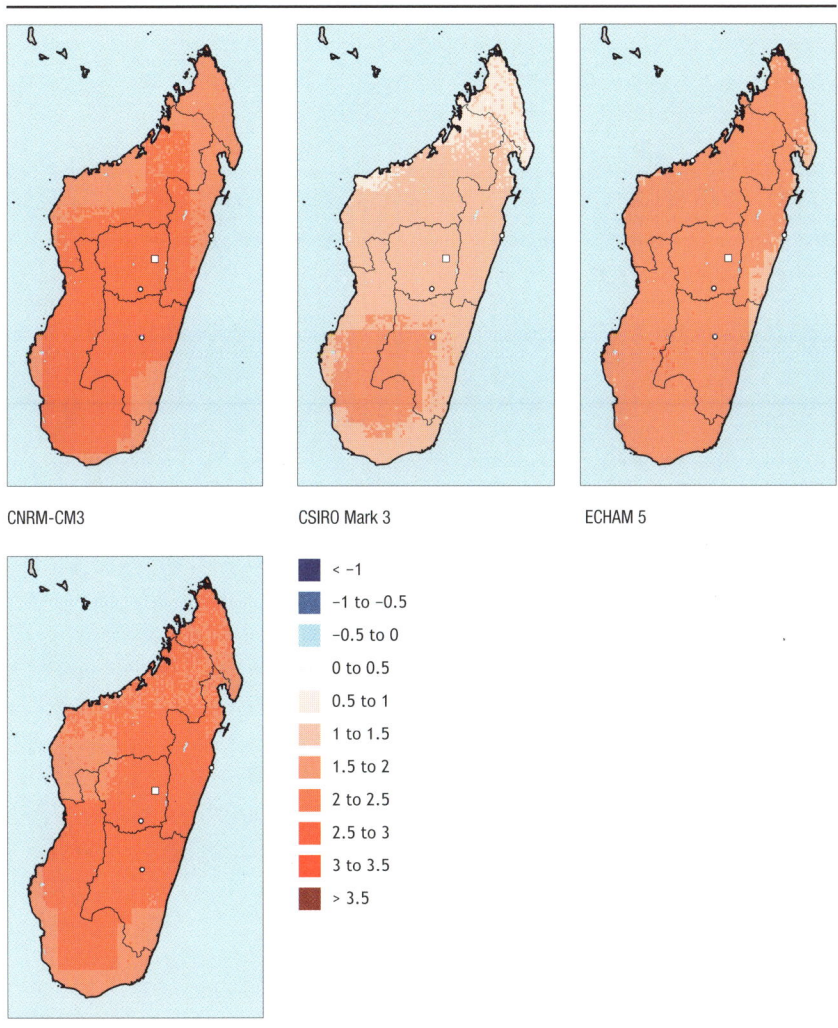

CNRM-CM3   CSIRO Mark 3   ECHAM 5

< −1
−1 to −0.5
−0.5 to 0
0 to 0.5
0.5 to 1
1 to 1.5
1.5 to 2
2 to 2.5
2.5 to 3
3 to 3.5
> 3.5

MIROC 3.2 medium-resolution

Source: Authors' calculations based on Jones, Thornton, and Heinke (2009).

Notes: A1B = greenhouse gas emissions scenario that assumes fast economic growth, a population that peaks midcentury, and the development of new and efficient technologies, along with a balanced use of energy sources; CNRM-CM3 = National Meteorological Research Center–Climate Model 3; CSIRO = climate model developed at the Australia Commonwealth Scientific and Industrial Research Organisation; ECHAM 5 = fifth-generation climate model developed at the Max Planck Institute for Meteorology (Hamburg); GCM = general circulation model; MIROC = Model for Interdisciplinary Research on Climate, developed by the University of Tokyo Center for Climate System Research.

MIROC 3.2 models.[3] Furthermore, the changes are not uniform spatially in most of the models. High temperatures promote evapotranspiration, thus reducing soil moisture and increasing soil degradation. High temperatures may also promote an increase in pests and diseases, as in the case of cassava mosaic disease, whose virus multiplies under higher temperatures.

Crop Models

The Decision Support System for Agrotechnology Transfer software was used to compute yields in the climate of 2000 and that of 2050. Results are shown in Figures 8.17 and 8.18.

Figure 8.17 shows remarkably similar changes across all general circulation models (GCMs). Each shows losses throughout the island. In the CSIRO Mark 3 model, most of the losses appear to be less than 25 percent, whereas in the other models the losses appear to be mostly greater than 25 percent. But there is a noticeable area of yield gain near Antananarivo, which our earlier maps showed had a high concentration of land devoted to irrigated rice. This particular area of gain is mixed, with part below 25 percent gain and part above. There is also a much smaller patch in the north that shows yield gain. Both of these patches of yield gain are in high elevations with colder temperatures. Rice yields are hampered by the cold, and what we see in these maps is that with climate change's warmer temperatures, rice will grow much better. Although it is difficult to say precisely whether the yield gains in these high-production areas will be sufficient to offset the yield losses, many of which will occur in low-production areas, we can see that there will likely be winners and losers among rice growers as a result of climate change, and there may be pressure for new settlements or more intensive cultivation in the areas that are projected to become more productive with climate change.

Figure 8.18 shows scattered gains in maize yield in all of the models, with most of the gains predicted to be greater than 25 percent. However, in all of these models the losses are more widespread and more divided between 5–25 percent yield loss and more than 25 percent yield loss. Maize, though a less-preferred crop in production and consumption than cassava and rice, adds to the food supplies available in the country.

---

3 CNRM-CM3 is National Meteorological Research Center–Climate Model 3. MIROC 3.2 is the Model for Interdisciplinary Research on Climate, developed at the University of Tokyo Center for Climate System Research. CSIRO Mark 3 is a climate model developed at the Australia Commonwealth Scientific and Industrial Research Organisation. ECHAM 5 is a fifth-generation climate model developed at the Max Planck Institute for Meteorology in Hamburg.

FIGURE 8.17   Yield change under climate change: Irrigated rice in Madagascar, 2000–2050, A1B scenario

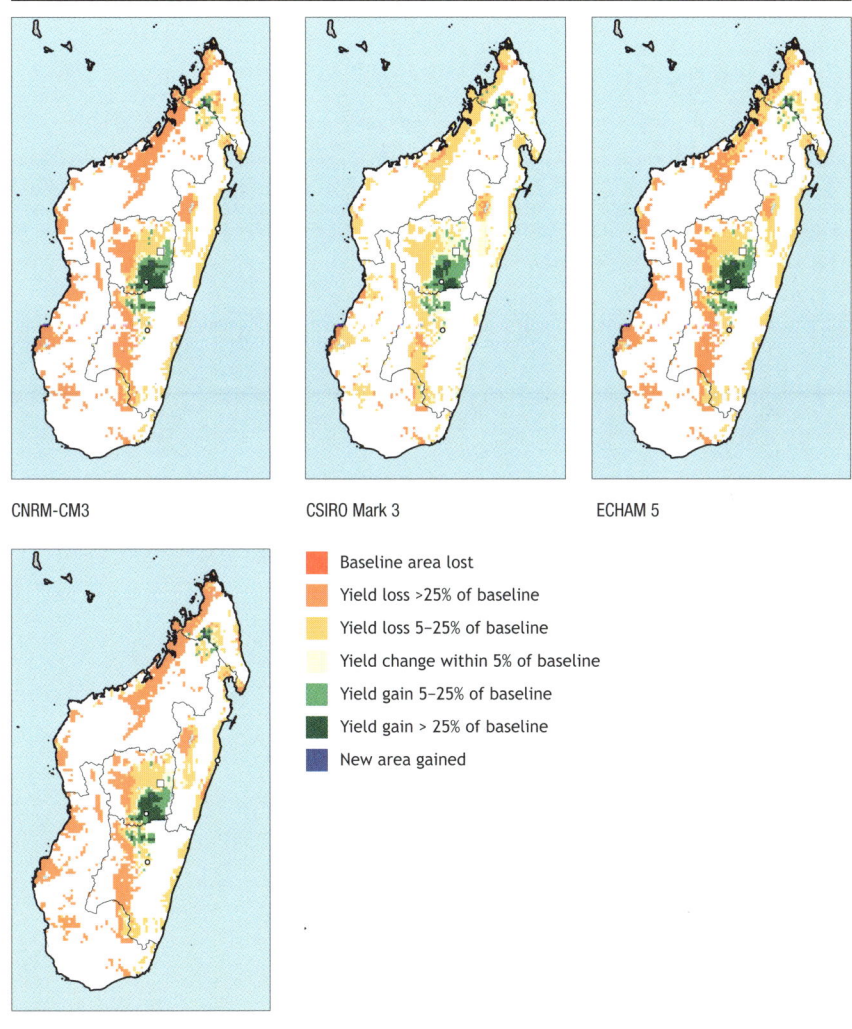

CNRM-CM3   CSIRO Mark 3   ECHAM 5

	Baseline area lost
	Yield loss >25% of baseline
	Yield loss 5–25% of baseline
	Yield change within 5% of baseline
	Yield gain 5–25% of baseline
	Yield gain > 25% of baseline
	New area gained

MIROC 3.2 medium-resolution

Source: Authors' calculations.

Notes: A1B = greenhouse gas emissions scenario that assumes fast economic growth, a population that peaks midcentury, and the development of new and efficient technologies, along with a balanced use of energy sources; CNRM-CM3 = National Meteorological Research Center–Climate Model 3; CSIRO = climate model developed at the Australia Commonwealth Scientific and Industrial Research Organisation; ECHAM 5 = fifth-generation climate model developed at the Max Planck Institute for Meteorology (Hamburg); GCM = general circulation model; MIROC = Model for Interdisciplinary Research on Climate, developed by the University of Tokyo Center for Climate System Research.

FIGURE 8.18 Yield change under climate change: Rainfed maize in Madagascar, 2000–2050, A1B scenario

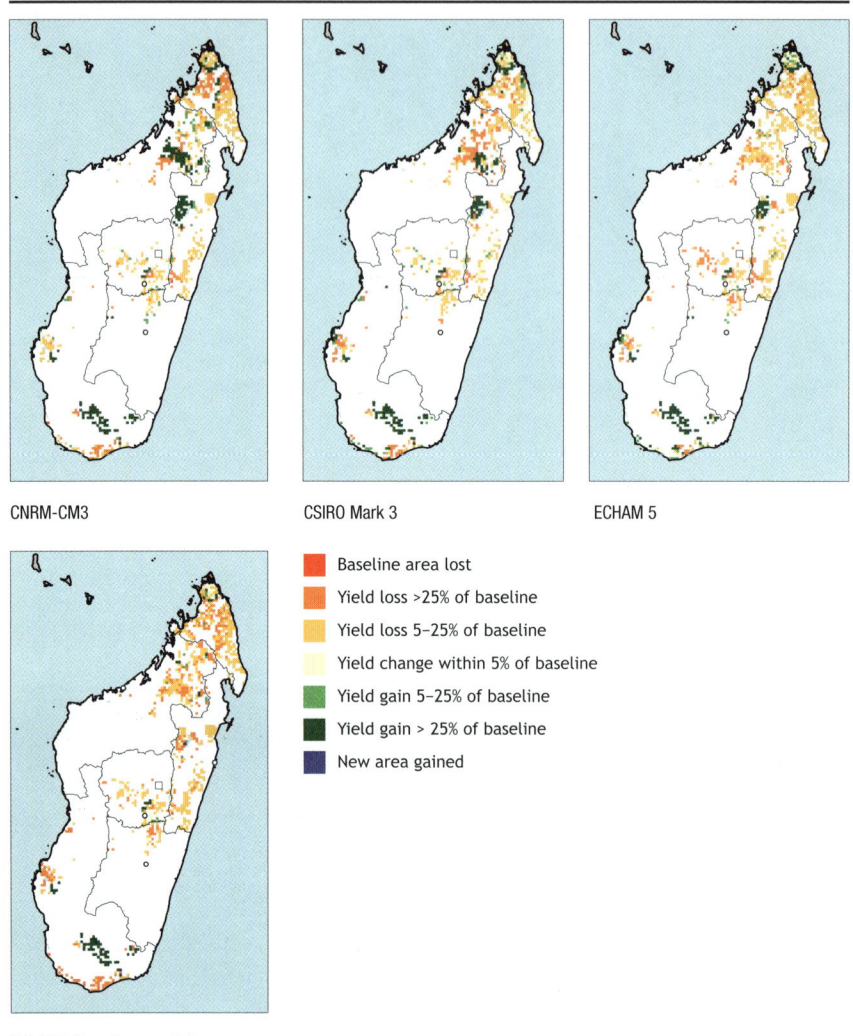

CNRM-CM3   CSIRO Mark 3   ECHAM 5

■ Baseline area lost
■ Yield loss >25% of baseline
■ Yield loss 5–25% of baseline
□ Yield change within 5% of baseline
■ Yield gain 5–25% of baseline
■ Yield gain > 25% of baseline
■ New area gained

MIROC 3.2 medium-resolution

Source: Authors' calculations.

Notes: A1B = greenhouse gas emissions scenario that assumes fast economic growth, a population that peaks midcentury, and the development of new and efficient technologies, along with a balanced use of energy sources; CNRM-CM3 = National Meteorological Research Center–Climate Model 3; CSIRO = climate model developed at the Australia Commonwealth Scientific and Industrial Research Organisation; ECHAM 5 = fifth-generation climate model developed at the Max Planck Institute for Meteorology (Hamburg); GCM = general circulation model; MIROC = Model for Interdisciplinary Research on Climate, developed by the University of Tokyo Center for Climate System Research.

Vulnerability

Figure 8.19 shows the impact of future GDP and population scenarios on the number of malnourished children under age five in Madagascar. Figure 8.20 shows the share of children who are malnourished. Figure 8.21 shows the kilocalories per capita available.

As the kilocalories available per capita increase, malnutrition of children under age five declines. By 2050, in the optimistic and baseline scenarios, the population in Madagascar will be less vulnerable. But the story is not as good for the pessimistic scenario, in which we see the kilocalories per capita barely getting back to the 2010 levels by 2050 after falling off by 11 percent. In part, this reflects the negative price effects dampening the positive income effects. But it likely represents an overdampening due to the own-price elasticities for staple crops being set too high in the International Model for Policy Analysis of Agricultural Commodities and Trade (IMPACT).

The number of malnourished children becomes slightly worse before it gets better in both the optimistic and the baseline scenarios, but the numbers increased more steeply in the pessimistic scenario and did not get down to the 2010 levels by 2050. With population increase, the rates of malnutrition will decline.

FIGURE 8.19   Number of malnourished children under five years of age in Madagascar in multiple income and climate scenarios, 2010–2050

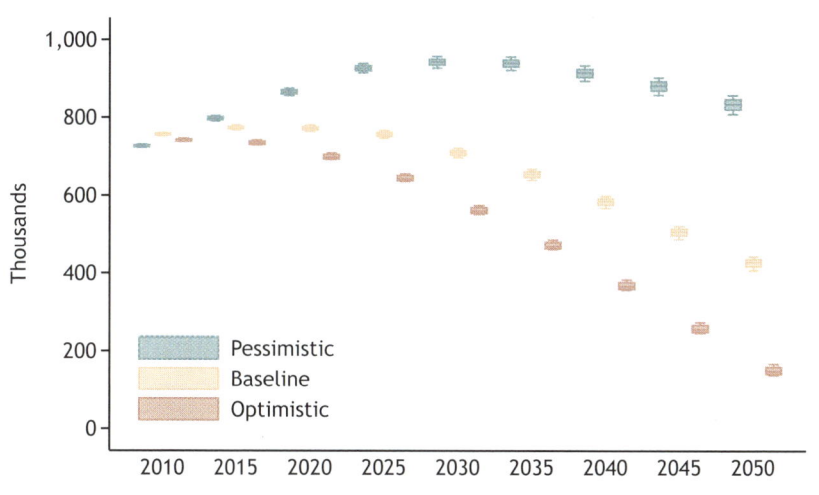

Source: Based on analysis conducted for Nelson et al. (2010).

Note: The box and whiskers plot for each socioeconomic scenario shows the range of effects from the four future climate scenarios.

FIGURE 8.20   Share of malnourished children under five years of age in Madagascar in multiple income and climate scenarios, 2010–2050

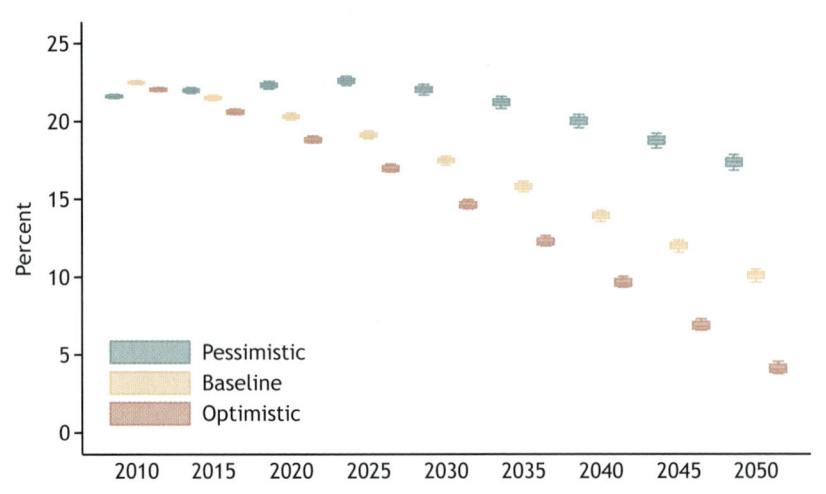

Source: Based on analysis conducted for Nelson et al. (2010).

Note: The box and whiskers plot for each socioeconomic scenario shows the range of effects from the four future climate scenarios.

FIGURE 8.21   Kilocalories per capita in Madagascar in multiple income and climate scenarios, 2010–2050

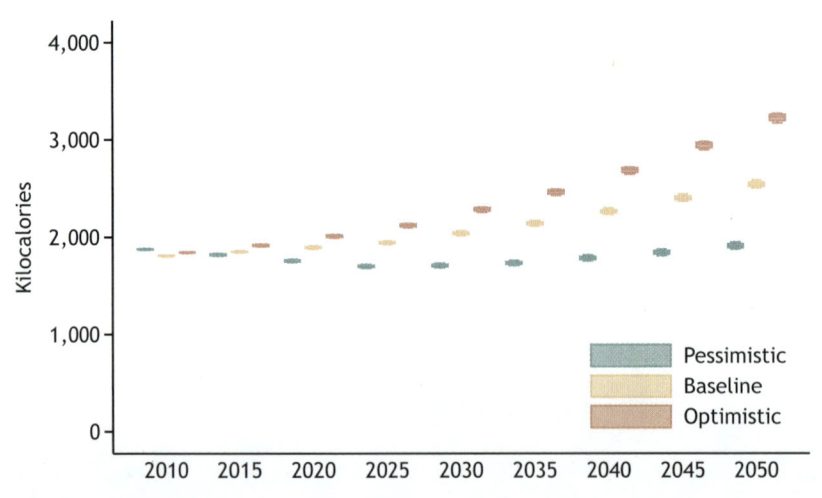

Source: Based on analysis conducted for Nelson et al. (2010).

Note: The box and whiskers plot for each socioeconomic scenario shows the range of effects from the four future climate scenarios.

FIGURE 8.22 Impact of changes in GDP and population on rice in Madagascar, 2010–2050

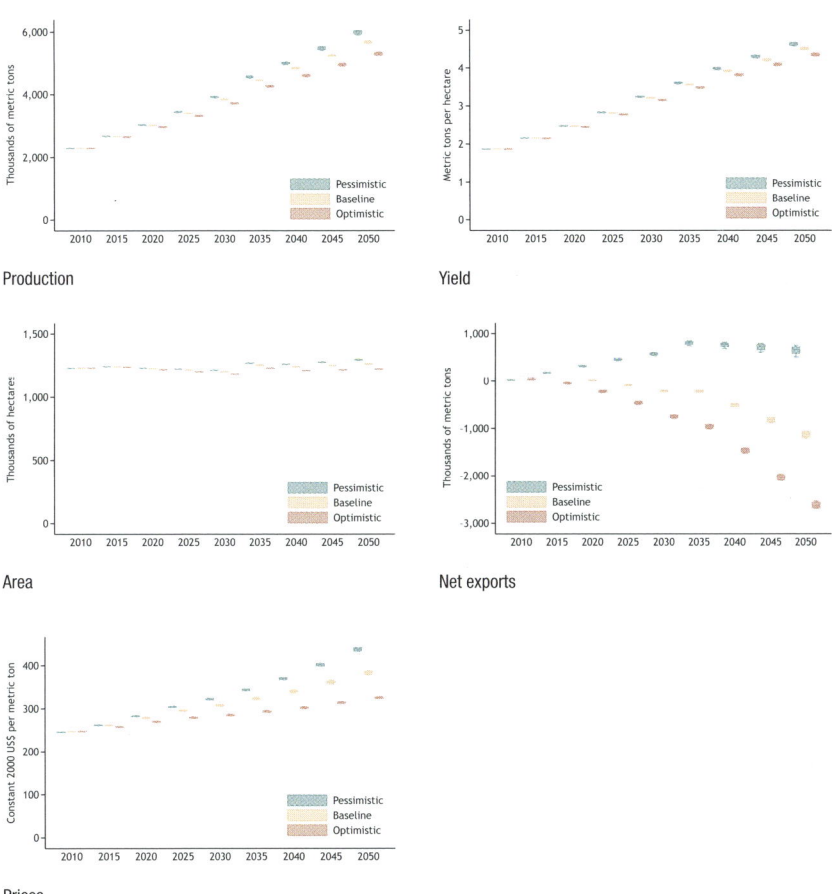

Production

Yield

Area

Net exports

Prices

Source: Based on analysis conducted for Nelson et al. (2010).

Notes: The box and whiskers plot for each socioeconomic scenario shows the range of effects from the four future climate scenarios. GDP = gross domestic product; US$ = US dollars.

Agricultural Outcomes

Figures 8.22–8.24 show simulation results from the IMPACT model associated with key agricultural crops in Madagascar. Each featured crop has five graphs, for production, yield, area, net exports, and world prices.

Figure 8.22 shows an increase in rice production, yields, and world market price across four decades. The harvested area remains more or less unchanged. Export trends appear less favorable over the same time period: the pessimistic

FIGURE 8.23   Impact of changes in GDP and population on cassava in Madagascar, 2010–2050

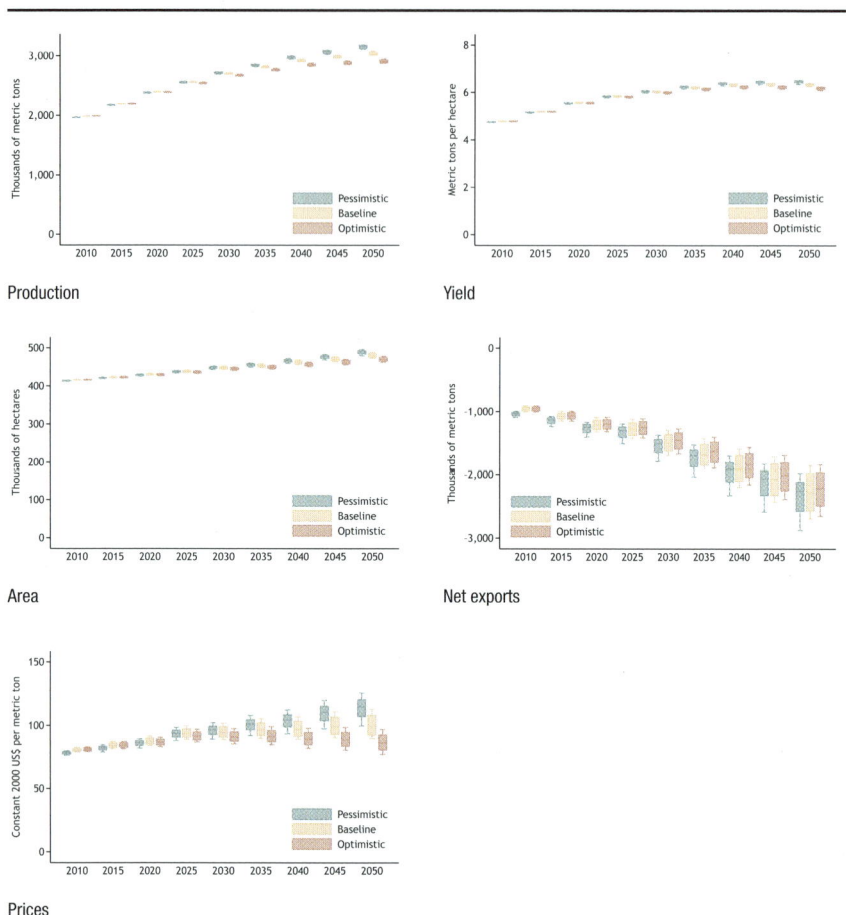

Production

Yield

Area

Net exports

Prices

Source: Based on analysis conducted for Nelson et al. (2010).

Notes: The box and whiskers plot for each socioeconomic scenario shows the range of effects from the four future climate scenarios. GDP = gross domestic product; US$ = US dollars.

scenario shows increasing exports until 2040 and a decline thereafter, whereas the baseline and optimistic scenarios show exports decreasing from zero. Currently the country is a net importer of rice. Consumers will pay higher rice prices in all scenarios; rising internal demand for the crop will cause increased imports in the baseline and optimistic scenarios.

For cassava, the model results show increased production, harvest area, and yields, as well as rising cassava prices (see Figure 8.23). However, there is also a clear trend for increasing imports from 2010 through 2050.

FIGURE 8.24 Impact of changes in GDP and population on maize in Madagascar, 2010–2050

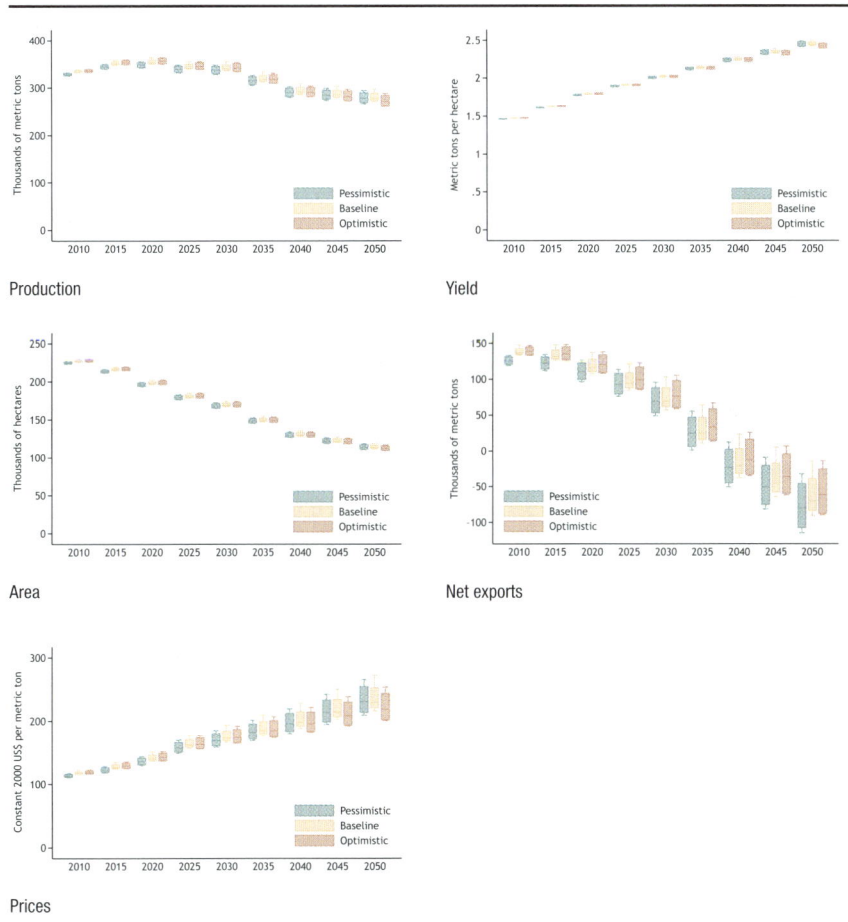

Source: Based on analysis conducted for Nelson et al. (2010).

Notes: The box and whiskers plot for each socioeconomic scenario shows the range of effects from the four future climate scenarios. GDP = gross domestic product; US$ = US dollars.

Future patterns in the maize crop are different (see Figure 8.24). Although the maize yield is projected to rise by around 60 percent, the area planted is projected to drop by around 40 percent, resulting in slightly increased production to 2020 and then a falling off, with 2050 production around 10 percent lower than that in 2010. Much as in the case of cassava and rice, the country will experience declining net exports of maize and face higher prices by 2050. Madagascar is projected to go from an exporter of maize to an importer. Although the rising kilocalories per capita indicate that the income of a large portion of the population will be sufficient to purchase food, this food will come more and more from trade than from own production.

Conclusions and Policy Recommendations

Madagascar is highly vulnerable to climate change due to heavy reliance on rainfed agriculture, high proportions of poverty, and an underdeveloped economy. The degree to which the country is exposed to risks in climate change concerning agriculture is assessed here using climate data from several GCMs together with crop modeling software and a global food and agriculture partial equilibrium model. Without technological change (including the use of more fertilizers and improved seeds), changes in precipitation and temperature are likely to inflict substantial damage, especially on maize production.

The projected growth in population has implications for food security and agricultural resources. The agricultural land in the country is very limited, yet a majority of the population depends on agriculture for subsistence. The population is predicted to more than double by 2050, increasing demand for land and other resources. Intense land pressure is most likely near urban centers, where populations are rising much faster than in rural areas. Although the country produces substantial amounts of food, the levels are already insufficient to meet domestic demand, especially for rice. Strategies should be developed to increase the awareness, access to, and use of family planning methods as well as increased farm productivity.

In this chapter we have considered three economic scenarios. In the optimistic and baseline scenarios we saw that by 2050 the population should be more resilient to the effects of climate change due to higher levels of per capita GDP and per capita kilocalories, along with reduced levels of malnutrition. However, to avoid the projected slow growth of GDP per capita in the pessimistic scenario, more investment is required in existing sectors to ensure a more robust economy and diversification in income-earning opportunities.

In addition, climate models show warmer temperatures and areas of reduced rainfall, which could potentially affect soil and crop productivity as well as increase the incidence of pests and disease. Beyond our ability to analyze with our models, some researchers suggest that unless a solution is found, cassava production will be most affected due to cassava mosaic disease, which favors such changes in climate. It is important that farmers be educated on integrated pest and disease management. Policies that encourage irrigation and develop drought-resistant varieties will also be critical.

In this chapter we have identified parts of Madagascar in which rice pro-duction would be aided by climate change due to high temperatures in areas that currently have yields limited by the cold. Additional yields for rice may be obtained by farming additional land or cultivating more intensively on the land currently used for rice, both of which could have potential detrimental effects on natural vegetation and climate mitigation capacities.

In order for farmers to be able to better adapt to climate changes, they will need the support of agricultural research and extension institutions to develop, test, and promulgate crop varieties that are better suited to the future climate conditions. In addition to simply developing new varieties, new crops (or crops that are not currently as widely used) might need to be introduced. With maize stressed by higher temperatures and by drier conditions in the south, substitute grains that are more resilient in hot, dry conditions could be tested—crops such as sorghum or millet.

Finally, Madagascar is frequently hit by extreme weather events such as cyclones and storms, damaging infrastructure, crops, livestock, and human lives. There are severe impacts on food security, water quality and supply, and public health. The country has only 100 weather stations, with limited capacity to forecast extreme weather events with accuracy. There is a need to improve the existing weather forecast information systems and to promote international cooperation on meteorological issues.

References

Bartholome, E., and A. S. Belward, 2005. "GLC2000: A New Approach to Global Land Cover Mapping from Earth Observation Data." *International Journal of Remote Sensing* 26 (9–10): 1959–1977.

CIESIN (Center for International Earth Science Information Network), Columbia University, IFPRI (International Food Policy Research Institute), World Bank, and CIAT (Centro Internacional de Agricultura Tropical). 2004. *Global Rural–Urban Mapping Project (GRUMP), Alpha Version: Population Density Grids.* Palisades, NY, US: Socioeconomic Data and Applications Center (SEDAC), Columbia University. http://sedac.ciesin.columbia.edu/gpw.

Dostie, B., J. Randriamamonjy, and L. Rabenasolo. 1999. "Cassava Production and Marketing Chains: The Forgotten Shock Absorber for the Vulnerable." Accessed September 7, 2010. www.cfnpp .cornell.edu/images /wp100engl.pdf.

Erdmann, T. K. 2003. "The Dilemma of Reducing Shifting Cultivation." In *The Natural History of Madagascar,* edited by S. M. Goodman and J. P. Benstead, 134–139. Chicago: University of Chicago Press.

FAO (Food and Agriculture Organization of the United Nations). 2000. "Special Report: FAO/ WFP Mission to Assess the Impact of Cyclones and Drought on the Food Supply Situation in Madagascar: FAO Global Information and Early Warning System on Food and Agriculture World Food Programme." Accessed March 5, 2011. www.fao.org/docrep/004/x7379e/ x7379e00.htm#P68_10450.

————. 2010a. FAOSTAT. Rome. http://faostat.fao.org.

————. 2010b. "Locust Swarms Threaten Agriculture in Madagascar." Media Center. Accessed September 8. www.fao.org/news/story/en/item/44696/icode/.

FINTRAC (Financial Transactions Reports Analysis Centre of Canada). 2008. *Best Analysis— Madagascar Bellmon Estimation Studies for Title Ii (Best) Project.* Washington, DC.

IMF (International Monetary Fund). 2007. "Republic of Madagascar: Poverty Reduction Strategy Paper." IMF Country Report 07/59. www.imf.org/external/pubs/ft/scr/2007/cr0759.pdf.

Jones, P. G., P. K. Thornton, and J. Heinke. 2009. *Generating Characteristic Daily Weather Data Using Downscaled Climate Model Data from the IPCC's Fourth Assessment.* Project report for the International Livestock Research Institute. Geneva: International Panel on Climate Change.

Lehner, B., and P. Döll. 2004. "Development and Validation of a Global Database of Lakes, Reservoirs, and Wetlands." *Journal of Hydrology* 296 (1–4): 1–22.

Madagascar, Ministère de l'Economie et de l'Industrie. 2011. *Enquête périodique auprès des ménages 2010.* Rapport principal. Antananarivo.

Madagascar, Ministère de l'Environnement et des Forêts. 2010. "Deuxième Communication Nationale sur le Changement Climatique." Antananarivo. Accessed September 6, 2011. http:// unfccc.int/essential_background/library/items/3599.php?rec=j&priref=7326#beg.

Madagascar, Ministère de l'Environnement, des Eaux et Forêts, Direction Generale de l'Environnement. 2006. *Programme d'action national d'adaptation au changement climatique.* Antananarivo.

Millennium Ecosystem Assessment. 2005. *Ecosystems and Human Well-being: Synthesis.* Washington, DC: Island Press. www.maweb.org/en/Global.aspx.

Nelson, G. C., M. W. Rosegrant, A. Palazzo, I. Gray, C. Ingersoll, R. Robertson, S. Tokgoz, et al. 2010. *Food Security, Farming, and Climate Change to 2050: Scenarios, Results, Policy Options.* Washington, DC: International Food Policy Research Institute.

PNUD (Programme des Nations Unies pour le Développement). 2010. Cinquième Rapport National sur le Développement Humain—Madagascar. Accessed September 7, 2011. http://hdr.undp .org/fr/rapports/national/afrique/madagascar/Madagascar_RNDH_2010_FR.pdf.

Preventionweb. 2011. "Madagascar's Risk Profile." Accessed August 31. www.preventionweb.net/
english/countries/statistics/risk.php?iso=mdg.

Ranomenjanahary, S., J. Ramelison, and J. Seruwagi. 2005. "Madagascar." In *Whitefly and Whitefly-
Borne Viruses in the Tropics: Building a Knowledge Base for Global Action,* edited by P. Anderson
and F. J. Morales. Cali, Colombia: International Center for Tropical Agriculture (CIAT).

Rapport National d'Investissement Madagascar. 2008. "L'eaupoy l'agriculture et l'energie en Afrique:
Les defis du changement climatique." Paper presented at the Conference de Haut Niveau Sur,
Syrte, Jamahiriya, Arab Libya, December 15–17.

Reliefweb. 2000. "Madagascar: Cyclones and Floods, Preliminary Appeal 06/2000." Accessed August
31, 2011. http://reliefweb.int/node/61234.

Sharp, M., and I. Kruse. 2011. *Health, Nutrition and Population in Madagascar.* Washington, DC:
World Bank.

UNEP (United Nations Environment Programme) and IUCN (International Union for the
Conservation of Nature). 2009. World Database on Protected Areas (WDPA): Annual Release.
No longer available online.

UNPOP (United Nations Secretariat, Department of Economic and Social Affairs, Population
Division). 2009. *World Population Prospects: The 2008 Revision.* Accessed April 06, 2010.
http://esa.un.org/unpp.

USDoS (United States Department of State). 2011. "Background Note: Madagascar." Accessed May 7.
www.state.gov/r/pa/ei/bgn/5460.htm.

Vision 2025 Madagascar. 2007. *Rapport national de suivi des OMD.* Antananarivo.

Wood, S., G. Hyman, U. Deichmann, E. Barona, R. Tenorio, Z. Guo, S. Castano, O. Rivera, E. Diaz, J.
Marin, 2010. "Sub-national poverty maps for the developing world using international poverty
lines: Preliminary data release." Accessed May 6, 2010. http://povertymap.info.

World Bank. 2009. *World Development Indicators.* Accessed May 2011. http://data.worldbank.org/
data-catalog/world-development-indicators.

———. 2010a. *Economics of Adaptation to Climate Change: Synthesis Report.* Washington, DC.
http://climatechange.worldbank.org/content/economics-adaptation-climate-change-study
-homepage.

———. 2010b. "Madagascar vers un agenda de relance économique." Accessed March 22, 2011.
http://documents.worldbank.org/curated/en/2010/06/14895586/madagascar-vers-un-agenda
-de-relance-economique.

You, L., and S. Wood. 2006. "An Entropy Approach to Spatial Disaggregation of Agricultural
Production." *Agricultural Systems* 90 (1–3): 329–347.

You, L., S. Wood, and U. Wood-Sichra. 2006. "Generating Global Crop Distribution Maps: From Census to Grid." Paper presented at the International Association of Agricultural Economists Conference, Brisbane, Australia, August 11–18.

———. 2009. "Generating Plausible Crop Distribution and Performance Maps for Sub-Saharan Africa Using a Spatially Disaggregated Data Fusion and Optimization Approach." *Agricultural Systems* 99 (2–3): 126–140.

Chapter 9

RWANDA

Ngoga G. Tenge, Mutabazi Alphonse, and Timothy S. Thomas

Rwanda is located 2° south of the equator in central Africa and covers a surface area of 26,338 square kilometers. The country is primarily mountainous, with the average altitude ranging from 900 meters in the southeast to 4,500 meters in the regions of the Congo–Nile crest. The countryside is covered by grasslands. The eastern slopes are more moderate, with rolling hills extending across central uplands gradually reducing in altitudes to the plains, swamps, and lakes of the eastern border region. Due to the high altitude, the country experiences average annual temperatures ranging from 16° to 20°C (REMA 2009).

The Rwandan economy is primarily based on rainfed agriculture, with coffee and tea the major cash crops. Farms are small, fragmented, and semisubsistence oriented. The country has few natural resources to exploit and a small, uncompetitive industrial sector. In 2009 its gross domestic product (GDP) was estimated to be $5.1 billion. Agriculture accounted for 36 percent, industry 14.2 percent, and services 43.7 percent of GDP. The Government of Rwanda remains dedicated to a strong and enduring economic climate for the country, focusing on poverty reduction, infrastructure development, privatization of government-owned assets, expansion of the export base, and trade liberalization. Agricultural reforms have improved crop yields, the national food supply, and farming methods and have increased the use of fertilizers. In addition, the government is pursuing educational and healthcare programs that bode well for the long-term quality of Rwanda's human resource skills base (REMA 2009; USDoS 2011).

However, several challenges remain for Rwanda. The country relies heavily on foreign aid. Exports continue to lag far behind imports, affecting the health of the economy. The persistent lack of economic diversification beyond the production of tea, coffee, and minerals keeps the country vulnerable to market fluctuations. Expensive electricity and limited transportation impede private-sector development and raise the costs of imports and exports (USDoS 2011).

In addition, recent climate change patterns have also threatened ongoing economic improvements. Climate-related events like heavy rainfall or too little rainfall occur more frequently than in years past and are affecting human well-being. Droughts are often responsible for famine, food shortages, a reduction in plant and animal species, and displacement of people in search of food and pasture. At times these have led to conflicts over different land resources, as was the case for protected areas. In the past 10 years, these disasters have occurred throughout the country, exacerbated by poor farming practices, deforestation, and environmental degradation. These climatic events have affected health, water quality, transportation, and agriculture, leaving the country drained of its wealth and increasing the level of poverty. The poor are most vulnerable to climatic effects. Although the Rwandan government has drafted policies to address these disasters, assessing their impact is vital to the management, reduction, and mitigation of potential risks (REMA 2009, 2010; SEI 2009).

Review of the Current Situation and Trends

Economic and Demographic Indicators

Population

Figure 9.1 shows the total and rural population counts (left axis) and the share of the population that is urban (right axis). The Rwandan development indicators published by the National Institute of Statistics of Rwanda (2008) listed the total population of the country as an estimated 8.9 million in 2005. The figure shows a clear decline of the total and rural populations from about 1990 to 1995, which coincides with the war and genocide that led to mass migrations and deaths. The urban population as a percentage of the total began to increase rapidly during that time and has slowed only slightly since around 2005.

The increases in urban population resulted from the rural exodus, the return of refugees who lived in foreign countries after the 1994 genocide, and improvements in health and well-being. These events increased the percentage of the population in urban areas from 5.5 percent to 16.7 percent between 1991 and 2002. In its Vision 2020 plan, Rwanda recognizes the productive management of land and effective basic infrastructure as two of the pillars. The vision stipulates that Rwandan towns will have the tools to achieve sustainable urban management and planning and that the percentage of the population living in urban areas will grow to 30 percent by 2020. Furthermore, the country's Poverty Reduction Strategy Paper (Rwanda, Ministry of Finance and Economic Planning 2007) also advocates mainstreaming capacity building for the decentralized entities to improve urban and habitat management. In the

FIGURE 9.1 Population trends in Rwanda: Total population, rural population, and percent urban, 1960–2008

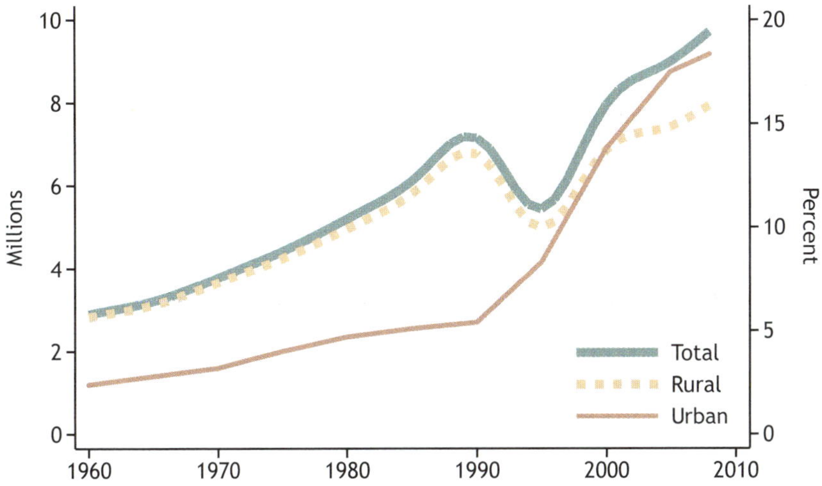

Source: *World Development Indicators* (World Bank 2009).

urban areas, a large percentage of the population lives below the poverty line, and more than 80 percent live in slums with no access to land.

Table 9.1 provides additional information concerning rates of population growth between 1960 and 2008. The total population growth for the first three decades was 2.6, 3.2, and 3.7 percent, respectively; in the decade 1990–1999, in which the genocide was perpetrated, the growth rate was only –0.1 percent, and the growth rate during the first decade of the 21st century was 2.3 percent (through 2008). These trends in growth rates line up with the tendencies observed in Figure 9.1.

TABLE 9.1 Population growth rates in Rwanda, 1960–2008 (percent)

Decade	Total growth rate	Rural growth rate	Urban growth rate
1960–1969	2.6	2.5	5.5
1970–1979	3.2	3.0	7.1
1980–1989	3.7	3.7	5.2
1990–1999	–0.1	–0.9	9.4
2000–2008	2.3	1.6	5.9

Source: Authors' calculations based on *World Development Indicators* (World Bank 2009).

FIGURE 9.2   Population distribution in Rwanda, 2000 (persons per square kilometer)

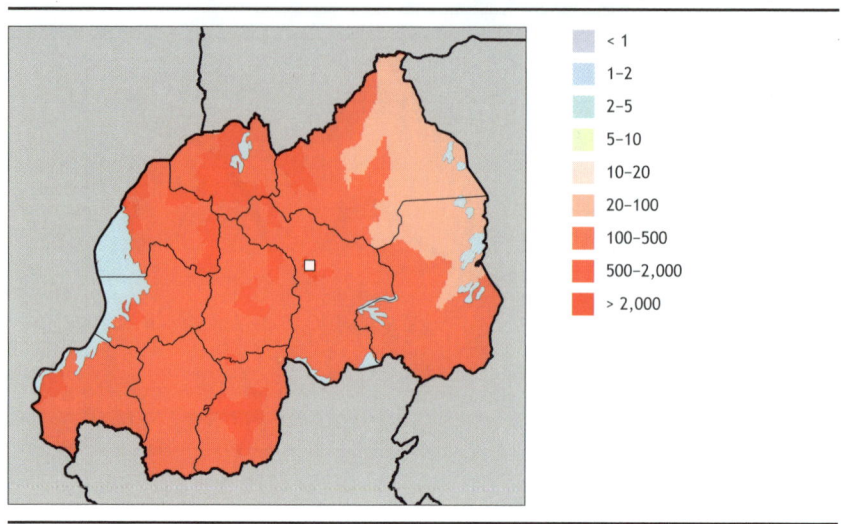

Source: CIESIN et al. (2004).

Figure 9.2 shows the geographic distribution of the population in Rwanda. Generally the country's recorded average population density of 321 persons per square kilometer is one of the highest of any country in the world, particularly one that relies on agriculture. The maps show that the country is densely populated except in places occupied by bodies of water located toward the borders with Democratic Republic of Congo and Tanzania.

Income

Figure 9.3 shows trends in GDP per capita and the proportion of GDP from agriculture in Rwanda. Recent information from the National Institute of Statistics of Rwanda (2008) shows that GDP per capita has increased considerably during the past five years, reaching US$465 (current US dollars, whereas the figure shows constant 2000 US dollars). Agriculture, forestry, and fishing activities contributed 39 percent of GDP, industry 14 percent, and services 41 percent.

Agriculture's contribution to the economy was nearly 80 percent of GDP at the beginning of the 1960s but is less than 40 percent today, which suggests the rising contribution of other sectors. Although the agricultural sector's proportion of GDP has declined, its importance in employment has changed little. The agricultural sector still employs 90 percent of the total labor population. The development of the agroindustry and appropriate post harvest technologies for each crop could increase employment in the industry and service sectors, thus enhancing the country's resilience to the threats of climate change.

FIGURE 9.3   Per capita GDP in Rwanda (constant 2000 US$) and share of GDP from agriculture (percent), 1960–2008

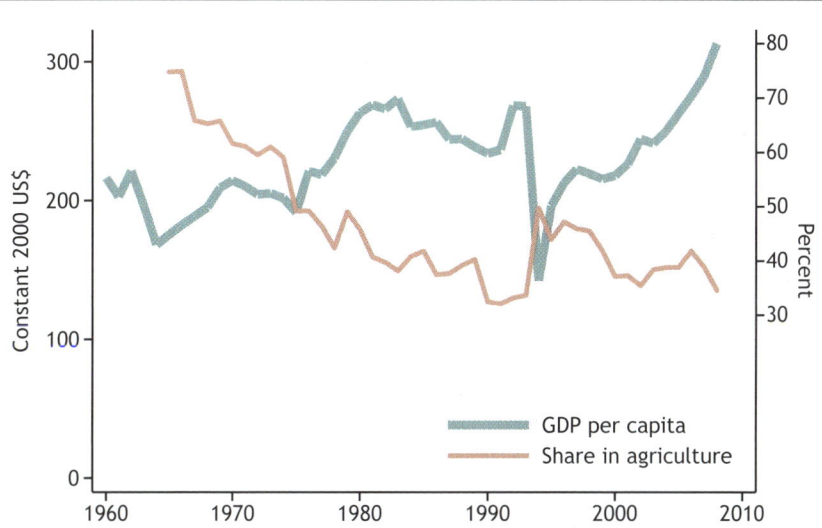

Source: *World Development Indicators* (World Bank 2009).
Note: GDP = gross domestic product; US$ = US dollars.

Vulnerability to Climate Change

Table 9.2 provides some data on additional indicators of vulnerability and resiliency to economic shocks: the education level of the population, the literacy rate, and the concentration of labor in poorer or less dynamic sectors. The high levels of primary education and adult literacy signify that much of the population has the potential to adapt to climate change through learning strategies, which would reduce their vulnerability to its adverse effects. However, the low levels of secondary education signify the low capacity of the population to take advantage of various employment opportunities. Hence, as observed in Table 9.2, most people remain engaged in agriculture—crop production, livestock rearing, fishing, and related services. Accordingly, many continue to be vulnerable to the impacts of climate change.

Figure 9.4 shows two noneconomic correlates of poverty, life expectancy and under-five mortality. Life expectancy at birth indicates the number of years a newborn infant would live if prevailing patterns of mortality at the time of its birth were to stay the same throughout its life.

Generally the graph shows very small improvements in under-five mortality and life expectancy trends in Rwanda. The sharp fall in life expectancy about 1991–1992 can be attributed to the genocide being carried out at that time. Life

TABLE 9.2   Education and labor statistics for Rwanda, 1990s and 2000s

Indicator	Year	Percent
Primary school enrollment: Percent gross (three-year average)	2007	147.4
Secondary school enrollment: Percent gross (three-year average)	2007	18.1
Adult literacy rate	2000	64.9
Percent employed in agriculture	1989	90.1
Percent with vulnerable employment (in agriculture on own farm or as a day laborer)	1996	92.5
Under-five malnutrition (weight for age)	2005	18.0

Source: *World Development Indicators* (World Bank 2009).

expectancy and child mortality in Rwanda are affected by a wide range of factors such as malaria, diarrhea, injuries, and HIV/AIDS (WHO 2006). Several environmental conditions contribute to these factors, such as floods and droughts. Future advancements in these health indicators may be compromised by climate change, which often acts to alter these conditions. Climate change may cause droughts and flooding that will affect food security, eventually leading to malnutrition, morbidity, or death. Rwanda, for instance, has experienced four major floods in the past seven years—in 2006, 2007, 2008, and 2009—which resulted

FIGURE 9.4   Well-being indicators in Rwanda, 1960–2008

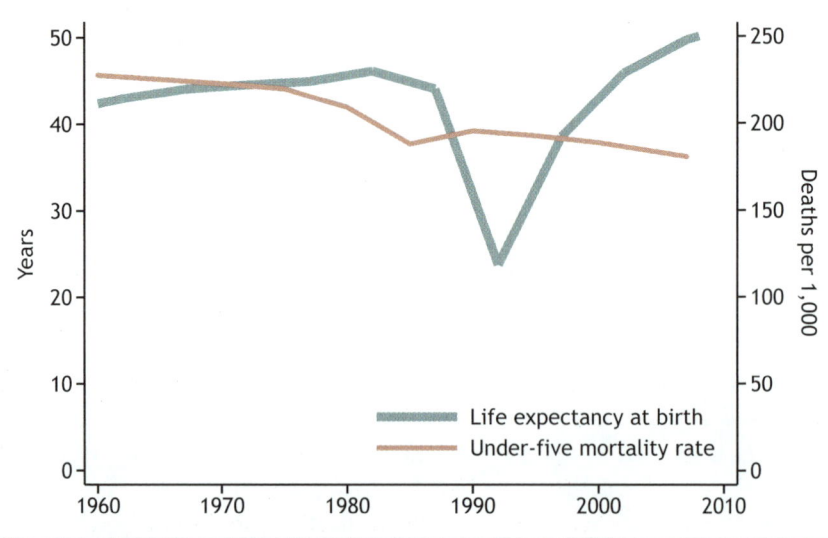

Source: *World Development Indicators* (World Bank 2009).

in fatalities, injuries, water contamination, and destruction of agriculture and infrastructure, among other damages. These effects have had economic consequences including increased direct medical costs, health protection costs, time lost at work, and welfare change. For example, the economic cost associated with the 2007 flood alone was about $22 million, of which the health economic costs were estimated to be between $1.6 million and $18.0 million. These events place a greater burden on the country's GDP, further increasing poverty (SEI 2009). Wood et al. (2010) report that in Rwanda, 85–95 percent of the people are living on less than $2 a day.

As a consequence of the traumatic upheaval of the genocide in the early 1990s, the baseline year for monitoring progress toward the Millennium Development Goal for poverty reduction in Rwanda was set to 2000. Progress in reducing poverty has been very slow in spite of strong economic growth. In 2006, 57 percent of the population lived below the national poverty line, and 37 percent of those people were regarded as extremely poor because they could not obtain the minimum calorie requirement on their income. Poverty declined most in urban areas in proportionate terms. In rural areas, poverty declined from 66.1 to 62.5 percent between 2000 and 2006. Improving rural incomes is critical to reducing poverty given that 90 percent of the population still lives in rural areas. A majority of rural income is generated through agriculture, which is affected by a number of input factors, including climatic patterns. Climate fluctuations will alter agricultural output directly and indirectly through change in input factors such as labor. With low incomes and little education, the rural poor are limited in their options for coping with the effects of climate change.

Review of Land Use, Potential, and Limitations

Land Use Overview

Figure 9.5 shows land cover and use in Rwanda as of 2000. Most of the country consists of cultivated and managed areas and shrub cover. The largest forested area is situated in the southwestern corner of the country, where Nyungwe National Park is located. Because agricultural land is taking more space over time as the forest cover diminishes, agroforestry is becoming a good option for poor households in the rural areas, providing not only forest products like firewood but also many other benefits, such as biomass for soil fertility, fodder for livestock, staking material for climbing beans, and erosion control. Agroforestry and agricultural intensification should be given more emphasis because agriculture puts a great amount of pressure on the land.

FIGURE 9.5   Land cover and land use in Rwanda, 2000

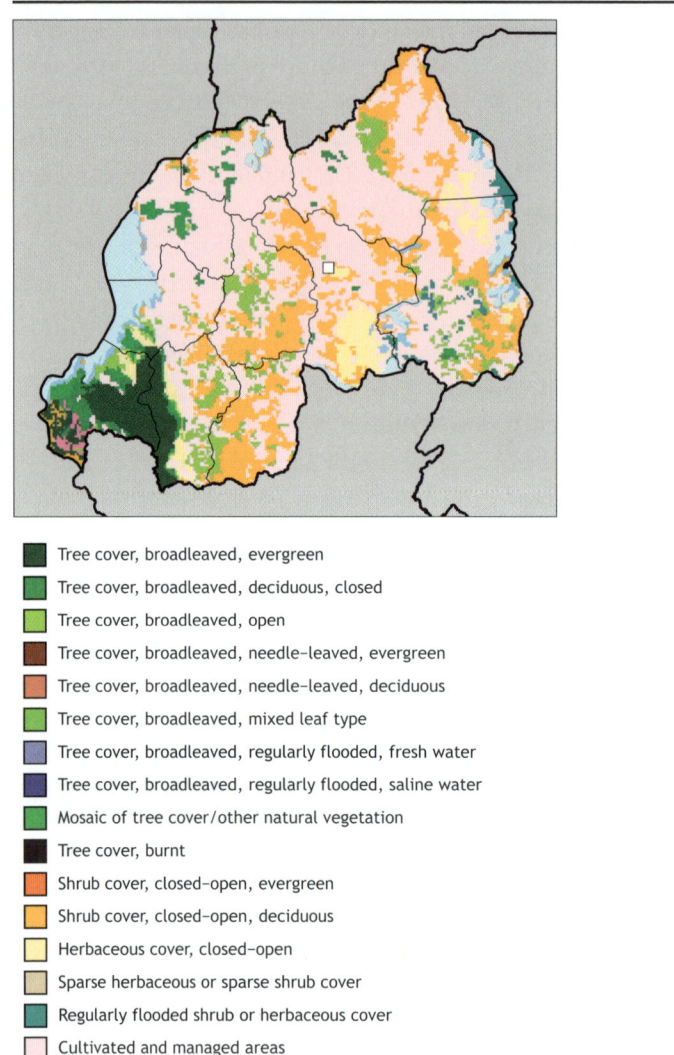

■ Tree cover, broadleaved, evergreen

■ Tree cover, broadleaved, deciduous, closed

■ Tree cover, broadleaved, open

■ Tree cover, broadleaved, needle-leaved, evergreen

■ Tree cover, broadleaved, needle-leaved, deciduous

■ Tree cover, broadleaved, mixed leaf type

■ Tree cover, broadleaved, regularly flooded, fresh water

■ Tree cover, broadleaved, regularly flooded, saline water

■ Mosaic of tree cover/other natural vegetation

■ Tree cover, burnt

■ Shrub cover, closed-open, evergreen

■ Shrub cover, closed-open, deciduous

■ Herbaceous cover, closed-open

■ Sparse herbaceous or sparse shrub cover

■ Regularly flooded shrub or herbaceous cover

■ Cultivated and managed areas

■ Mosaic of cropland/tree cover/other natural vegetation

■ Mosaic of cropland/shrub/grass cover

■ Bare areas

■ Water bodies

□ Snow and ice

■ Artificial surfaces and associated areas

□ No data

Source: GLC2000 (Bartholome and Belward 2005).

FIGURE 9.6 Protected areas in Rwanda, 2009

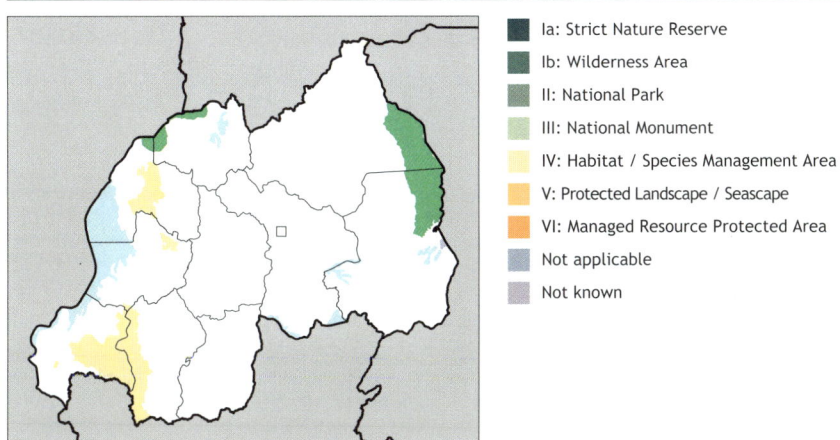

Ia: Strict Nature Reserve
Ib: Wilderness Area
II: National Park
III: National Monument
IV: Habitat / Species Management Area
V: Protected Landscape / Seascape
VI: Managed Resource Protected Area
Not applicable
Not known

Sources: Protected areas are from the World Database on Protected Areas (UNEP and IUCN 2009). Water bodies are from the World Wildlife Fund's Global Lakes and Wetlands Database (Lehner and Döll 2004).

According to the *Rwanda State of Environment and Outlook Report* (REMA 2009), protected areas in Rwanda include national parks (Akagera, 108,500 hectares; Nyungwe, 101,900 hectares; and Volcanoes National Park, 16,000 hectares); forest reserves (Gishwati, 700 hectares; Iwawa Island and Mukura, 1,933 hectares); forests of cultural importance (Buhanga Forest); and wetlands of global importance (the Rugezi–Bulera–Ruhondo wetland complex).

Figure 9.6 shows the locations of protected areas, including parks and reserves, that safeguard fragile environmental areas, conserve biodiversity, enhance water catchment and purification, mitigate climate change, and serve as sources of energy, construction materials, food, and health products. Protected areas (particularly forests) indirectly support ecological systems of agriculture. These areas are also important for the tourism industry and economic growth. The tourism sector has experienced considerably enhanced performance since 2002. In 2007, for example, gorilla tourism attracted 16,000 visits, generating $42 million, making it the largest foreign exchange earner in the country. In spite of their contributions, protected areas are threatened by several problems, including governance issues, an inadequate legal framework, and population pressures that have led to encroachment and deforestation. Protected forest areas were reduced by 64 percent between 1960 and 2007. The end result is increased vulnerability to the effects of climate change and soil degradation (REMA 2009).

Figure 9.7 shows travel times to cities and towns of various sizes in Rwanda. These maps help us better understand how far rural areas are from potential

FIGURE 9.7   Travel time to urban areas of various sizes in Rwanda, circa 2000

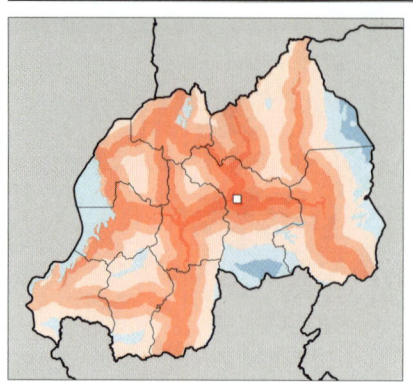

To cities of 500,000 or more people

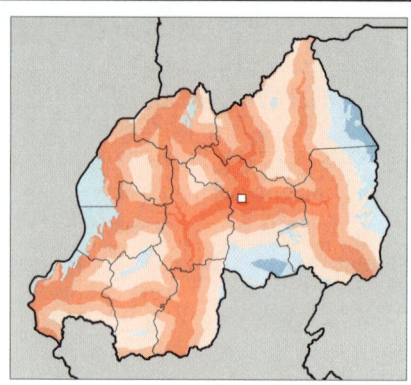

To cities of 100,000 or more people

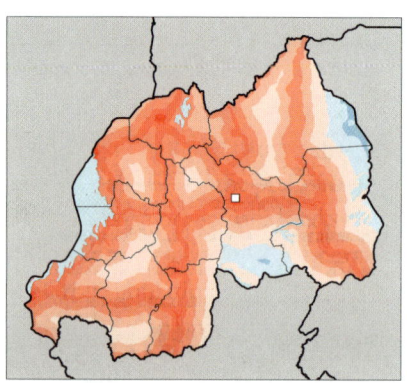

To towns and cities of 25,000 or more people

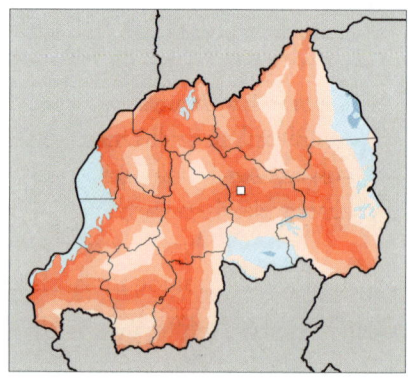

To towns and cities of 10,000 or more people

 Urban location
< 1 hour
1–3 hours
3–5 hours
5–8 hours
8–11 hours
11–16 hours
16–26 hours
> 26 hours

Source: Authors' calculations.

markets for agricultural products, as well as places for farmers to purchase inputs and consumer goods. Accessibility to markets in Rwandan cities is fairly easy. Rwanda is a small country with a relatively good road network that is continuously improved by the Ministry of Infrastructure in collaboration with local governments. The maps in Figure 9.7 show that most of the urban centers can be reached within 1–3 hours except when one is traveling from protected areas, which are more remote than other parts of the country.

Agriculture

Tables 9.3–9.5 show key agricultural commodities in terms of area harvested, value of the harvest, and consumption (ranked by weight). All values are based on three-year averages. Plantains and bananas appear to be the most important staple crops grown and consumed in the country. Rwanda's daily consumption of these crops, recorded at 250 grams per capita, is higher than anywhere else in the world (Mpawenimana 2005).

Figures 9.8–9.12 show the estimated yields and growing areas for key crops in Rwanda: bananas, beans, potatoes, sorghum, and cassava. The maps

TABLE 9.3   Harvest area of leading agricultural commodities in Rwanda, 2005–2007 (thousands of hectares)

Rank	Crop	Percent of total	Harvest area
	Total	100.0	1,692
1	Bananas and plantains	22.2	375
2	Beans	21.2	359
3	Sorghum	10.2	172
4	Potatoes	8.3	140
5	Sweet potatoes	8.2	139
6	Cassava	6.7	113
7	Maize	6.6	112
8	Pumpkins, squash, and gourds	2.5	43
9	Soybeans	2.5	43
10	Coffee	2.0	34

Source: FAOSTAT (FAO 2010).
Note: All values are based on the three-year average for 2005–2007.

TABLE 9.4   Value of production for leading agricultural commodities in Rwanda, 2005–2007 (millions of constant 2000 US$)

Rank	Crop	Percent of total	Value of production
	Total	100.0	603.4
1	Bananas and plantains	29.2	176.2
2	Potatoes	21.5	129.5
3	Beans	8.9	53.9
4	Cassava	7.4	44.9
5	Rice	5.4	32.7
6	Sorghum	5.0	30.1
7	Sweet potatoes	4.5	27.3
8	Tea	4.0	24.1
9	Taro cocoyams	3.4	20.3
10	Other fresh fruit	1.8	11.1

Source: FAOSTAT (FAO 2010).
Notes: All values are based on the three-year average for 2005–2007. US$ = US dollars.

TABLE 9.5   Consumption of leading food commodities in Rwanda, 2003–2005 (thousands of metric tons)

Rank	Crop	Percent of total	Food consumption
	Total	100.0	5,526
1	Bananas and plantains	22.5	1,245
2	Potatoes	18.7	1,033
3	Cassava	15.3	847
4	Sweet potatoes	15.1	834
5	Fermented beverages	7.3	406
6	Other vegetables	4.4	241
7	Beans	3.6	199
8	Sorghum	2.7	150
9	Other roots and tubers	2.2	124
10	Maize	1.9	106

Source: FAOSTAT (FAO 2010).
Note: All values are based on the three-year average for 2003–2005.

FIGURE 9.8 Yield (metric tons per hectare) and harvest area density (hectares) for rainfed plantains and bananas in Rwanda, 2000

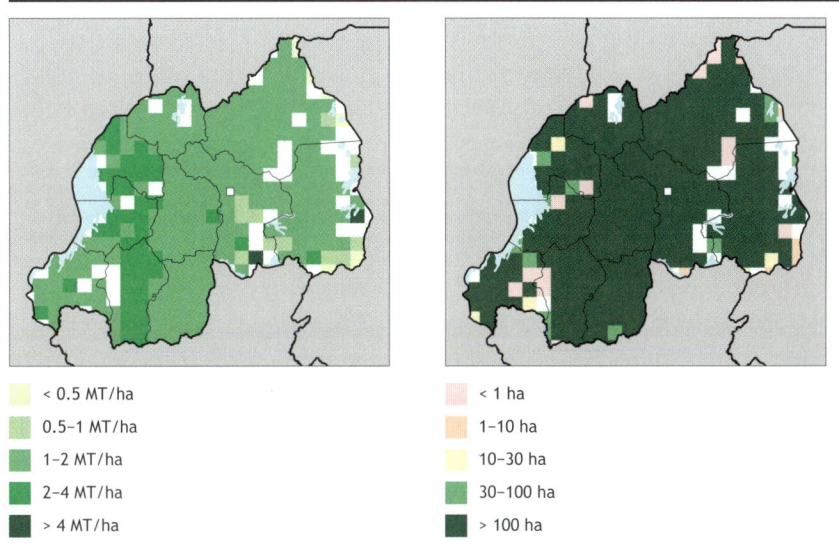

< 0.5 MT/ha	< 1 ha
0.5–1 MT/ha	1–10 ha
1–2 MT/ha	10–30 ha
2–4 MT/ha	30–100 ha
> 4 MT/ha	> 100 ha

Source: SPAM (Spatial Production Allocation Model) (You and Wood 2006; You, Wood, and Wood-Sichra 2006, 2009).
Note: ha = hectare; MT/ha = metric tons per hectare.

FIGURE 9.9 Yield (metric tons per hectare) and harvest area density (hectares) for rainfed beans in Rwanda, 2000

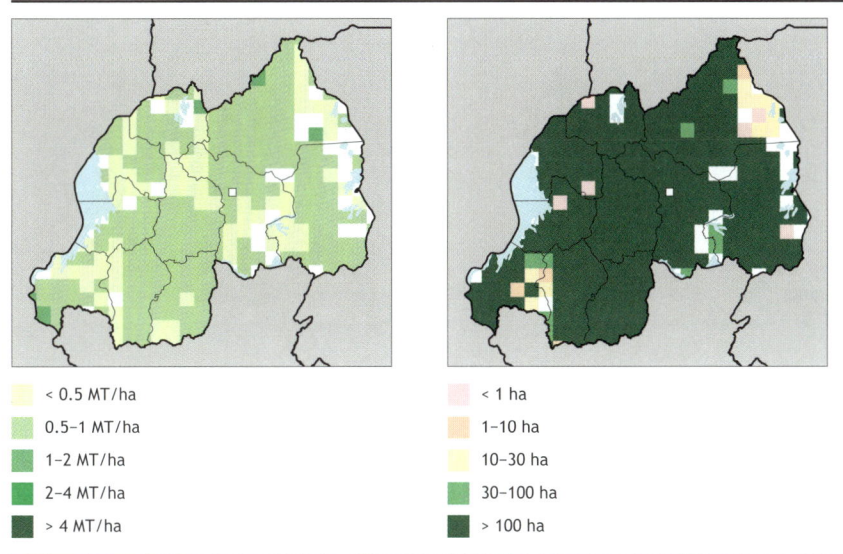

< 0.5 MT/ha	< 1 ha
0.5–1 MT/ha	1–10 ha
1–2 MT/ha	10–30 ha
2–4 MT/ha	30–100 ha
> 4 MT/ha	> 100 ha

Source: SPAM (Spatial Production Allocation Model) (You and Wood 2006; You, Wood, and Wood-Sichra 2006, 2009).
Note: ha = hectare; MT/ha = metric tons per hectare.

show that the crops are produced throughout the country. Banana production occupies 35 percent of the total cultivated land and is usually done by small-scale farmers. The crop is used mainly for food security and income generation, especially by peasant farmers. Bananas grow well in tropical temperatures of about 15°–30°C, rainfall of about 1,000–1,200 millimeters, and deep soils (Rwanda, Ministry of Agriculture and Animal Resources 2005). Like bananas, the other four crops are also grown for food and income and need similar conditions to grow. However, beans are grown at a much higher altitude of about 1,700 meters above sea level, where temperatures are cooler (14°–18°C). Bean production covers an estimated 22–30 percent of cultivated land and yields about 200,000–300,000 tons annually (Rwanda Agricultural Research Institute 2010). Other factors that may constrain crop yields are pests and diseases, lack of agricultural inputs, poor varieties, and weak extension systems. More effort is needed to make available improved seed varieties and proper integrated soil fertility management technologies to raise the yield for beans and sorghum to 4 tons per hectare and that for potatoes and cassava above 10 tons per hectare. More effort is also needed in the production and transformation of crops for value addition.

FIGURE 9.10   Yield (metric tons per hectare) and harvest area density (hectares) for rainfed potatoes in Rwanda, 2000

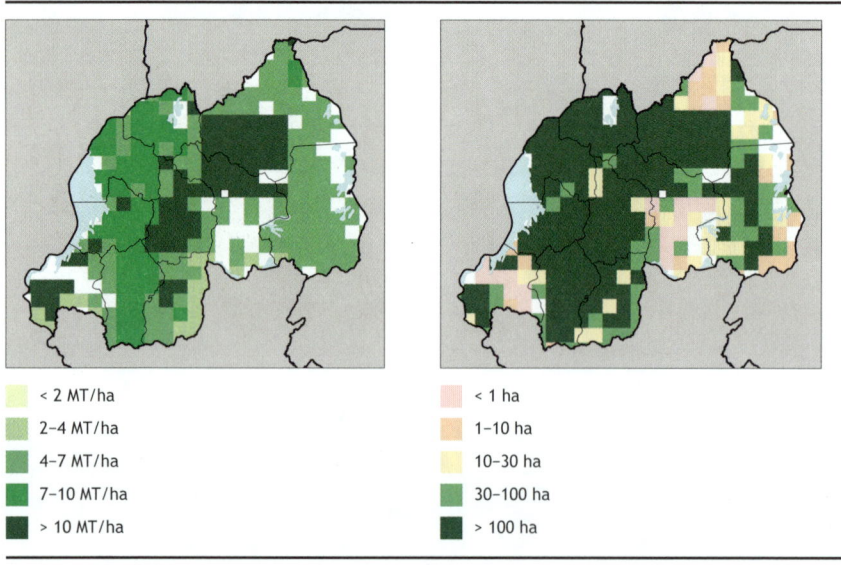

< 2 MT/ha	< 1 ha
2–4 MT/ha	1–10 ha
4–7 MT/ha	10–30 ha
7–10 MT/ha	30–100 ha
> 10 MT/ha	> 100 ha

Source: SPAM (Spatial Production Allocation Model) (You and Wood 2006; You, Wood, and Wood-Sichra 2006, 2009).
Note: ha = hectare; MT/ha = metric tons per hectare.

FIGURE 9.11   Yield (metric tons per hectare) and harvest area density (hectares) for rainfed sorghum in Rwanda, 2000

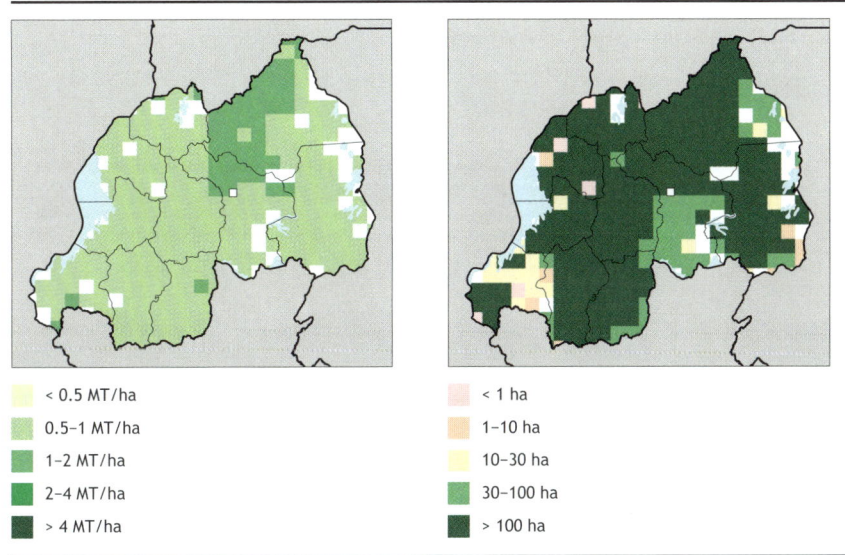

< 0.5 MT/ha	< 1 ha
0.5–1 MT/ha	1–10 ha
1–2 MT/ha	10–30 ha
2–4 MT/ha	30–100 ha
> 4 MT/ha	> 100 ha

Source: SPAM (Spatial Production Allocation Model) (You and Wood 2006; You, Wood, and Wood-Sichra 2006, 2009).
Note: ha = hectare; MT/ha = metric tons per hectare.

FIGURE 9.12   Yield (metric tons per hectare) and harvest area density (hectares) for rainfed cassava in Rwanda, 2000

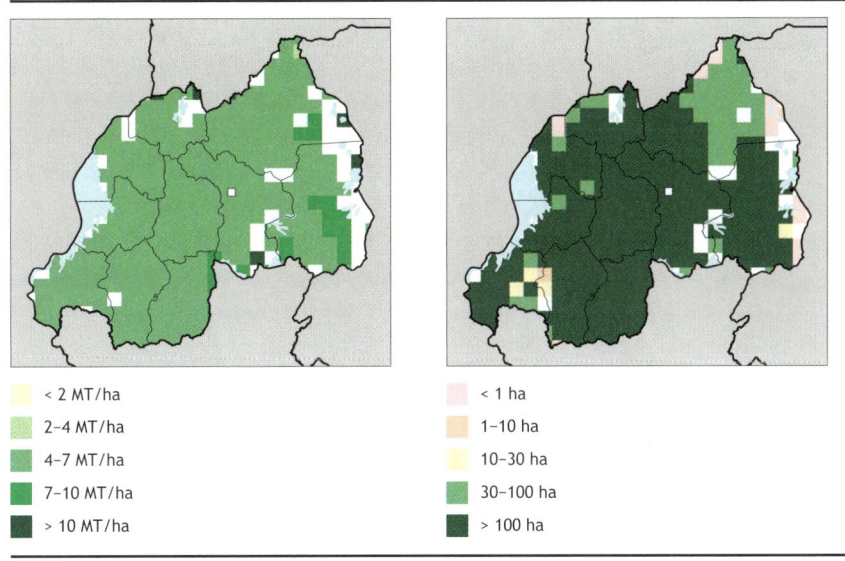

< 2 MT/ha	< 1 ha
2–4 MT/ha	1–10 ha
4–7 MT/ha	10–30 ha
7–10 MT/ha	30–100 ha
> 10 MT/ha	> 100 ha

Source: SPAM (Spatial Production Allocation Model) (You and Wood 2006; You, Wood, and Wood-Sichra 2006, 2009).
Note: ha = hectare; MT/ha = metric tons per hectare.

Scenarios for the Future

Economic and Demographic Indicators

Population

Figure 9.13 shows population projections by the UN Population Division through 2050. The low variant of the UN population projections (UNPOP 2009) estimates that the Rwandan population will almost double by 2050. The low variant is characterized by a declining fertility rate, which the country has been experiencing in the past decade (Indexmundi 2011). Attaining the low-variant projection would require appropriate education, family planning programs, and increased agricultural productivity to meet the food demands of the growing population. Given the already high population density in Rwanda, additional numbers of persons are likely to increase population pressure and land degradation. Therefore, emphasis should be placed on well-planned settlements in both rural and urban areas to reduce potential risks. The high-variant scenario is even worse, with the population in 2050 projected to be almost two and a half times that of 2010.

Income

Figure 9.14 shows the three scenarios for GDP per capita used for this study. These are the result of combining three GDP projections with the three population projections of Figure 9.14 from the United Nations Population Division. The optimistic scenario combines high GDP with low population. The baseline scenario combines the medium GDP projection with the medium population projection. Finally, the pessimistic scenario combines the low GDP projection with the high population projection.

GDP per capita rose by 44 percent between 2000 and 2008 (World Bank 2009). If growth between 2010 and 2050 were to maintain that pace, we would find Rwanda's GDP per capita about midway between the baseline scenario and the optimistic scenario. In the optimistic scenario GDP per capita will reach just under $2,300 by 2050, while in the baseline scenario GDP per capita will be just under $1,600 by 2050. Attaining either of these levels would require improving agricultural productivity, which is a means of livelihood for 90 percent of the population. Adopting high-yielding and pest-resistant varieties and appropriate soil management techniques, improving markets and infrastructure, and providing extension services will be key to increasing productivity, rural incomes, and GDP. Agricultural development should be coupled with adequate measures for adaptation to climate change.

FIGURE 9.13   Population projections for Rwanda, 2010–2050

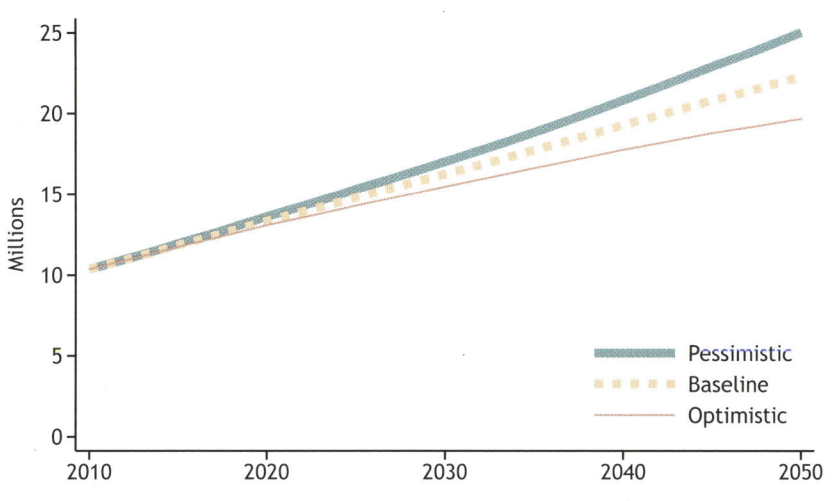

Source: UNPOP (2009).

FIGURE 9.14   Gross domestic product (GDP) per capita in Rwanda, future scenarios, 2010–2050

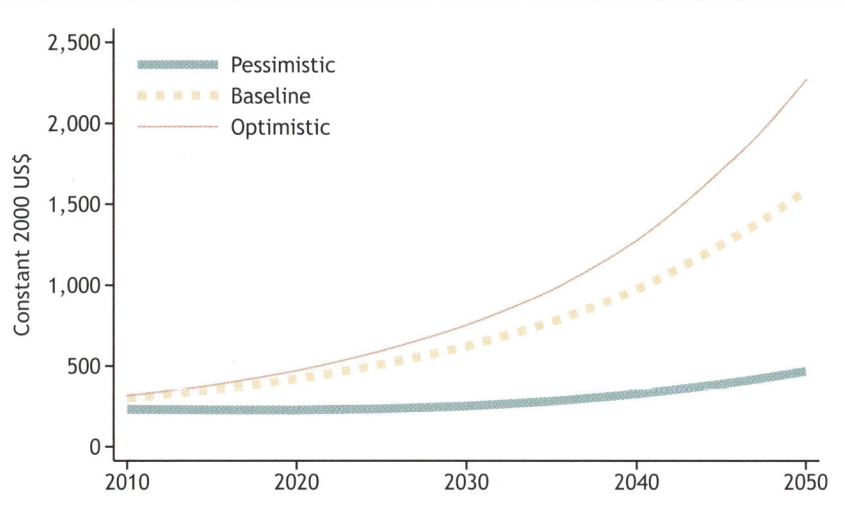

Sources: Computed from GDP data from the World Bank Economic Adaptation to Climate Change project (World Bank 2010), from the Millennium Ecosystem Assessment (2005) reports, and from population data from the United Nations (UNPOP 2009).

Note: US$ = US dollars.

Biophysical Analysis

Climate Models

Figure 9.15 shows projected precipitation changes under the four downscaled climate models (general circulation models or GCMs) used in this chapter with the A1B scenario.[1] The results from the CNRM-CM3 and ECHAM 5 GCMs predict neither an increase nor a decrease in rainfall in Rwanda.[2] But the CSIRO Mark 3 GCM indicates a drier future, and the MIROC 3.2 GCM indicates a wetter future. This kind of inconsistency is consistent with the type of information provided by climate models and tells us that any plans and policies made now would have to be sufficiently flexible either to benefit farmers under all three types of rainfall outcomes or to be changed quickly as the years go by and we are able to see what kind of climate is actually being experienced. One other study found that Rwanda is likely to have a wetter climate with more intense wet seasons and less severe droughts (Shongwe et al. 2010).

Figure 9.16 shows changes in normal mean daily maximum temperature for the month with the highest mean daily maximum temperature. All the models demonstrate a maximum temperature increase ranging from 1° to 2.5°C. There are some significant differences among the models. The CSIRO Mark 3 GCM has temperatures that range from 1° to 1.5°C, whereas the CNRM-CM3 model (and the only slightly cooler ECHAM 5 model) have temperatures that range mostly or entirely from 2° to 2.5°C. Higher temperatures are likely to affect crop yields, especially those of beans, which grow in cooler temperatures at high altitudes. Breeding for varieties adapted to relatively warmer temperatures is likely to keep the level of productivity high. High temperatures may also promote plant pest and disease multiplication as well as increased transmission of human diseases, particularly malaria. Rwandans can likely counter the effects of temperature increases by using appropriate pesticides and pest-resistant plant varieties and setting up appropriate health services.

---

1   The A1B scenario is a greenhouse gas emissions scenario that assumes fast economic growth, a population that peaks midcentury, and the development of new and efficient technologies, along with a balanced use of energy sources.

2   CNRM-CM3 is National Meteorological Research Center–Climate Model 3. MIROC 3.2 is the Model for Interdisciplinary Research on Climate, developed at the University of Tokyo Center for Climate System Research. CSIRO Mark 3 is a climate model developed at the Australia Commonwealth Scientific and Industrial Research Organisation. ECHAM 5 is a fifth-generation climate model developed at the Max Planck Institute for Meteorology in Hamburg.

FIGURE 9.15   Changes in mean annual precipitation in Rwanda, 2000–2050, A1B scenario (millimeters)

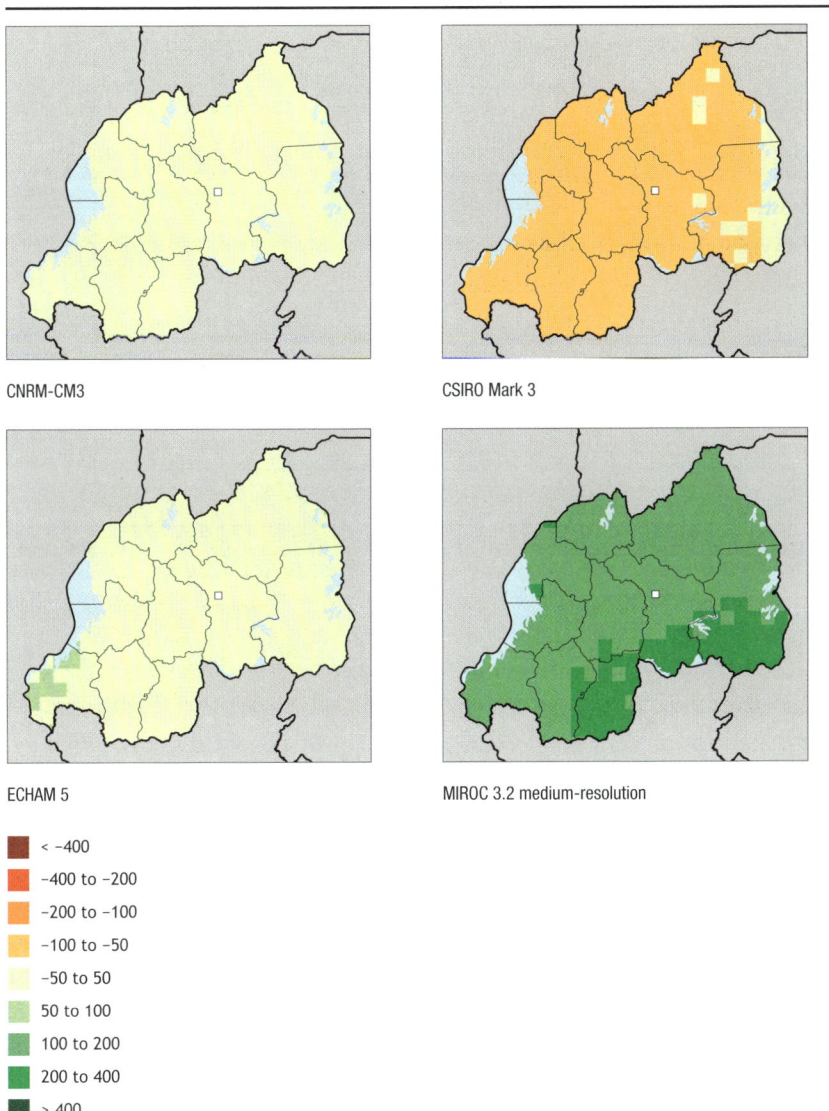

CNRM-CM3

CSIRO Mark 3

ECHAM 5

MIROC 3.2 medium-resolution

- < −400
- −400 to −200
- −200 to −100
- −100 to −50
- −50 to 50
- 50 to 100
- 100 to 200
- 200 to 400
- > 400

Source: Authors' calculations based on Jones, Thornton, and Heinke (2009).

Notes: A1B = greenhouse gas emissions scenario that assumes fast economic growth, a population that peaks midcentury, and the development of new and efficient technologies, along with a balanced use of energy sources; CNRM-CM3 = National Meteorological Research Center–Climate Model 3; CSIRO = climate model developed at the Australia Commonwealth Scientific and Industrial Research Organisation; ECHAM 5 = fifth-generation climate model developed at the Max Planck Institute for Meteorology (Hamburg); GCM = general circulation model; MIROC = Model for Interdisciplinary Research on Climate, developed by the University of Tokyo Center for Climate System Research.

FIGURE 9.16   Changes in monthly mean maximum daily temperature in Rwanda for the warmest month, 2000–2050, A1B scenario (°C)

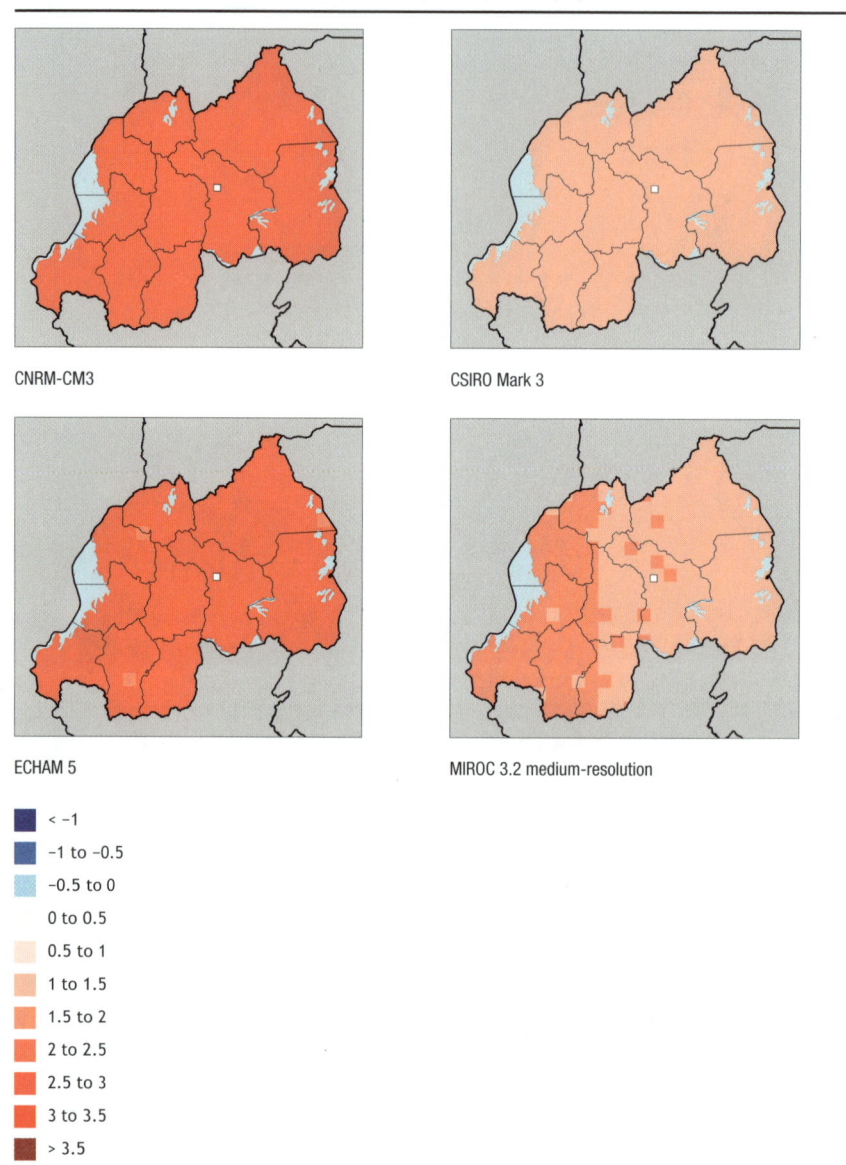

CNRM-CM3

CSIRO Mark 3

ECHAM 5

MIROC 3.2 medium-resolution

■	< −1
■	−1 to −0.5
■	−0.5 to 0
	0 to 0.5
■	0.5 to 1
■	1 to 1.5
■	1.5 to 2
■	2 to 2.5
■	2.5 to 3
■	3 to 3.5
■	> 3.5

Source: Authors' calculations based on Jones, Thornton, and Heinke (2009).

Notes: A1B = greenhouse gas emissions scenario that assumes fast economic growth, a population that peaks midcentury, and the development of new and efficient technologies, along with a balanced use of energy sources; CNRM-CM3 = National Meteorological Research Center–Climate Model 3; CSIRO = climate model developed at the Australia Commonwealth Scientific and Industrial Research Organisation; ECHAM 5 = fifth-generation climate model developed at the Max Planck Institute for Meteorology (Hamburg); GCM = general circulation model; MIROC = Model for Interdisciplinary Research on Climate, developed by the University of Tokyo Center for Climate System Research.

FIGURE 9.17 Yield change under climate change: Rainfed sorghum in Rwanda, 2000–2050, A1B scenario

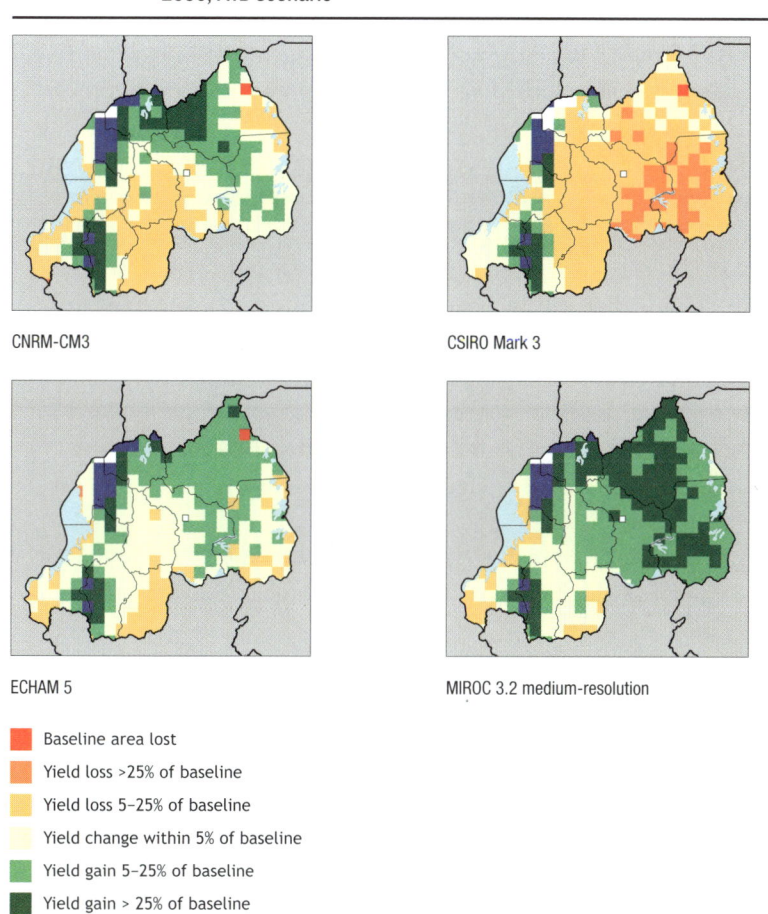

CNRM-CM3

CSIRO Mark 3

ECHAM 5

MIROC 3.2 medium-resolution

- ■ Baseline area lost
- ■ Yield loss >25% of baseline
- ■ Yield loss 5–25% of baseline
- □ Yield change within 5% of baseline
- ■ Yield gain 5–25% of baseline
- ■ Yield gain > 25% of baseline
- ■ New area gained

Source: Authors' calculations.
Notes: A1B = greenhouse gas emissions scenario that assumes fast economic growth, a population that peaks midcentury, and the development of new and efficient technologies, along with a balanced use of energy sources; CNRM-CM3 = National Meteorological Research Center–Climate Model 3; CSIRO = climate model developed at the Australia Commonwealth Scientific and Industrial Research Organisation; ECHAM 5 = fifth-generation climate model developed at the Max Planck Institute for Meteorology (Hamburg); GCM = general circulation model; MIROC = Model for Interdisciplinary Research on Climate, developed by the University of Tokyo Center for Climate System Research.

Crop Models

The Decision Support System for Agrotechnology Transfer software was used to compute baseline and future crop yields with current temperatures and precipitation. The results for sorghum are compared for the climate in the year 2000 and that of 2050 in Figure 9.17. There are significant differences among the climate models. The MIROC 3.2 model, for example, predicts possible gains of more than 25 percent in most of the eastern half of the country. The MIROC 3.2 model has only modest temperature increases and significant rainfall increases. The CSIRO Mark 3 model predicts losses for almost the entire country, though most of the losses range from 5 to 25 percent. The CSIRO Mark 3 model predicts modest temperature increases but lower rainfall. A consistent result across all maps is that some new areas in the western part of Rwanda will be able to cultivate sorghum that were previously unable to cultivate it. These happen to be higher-elevation areas that are currently too cold for sorghum but will become warm enough with climate change. Increasing sorghum production is important for improving the nutrition of mothers and children as well as generating incomes for families and reducing poverty. However, cultivating new lands may damage natural resources, especially because the land portions that will be better for sorghum appear to be protected areas displayed in Figure 9.6. Increasing productivity on the already cultivated land through applying appropriate agricultural technologies (for example, improved high-yielding plant varieties and good soil management techniques) will strike a balance between higher food production and conservation.

Vulnerability

Figure 9.18 shows the impact of future GDP and population scenarios on the number of malnourished children under age five in Rwanda. Figure 9.19 shows the share of children who are malnourished. In the baseline and optimistic scenarios, the number of children who are malnourished rises until 2015. Thereafter, they decrease over the next three and a half decades. However, the pessimistic scenario predicts a rise in the number of children who are malnourished throughout the next four decades. The malnutrition rate declines slightly during the same period.

In the optimistic scenario, greater per capita incomes will give families easier access to food (increasing caloric availability), education, and health services, consequently lowering malnutrition rates. This will lead to a decrease in the number of malnourished children under age five to fewer than 380,000 by 2050. It is important to note that changes in other factors may bring about a situation far from the optimistic scenario.

Figure 9.20 shows the kilocalories per capita available to each person in Rwanda. Even in the case of moderate increases in income represented by the baseline scenario, the number of kilocalories available per capita will increase

FIGURE 9.18   Number of malnourished children under five years of age in Rwanda in multiple income and climate scenarios, 2010–2050

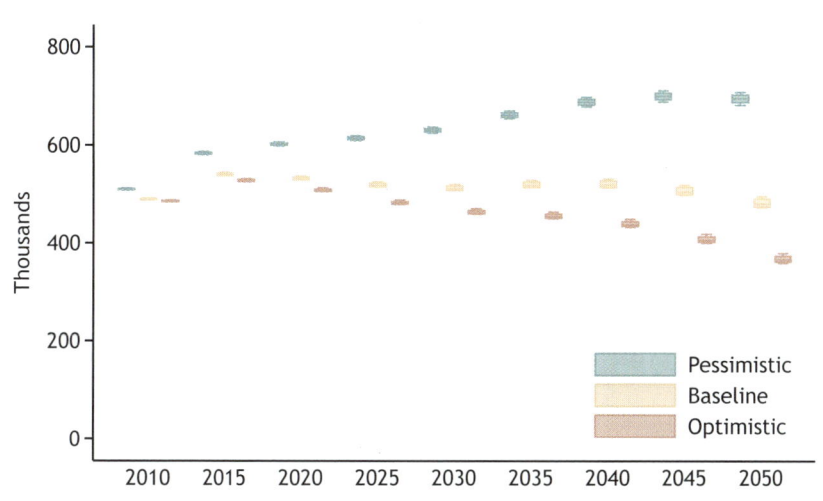

Source: Based on analysis conducted for Nelson et al. (2010).

Note: The box and whiskers plot for each socioeconomic scenario shows the range of effects from the four future climate scenarios.

FIGURE 9.19   Share of malnourished children under five years of age in Rwanda in multiple income and climate scenarios, 2010–2050

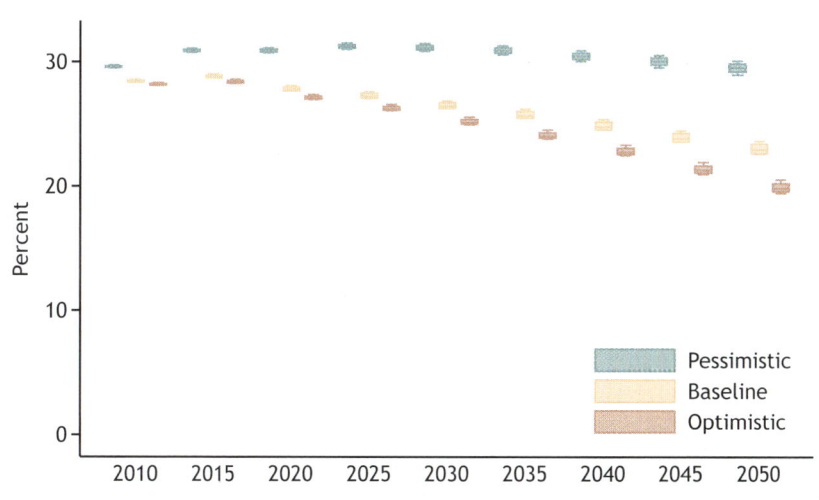

Source: Based on analysis conducted for Nelson et al. (2010).

Note: The box and whiskers plot for each socioeconomic scenario shows the range of effects from the four future climate scenarios.

FIGURE 9.20   Kilocalories per capita in Rwanda in multiple income and climate scenarios, 2010–2050

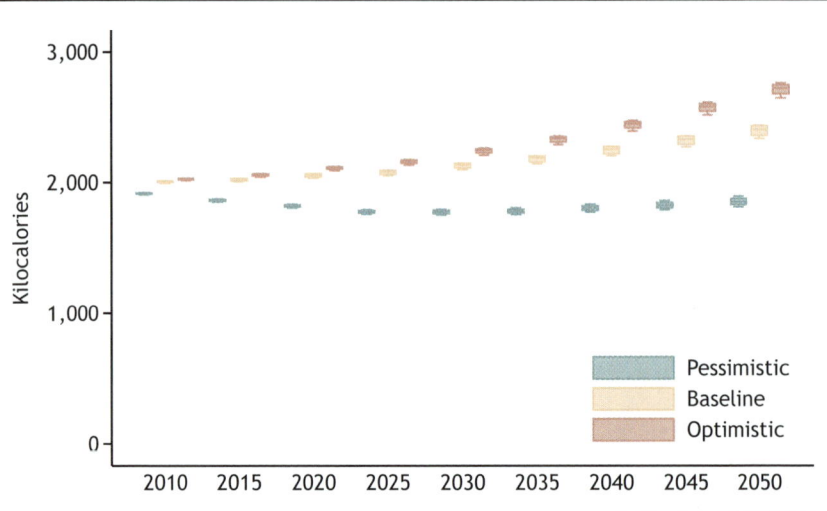

Source: Based on analysis conducted for Nelson et al. (2010).

Note: The box and whiskers plot for each socioeconomic scenario shows the range of effects from the four future climate scenarios.

to as many as 2,400 by 2050 given an ideal climate (baseline scenario). Higher income will result in greater per capita kilocalorie availability—up to 2,700 by 2050 (optimistic scenario). However, a low rate of growth in income will result in the availability of a lower number of kilocalories per capita by 2050 (pessimistic scenario), reflecting how the price effects negate gains from the income effects.

Agricultural Outcomes

Figures 9.21–9.23 show simulation results from the International Model for Policy Analysis of Agricultural Commodities and Trade associated with key agricultural crops in Rwanda. Each featured crop has five graphs, one each showing production, yield, harvested area, net exports, and world price.

The results in Figure 9.21 show an improved food security situation concerning potatoes given the expected changes in climate. All scenarios point to increased production, yield, harvested area, net exports, and price of potatoes by 2050. Increased net exports implies that by 2050 Rwanda will be able to meet all its domestic demand for potatoes and have supplies left to meet global demand.

Unlike in many other countries, the impact of climate on yield (and hence on production and net exports) will be quite large in Rwanda by 2050. That is,

FIGURE 9.21   Impact of changes in GDP and population on potatoes in Rwanda, 2010–2050

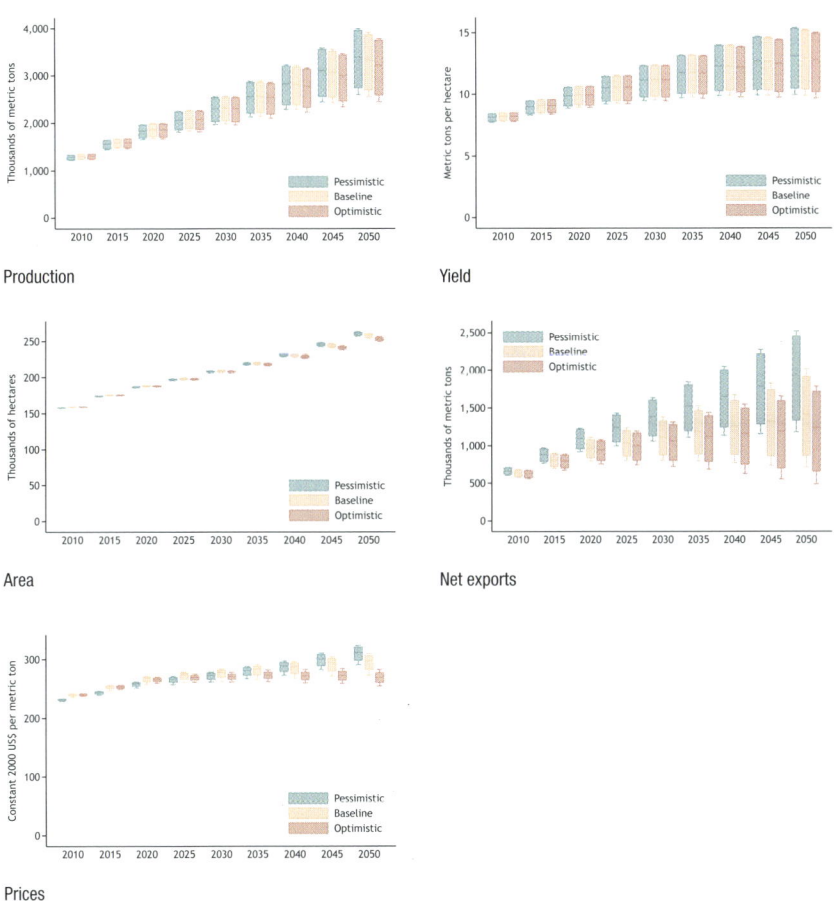

Source: Based on analysis conducted for Nelson et al. (2010).
Notes: The box and whiskers plot for each socioeconomic scenario shows the range of effects from the four future climate scenarios. GDP = gross domestic product; US$ = US dollars.

the difference for any scenario between the top whisker for 2050 and the bottom whisker for 2050 is quite large; most countries will have less than half of the range of Rwanda. This means that it is more difficult to generalize the results for Rwanda. The bottom whisker for the yield of potatoes indicates that there will be a gain in yield of only around 25 percent between 2010 and 2050, while the top whisker indicates that there will be a gain in yield of about 90 percent.

Figure 9.22 shows increased production, yield, and harvested areas of soghum, with yields doubling on average. Although the price for sorghum will

FIGURE 9.22   Impact of changes in GDP and population on sorghum in Rwanda, 2010–2050

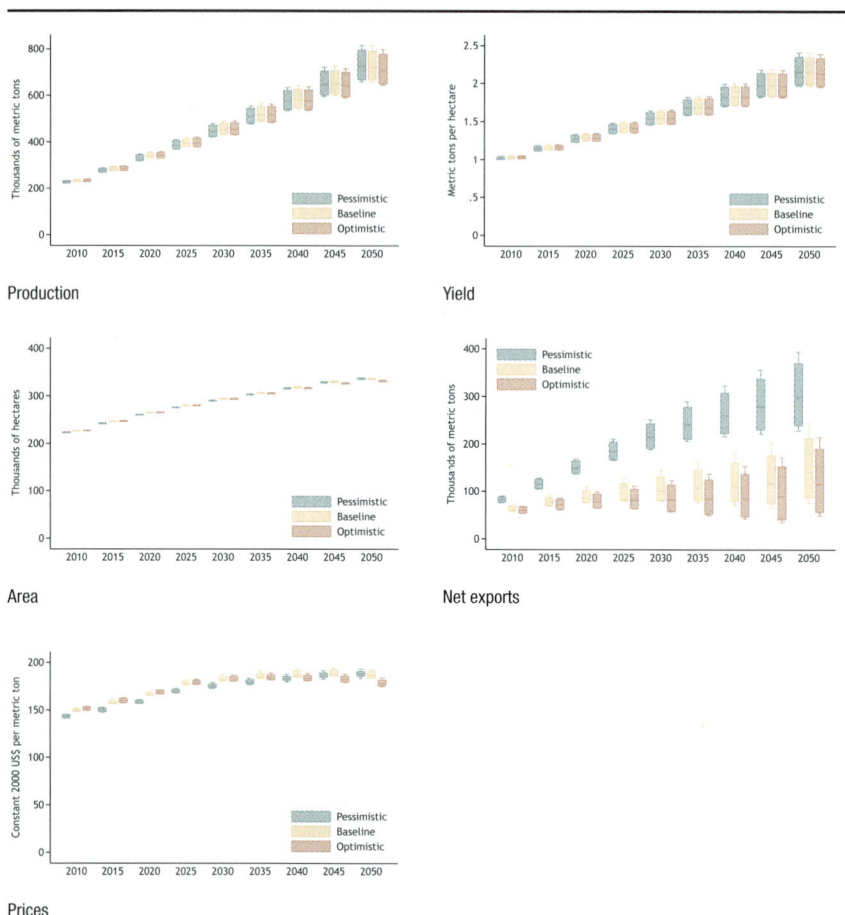

Production

Yield

Area

Net exports

Prices

Source: Based on analysis conducted for Nelson et al. (2010).

Notes: The box and whiskers plot for each socioeconomic scenario shows the range of effects from the four future climate scenarios. GDP = gross domestic product; US$ = US dollars.

be roughly 50 percent higher by 2050, even in the pessimistic scenario per capita income will have doubled, making higher food prices easier to bear. Net exports almost always show the greatest variance of the five graphs in each figure because they represent the difference between production and consumption, and both of these have their own variances. We generally observe that exports of sorghum will hold constant or grow, and in the few cases in which they will fall, the drop will be small.

FIGURE 9.23 Impact of changes in GDP and population on cassava in Rwanda, 2010–2050

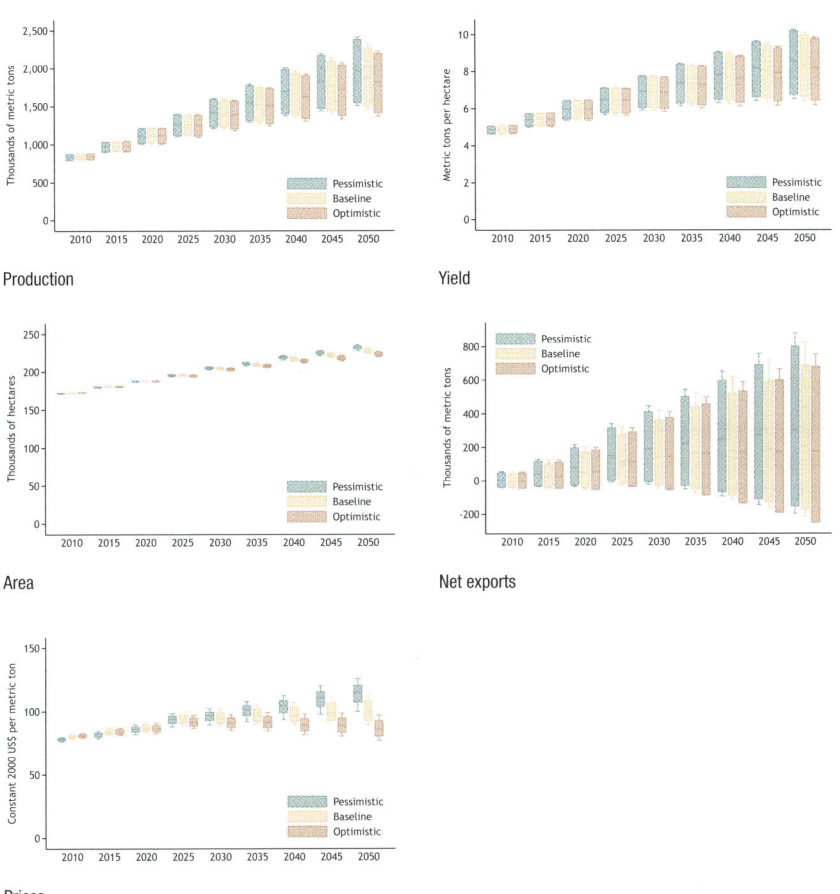

Production

Yield

Area

Net exports

Prices

Source: Based on analysis conducted for Nelson et al. (2010).

Notes: The box and whiskers plot for each socioeconomic scenario shows the range of effects from the four future climate scenarios. GDP = gross domestic product; US$ = US dollars.

Figure 9.23 illustrates findings somewhat similar to those in Figure 9.22. The graphs show increased yield and area, which will naturally result in increased production. Prices will rise slightly except in the optimistic scenario, in which they will rise and then fall, with the price in 2050 almost the same as in 2010. There is a high variance on net exports. It is hard to generalize, but more scenarios show exports rather than imports increasing. However, it would be improper to conclude that either outcome is definitive.

Conclusions and Policy Recommendations

In this chapter we set out to assess the vulnerability of Rwanda to the impacts of climate change. This evaluation was based on several economic, demographic, health, and food security indicators, and we used crop models and a global partial equilibrium model to determine the impact of climate change across diverse projections by climate models.

For a country that already has a high population density, one of the most prominent projections that went into the partial equilibrium model is that in the best-case scenario, which is the low population projection, there should be a doubling in the population by 2050. This suggests that policies should be created to reduce fertility rates. To ensure sustainable development, country strategies should include development of a strong education system, increased use of family planning methods, planned settlements, and increased productivity to feed the growing population.

If Rwanda is to attain the higher growth rates of either the baseline or the optimistic scenario, strategies that improve its agricultural productivity will be critical given that they will increase incomes for a majority of the population. Strengthening the education system beyond the primary level will also enhance individuals' capacity to maneuver between alternative forms of employment. In addition, reinforcing the service sector will boost livelihoods in the country given its increasing contribution to the economy and employment.

The four GCMs we used differed on the direction of change in annual rainfall, with one showing a general increase, one showing a general decrease, and two showing rainfall remaining unchanged. Although the GCMs clearly predict that the country is likely to experience warmer temperatures over the next four decades, they disagree as to how much, with two essentially suggesting that the increases will be only 1°–1.5°C, while the other two suggest that the increases will be 2°–2.5°C. The risks associated with greater increases include an increased number of plant and human pests and diseases as well as reduced productivity for crops acclimated to cooler temperatures. Policies supporting a stronger health system, adoption of integrated pest management, and the breeding of crop varieties suitable for warm temperatures will counter such effects.

Higher temperatures will permit the growing of some crops in places that are currently too cold. Using improved technology to increase the productivity of already cultivated areas will be far more beneficial in increasing yields and regulating climate change.

The ability of farmers to produce more than consumers demand will increase for some key crops such as potatoes but is less certain for other crops, for which the models show great variance, so it is too difficult to determine whether production gains will outpace increases in consumer demand. Setting up and improving institutions and infrastructure to coordinate and allow movement of foodstuffs and production inputs will further mitigate the food shortages brought about by change climate.

References

Bartholome, E., and A. S. Belward. 2005. "GLC2000: A New Approach to Global Land Cover Mapping from Earth Observation Data." *International Journal of Remote Sensing* 26 (9–10): 1959–1977.

CIESIN (Center for International Earth Science Information Network), Columbia University, IFPRI (International Food Policy Research Institute), World Bank, and CIAT (Centro Internacional de Agricultura Tropical). 2004. *Global Rural–Urban Mapping Project (GRUMP), Alpha Version: Population Density Grids.* Palisades, NY, US: Socioeconomic Data and Applications Center (SEDAC), Columbia University. http://sedac.ciesin.columbia.edu/gpw.

FAO (Food and Agriculture Organization of the United Nations). 2010. FAOSTAT. Rome. http://faostat.fao.org.

Indexmundi. 2011. "Rwanda GDP-per capita (PPP)." Accessed December 14. www.indexmundi.com/rwanda/gdp_per_capita_(ppp).html.

Jones, P. G., P. K. Thornton, and J. Heinke. 2009. *Generating Characteristic Daily Weather Data Using Downscaled Climate Model Data from the IPCC's Fourth Assessment.* Project report for the International Livestock Research Institute. Geneva: International Panel on Climate Change.

Lehner, B., and P. Döll. 2004. "Development and Validation of a Global Database of Lakes, Reservoirs, and Wetlands." *Journal of Hydrology* 296 (1–4): 1–22.

Millennium Ecosystem Assessment. 2005. *Ecosystems and Human Well-being: Synthesis.* Washington, DC: Island Press. www.maweb.org/en/Global.aspx.

Mpawenimana, J. 2005. "Analysis of Socio-economic Factors Affecting the Production of Bananas in Rwanda: A Case Study of Kanama District." Accessed December 16, 2011. www.unipv.eu/on-line/en/Home/InternationalRelations/CICOPS/PublicationsandMaterials.html.

National Institute of Statistics of Rwanda. 2008. *Rwanda Development Indicators, 2006 Edition.* Kigali.

Nelson, G. C., M. W. Rosegrant, A. Palazzo, I. Gray, C. Ingersoll, R. Robertson, S. Tokgoz, et al. 2010. *Food Security, Farming, and Climate Change to 2050: Scenarios, Results, Policy Options.* Washington, DC: International Food Policy Research Institute.

REMA (Rwanda Environment Management Authority). 2009. *Rwanda State of Environment and Outlook Report*. Kigali. www.rema.gov.rw/soe/full.pdf.

———. 2010. *Environment Sub-sector Strategic Plan, 2010–2015*. Kigali.

Rwanda, Ministry of Agriculture and Animal Resources. 2005. *Programme national pour le development de la banane*. Kigali.

Rwanda, Ministry of Finance and Economic Planning. 2007. *Economic Development and Poverty Reduction Strategy, 2008–2012*. Kigali.

Rwanda Agricultural Research Institute. 2010. "Banana Programme." Accessed August 15. www.isar.rw/spip.php?article45.

SEI (Stockholm Environmental Institute). 2009. "The Economics of Climate Change in Rwanda." Accessed August 31, 2011. www.rema.gov.rw/ccr/Final percent20report.pdf.

Shongwe, M. E., G. J. van Oldenborgh, B. van den Hurk, and M. van Aalst. 2010. "Projected Changes in Mean and Extreme Precipitation in Africa under Global Warming, Part 2: East Africa." *Journal of Climate* 24 (14): 3718–3733.

UNEP (United Nations Environment Programme) and IUCN (International Union for the Conservation of Nature). 2009. World Database on Protected Areas (WDPA): Annual Release. No longer available online.

UNPOP (United Nations Secretariat, Department of Economic and Social Affairs, Population Division). 2009. *World Population Prospects: The 2008 Revision*. Accessed April 06, 2010. http://esa.un.org/unpp.

USDoS (United States Department of State). 2011. "Background Note: Rwanda Bureau of African Affairs." Accessed August 31. www.state.gov/r/pa/ei/bgn/2861.htm.

WHO (World Health Organization). 2006. *The World Health Report 2006: Working Together for Health*. Accessed October 19, 2010. http://whqlibdoc.who.int/publications/2006/9241563176_eng.pdf.

Wood, S., G. Hyman, U. Deichmann, E. Barona, R. Tenorio, Z. Guo, S. Castano, O. Rivera, E. Diaz, and J. Marin. 2010. "Sub-national Poverty Maps for the Developing World Using International Poverty Lines: Preliminary Data Release." Accessed May 6. http://povertymap.info.

World Bank. 2009. *World Development Indicators*. Accessed May 2011. http://data.worldbank.org/data-catalog/world-development-indicators.

———. 2010. *Economics of Adaptation to Climate Change: Synthesis Report*. Washington, DC. http://climatechange.worldbank.org/content/economics-adaptation-climate-change-study-homepage.

You, L., and S. Wood. 2006. "An Entropy Approach to Spatial Disaggregation of Agricultural Production." *Agricultural Systems* 90 (1–3): 329–347.

You, L., S. Wood, and U. Wood-Sichra. 2006. "Generating Global Crop Distribution Maps: From Census to Grid." Paper presented at the International Association of Agricultural Economists Conference, Brisbane, Australia, August 11–18.

———. 2009. "Generating Plausible Crop Distribution and Performance Maps for Sub-Saharan Africa Using a Spatially Disaggregated Data Fusion and Optimization Approach." *Agricultural Systems* 99 (2–3): 126–140.

Chapter 10

SUDAN

Abdelmoneim Taha, Timothy S. Thomas, and Michael Waithaka

S udan is a country of fragile ecosystems, frequent droughts, and, as a result, pressing challenges to address the national priorities of food security, water supply, and public health. An examination of Sudan's ecological zones indicates that the majority of its land is quite vulnerable to changes in temperature and precipitation. More than half the country can be classified as desert or semidesert, with another quarter arid savanna. The country's inherent vulnerability is evident in the fact that food security in Sudan is mainly determined by rainfall, particularly in rural areas, where 70 percent of the population lives (Republic of the Sudan, MEPD 2003).

Sudan lies in the tropical zone, between latitudes 3° and 22°N and longitudes 22° and 38°E. Mean annual temperatures vary between 26° and 32°C across the country. Rainfall, which supports the overwhelming majority of the country's agricultural activity, is erratic and varies significantly from the northern to the southern regions of the country. Overall, the country's land and water resources can be classified into four major ecological regions:

1. Arid and semiarid ecosystems, which occur in the northern and central parts of the country and represent over 50 percent of total area with about 125 million hectares. Summer temperatures can often exceed 43°C, and sandstorms blow across the Sahara from April to September, with rainfall averaging about 200 millimeters per year and rarely exceeding 700 millimeters per year.

2. Savanna ecosystems (clay), which are typified by low rainfall and the prevalence of clay soils and represent about 5 percent of total area, with about 12 million hectares.

3. Savanna ecosystems (sand), which are typified by low rainfall and the prevalence of sandy soils. They represent about 3 percent of the country's total area, with about 8 million hectares.

This chapter was written prior to the independence of South Sudan, and it covers both the north and the south.

4. Southern flood-prone ecosystems, which are located below latitude 10°N and represent about 3 percent of the total area, with about 8.5 million hectares. Their climatic conditions are more equatorial, with an average annual temperature of about 29°C and average annual rainfall greater than 1,000 millimeters per year.

Traditional subsistence agriculture dominates the Sudanese economy, with over 70 percent of the population dependent for their livelihoods on crop production, livestock husbandry, or both. The agricultural sector is dominated by small-scale farmers—who typically live in conditions of persistent poverty and are reliant on rainfed and traditional agricultural practices. This combination renders them highly vulnerable to climate variability, as evidenced by the widespread suffering in rural areas during past droughts. Indeed, chronic drought is one of the most important climate risks facing Sudan. Recurring series of dry years have become a normal occurrence in the Sudan–Sahel region. Drought is threatening the existing cultivation of about 12 million hectares of rainfed mechanized farming and 6.6 million hectares of traditional rainfed cropland (Balgis-Osman et al. 2005). Pastoral and nomadic groups in the semiarid areas of Sudan are also affected. Poverty is deeply entrenched in rural areas, where more than 20 million people live on less than $1 a day. Sudan's diverse agroecological zones and abundant surface water offer the potential to produce a range of crops as well as livestock. Yet production consistently remains quite low, due in large part to an agricultural system that is not well adapted to rainfall variability and prolonged drought events.

Sudan also faces numerous other development challenges. For example, land degradation and desertification brought on by human land-use pressures and recurrent drought affect large areas and threaten already vulnerable arable zones. Depletion of forests—primarily for household fuel consumption— threatens biological diversity and human communities and reduces the other valuable products and services that forests provide.

The process involved in developing Sudan's National Adaptation Programme of Action (NAPA) examined each of these ecosystems as distinct zones, each meriting its own locally driven assessment of priority interventions to address looming climate risks. In arid and semiarid zones, frequent droughts exacerbate declining soil fertility, low agricultural productivity, and persistent food insecurity. Frequent droughts also afflict savanna areas, where they compound the problems of overgrazing, soil erosion, and outbreaks of public health epidemics such as malaria. In southern areas, chronic flooding and frequent malaria outbreaks impose great strains on communities and infrastructure, newly emerging from decades of civil strife.

A trend of decreasing annual rainfall is contributing to drought conditions in many parts of Sudan. Moreover, the coefficient of variability of rainfall shows an overall increasing trend, suggesting greater rainfall unreliability. The variability in rainfall is most serious in the arid northern parts of the country, where the average variability now exceeds 100 percent (Zaki-Eldeen 2009). The situation is less serious in the central parts of the country, where average rainfall variability ranges from 20 to 60 percent, and in the south, where it varies between 15 and 20 percent. At the national level, however, there is a trend toward greater rainfall variability, increasing at a rate of about 0.2 percent per year (FAO 2011). These rainfall patterns are associated with serious drought episodes throughout the country—even in the south. The hardest-hit areas are in the western and northern regions, in the semiarid portions of the Nile basin. A succession of dry years from 1978 to 1987 resulted in severe social and economic impacts, including many human and livestock fatalities; nearly 3 million people living along the Nile and in urban areas had to be resettled. Problems will increase if these trends continue without efforts to adapt.

Sudan has been actively seeking to mainstream adaptation to climate change in its development process by including climate and vulnerability in the sectoral and development policies that complement the climate change and environmental policies embodied in the 10-year Comprehensive National Strategy (1992–2002) and the 25-year Comprehensive National Strategy Outlines. Many ongoing national policy processes have aims parallel to climate change adaptation, such as the poverty reduction strategy (2004–2008), the malaria roll-back program, and water harvesting.

In an attempt to address climate change and related issues, Sudan has already completed several activities. It ratified the United Nations Framework Convention on Climate Change (UNFCCC) in 2003 and submitted its initial national communication the same year. The Higher Council for Environment and Natural Resources (HCENR), the government's national focal point for the UNFCCC, plays an advisory policymaking role with regard to climate-related initiatives. The HCENR is also the national executing agency for Sudan's NAPA, completed in 2007, which focuses on major climate impacts and vulnerabilities in five regions representing the different ecological settings across the country.

This chapter is meant to further the understanding of policymakers concerning the potential impact of climate change on agriculture in Sudan, along with various adaptation options. The chapter begins with a review of some national statistics that will help characterize the capacity of the population to adapt to climate change. We also look at Sudan's land base and current agricultural situation. After this we show the results of crop models that incorporate climate projections from recent global models. Finally, we incorporate those results into a global model for food and agricultural supply and demand that takes into account demographic and economic projections for Sudan.

Review of the Current Situation and Trends

Economic and Demographic Indicators

Population

Figure 10.1 shows the total and rural population of Sudan (left axis) as well as the share of the population that is urban (right axis). Figure 10.2 shows the geographic distribution of the population within Sudan. Both are estimates based on census data and other sources. Table 10.1 provides additional information concerning rates of population growth.

As Figure 10.1 shows, the population of Sudan was estimated at just over 41 million in 2008. Assessments of annual population growth rates vary from around 2.1 to 2.6 percent, placing Sudan's rate of population growth among the highest in the world (Republic of the Sudan, MEPD 2007). Its overall population density is about 10 people per square kilometer, but the density on arable land is considerably higher, at 63 people per square kilometer— and higher still on cultivated land, with about 370 people per square kilometer. Much of the population is clustered in central Sudan and along the Nile River (Republic of Sudan, MEPD 2007). Though annual rates of growth are projected to fall by 2050, the population is still projected to double over that period. Meanwhile, the proportion of urban residents is projected to increase from around 40.0 percent in 2005 to 60.7 percent by 2030, suggesting that almost all (94 percent) of the country's population growth will be accounted for by urban areas during this period. Historically, the population of Sudan has been highly mobile. On average, 40 percent of the population is believed to be on the move every year, especially in the semiarid regions of western Sudan. There are different systems of migration, but currently rural-to-urban is the most common pattern.

Climate variability, directly and indirectly, is a key driver of displacement and forced migration. There are an estimated 6 million internally displaced

FIGURE 10.1   Population trends in Sudan: Total population, rural population, and percent urban, 1960–2008

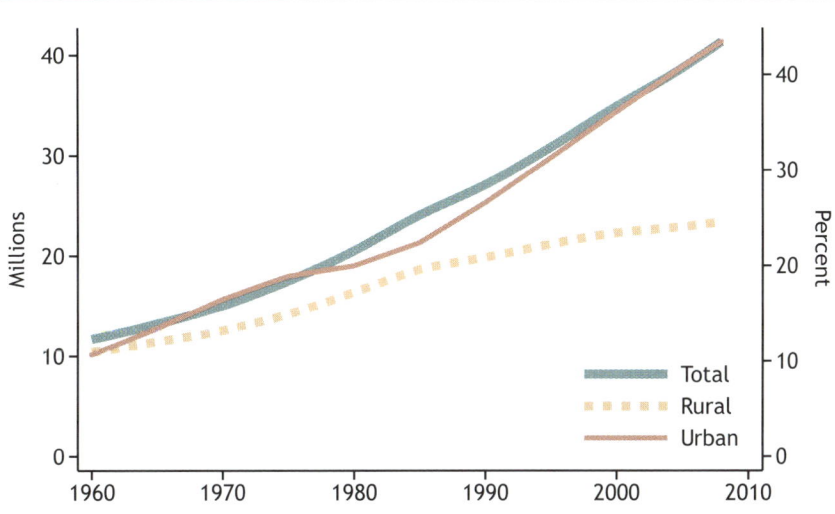

Source: *World Development Indicators* (World Bank 2009).

FIGURE 10.2   Population distribution in Sudan, 2000 (persons per square kilometer)

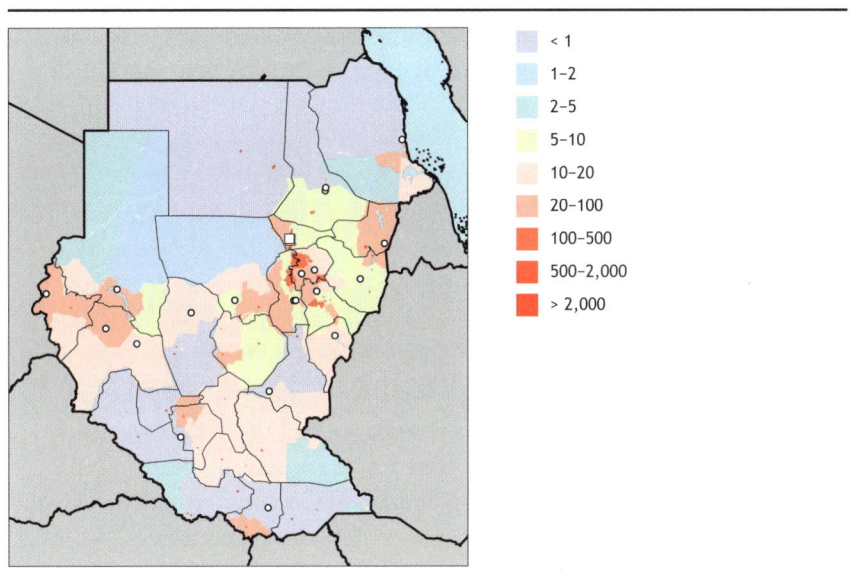

Source: CIESIN et al. (2004).

TABLE 10.1   Population growth rates in Sudan, 1960–2008 (percent)

Decade	Total growth rate	Rural growth rate	Urban growth rate
1960–1969	2.5	1.8	6.9
1970–1979	3.1	2.6	5.1
1980–1989	2.8	2.0	5.6
1990–1999	2.6	1.2	5.7
2000–2008	2.1	0.6	4.4

Source: Authors' calculations based on *World Development Indicators* (World Bank 2009).

persons in Sudan; they are displaced largely by drought, desertification, and famine in the north and by conflict, famine, and flood-induced epidemics in the south. The hardest-hit areas over the past three decades have been the western and northern regions in the semiarid portions of the Nile basin. A succession of dry years from 1978 to 1987, for example, resulted in the resettlement of almost 3 million people along the Nile Valley and urban peripheries, particularly Khartoum. Climate change is believed to be among the key factors contributing to the Darfur conflict (UNEP 2007).

Another factor is economic: uneven patterns of growth have resulted in great disparities between urban and rural areas and between regions. This has contributed to growing inequalities and an increasing urban informal sector, accounting for more than 60 percent of gross domestic product (GDP) (UNDP 2008).

Vulnerability to Climate Change

Table 10.2 provides some data on indicators of the population's vulnerability and resiliency to economic shocks: levels of education, literacy, and malnutrition and concentration of labor in poorer or less dynamic sectors.

The country is characterized by a moderately educated population, as is evidenced by the 61 percent literacy rate shown in Table 10.2. The level of primary school enrollment is relatively high due to the country's policy of free education for all children 6–13 years of age. The low level of secondary school education is due to the fact that males are required to enroll in military service before completing school (Library of Congress Federal Research Division 2004). Regional disparities exist, with most schools located in the prosperous north and urban centers, leaving the south and rural areas lagging behind in learning. Furthermore, girls' education levels fall behind those of boys. For example, at the primary school level there are 80 girls for every

TABLE 10.2 Education and labor statistics for Sudan, 2000s

Indicator	Year	Percent
Primary school enrollment: Percent gross (three-year average)	2007	66.4
	2009[a]	71.0
Secondary school enrollment: Percent gross (three-year average)	2007	33.4
Adult literacy rate	2000	60.9
Percent with vulnerable employment (in agriculture on own farm or as a day laborer)	2009[b]	20.0
Under-five malnutrition (weight for age)	2000	38.4

Source: *World Development Indicators* (World Bank 2009).
[a]See also Republic of Sudan, Ministry of Welfare and Social Security, NPC/GS (2010).
[b]See also Republic of Sudan, CBS (2010).

100 boys. Enrollment rates for girls are worse among those internally displaced by conflict and nomads. In spite of these challenges, the country has recorded improvements in the sector. The gross enrollment rate in basic education increased in northern Sudan from 65 percent in 2004 to 71 percent in 2009. The literacy rate for people between 15 and 24 years of age increased from 27.0 percent in 1990 to 69.0 percent in 2009 and to 72.5 percent in 2010 (UNDP 2011). The current education levels form a foundation for climate change advocacy through learning and will prove critical to the effectiveness of various mitigation strategies.

The level of child malnutrition in Sudan remains very serious. The recent UN Millennium Development Goals progress report for Sudan (UNDP 2011) assesses the nutrition situation in the country as characterized by high levels of underweight, chronic malnutrition, and acute malnutrition. Nationally, 31 percent of children under the age of five are moderately or severely underweight (UNDP 2011). An equal proportion (32.5 percent) suffers from moderate or severe chronic malnutrition. The proportions vary significantly between states. This points to the extreme low levels of caloric availability, which are not surprising given that the country has just recently emerged from civil wars that are associated with reduced agricultural productivity and incomes. The implication is that the population will have a low rate of productivity and be highly vulnerable to extreme events of climate change, which affects agricultural productivity and reduces food reserves.

Income

Over the decade 1999–2008, Sudan's economy witnessed its longest and strongest growth episode, driven by the discovery of oil in 1999 (Figure 10.3).

FIGURE 10.3   Per capita GDP in Sudan (constant 2000 US$) and share of GDP from
agriculture (percent), 1960–2008

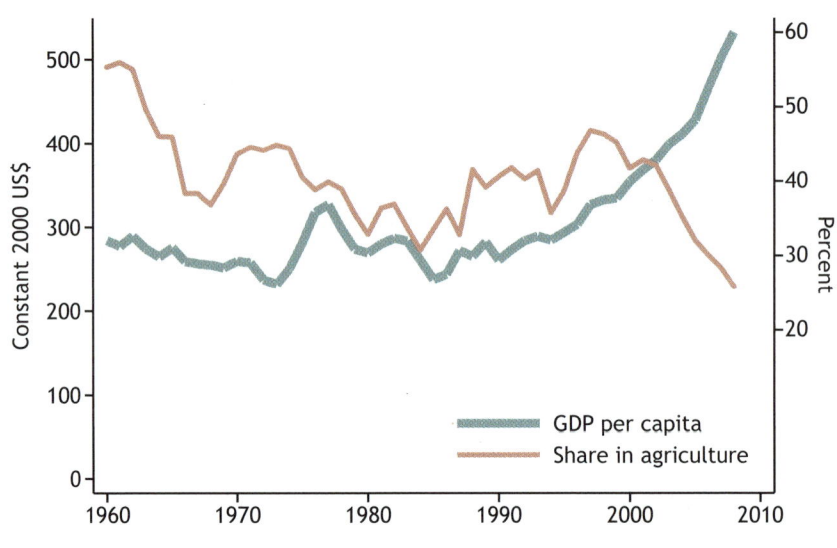

Source: *World Development Indicators* (World Bank 2009).
Note: GDP = gross domestic product; US$ = US dollars.

The size of its economy, measured by gross national product, has grown
fivefold—from $10 billion in 1999 to $53 billion in 2008. Per capita income,
a rough measure of the living standard of average citizens, increased from
$348 to $532 over the same period. In contrast, before oil was discovered, per
capita income hardly changed over four decades, remaining within the $200–
$300 range.

However, growth has not been sufficiently broad based, and there are significant disparities between urban and rural areas as well as between regions.
This contributes to growing inequalities and an increasing urban informal
sector, currently accounting for more than 60 percent of GDP. This state
of affairs has been encouraging a rural–urban migration that might weaken
agricultural productivity and deepen poverty in both urban and rural areas.
The agricultural sector has performed poorly in the past decade. Although it
remains an important sector, its share of GDP in the economy has declined,
the rate of growth of rural incomes has decreased, and the level of poverty in
rural areas remains high. The average annual growth rate of the agricultural

sector between 2000 and 2008 was 3.6 percent, substantially lower than the 10.8 percent during the previous decade (FAO 2011).

Poverty

A country's vulnerability and resiliency to economic shocks can be assessed from the status of its progress toward the United Nations Millennium Development Goals (MDGs). Sudan has made progress toward achieving the MDGs in the areas of education, infant and child mortality, and access to water and sanitation (UNDP 2011). However, poverty remains widespread throughout the country. In southern Sudan, approximately half of the population (50.6 percent) is below the official poverty line. The poverty headcount ratio in rural areas (55.4 percent) is more than double the ratio in urban areas (24.4 percent). Poverty rates vary significantly between states: three out of four people are poor in Northern Bahr el Ghazal state (75.6 percent), but only one in four people in Upper Nile state is poor (25.7 percent). In the northern states, the level of food deprivation varies significantly, with 44 percent in the Red Sea region and 15 percent in the Gezira and River Nile region. At the household level, the rate of food deprivation is higher in female-headed households (37 percent) than in male-headed households (31 percent), reflecting that males have more access to education and income. The rate of food deprivation also differs according to household size, ranging from 5 percent in households of one or two members to 49 percent in households with more than nine members (Faki et al. 2009; IFAD 2011).

Figure 10.4 shows the human poverty map of Sudan using the human poverty index (HPI) as a measure of poverty in different states. The HPI, as a composite measure of human development, combines three observable and measurable dimensions of human well-being or deprivation: the ability to live long (longevity), the ability to acquire knowledge (knowledge ability), and the ability to live comfortably (decent standard of living). See http://hdr.undp.org/en/statistics/indices/hpi/.

The nutrition situation in Sudan is poor, characterized by high levels of underweight, chronic malnutrition, and acute malnutrition. Nationally, one-third (31 percent) of children under the age of five are moderately or severely underweight, and a similar number (32.5 percent) suffer from moderate or severe chronic malnutrition (Republic of Sudan, Ministry of Welfare and Social Security, NPC/GS 2010). In fact, the national level of acute malnutrition is just below the internationally recognized standard indicating a nutrition emergency. These figures, too, vary significantly between states. Figure 10.5 nevertheless shows a gradual improvement in two noneconomic correlates of poverty—life expectancy and under-five mortality.

FIGURE 10.4   Poverty in Sudan, circa 2005 (percentage of population below US$2 per day)

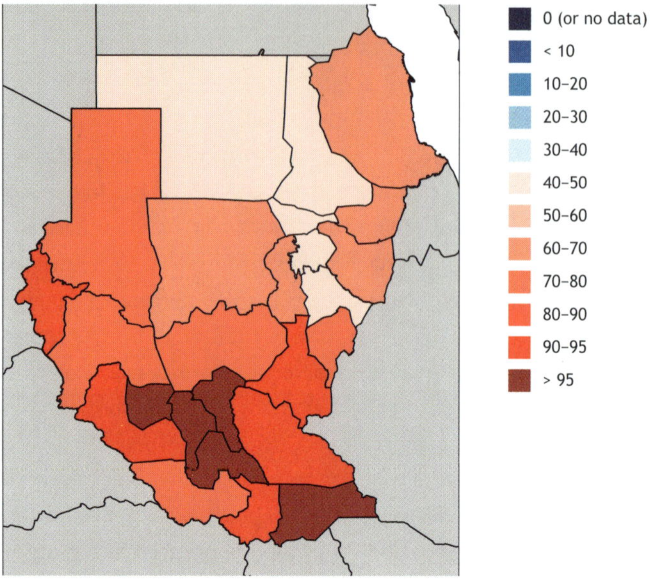

Source: Wood et al. (2010).
Note: Based on 2005 US$ (US dollars) and on purchasing power parity value.

Review of Land Use, Potential, and Limitations

Arable land (around 84 million hectares) constitutes about one-third of the total land area of Sudan, but only 21 percent of its arable land is cultivated (Metz 1991; FAO 2010). More than 40 percent of the total area of Sudan consists of pasture and forests (FAO 2010). Nearly all livestock are grazed on natural pasture. The annual production of animal feed is estimated at 78 million tons; production fluctuates from year to year, affected by varying rainfall and fire hazards. Forests and woodlands are used to meet the population's demand for wood products and are exposed to pressure from both agricultural expansion and demand for firewood.

Agriculture is the backbone of Sudan's economy. It is characterized by three major production systems: irrigated, rainfed semimechanized, and rainfed traditional agriculture. These farming systems are used for both crop and livestock production. Forestry is also an important subsector. Traditional

FIGURE 10.5   Well-being indicators in Sudan, 1960–2008

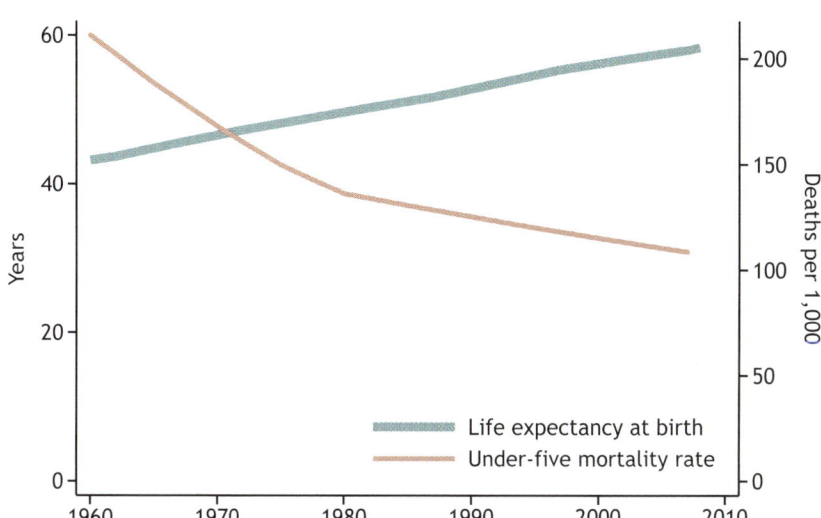

Source: *World Development Indicators* (World Bank 2009).

subsistence agriculture is a common practice of more than 70 percent of the population who depend on it for their livelihoods. Small-scale farmers dominate the sector and typically live in persistent poverty, which renders them vulnerable to climate variability, as evidenced by the widespread suffering in rural areas during past droughts.

Figure 10.7 shows the locations of protected areas, including parks and reserves. These locations provide important protection for fragile environmental areas, which may also be important for tourism. Sudan has about 27 protected areas, amounting to more than 7 percent of the country's total area. There are 3,246 reserved forests (not reflected in the current map), with a total area of 12.3 million hectares.

Figure 10.8 shows travel times to urban areas. There has been a notable improvement in the road network, which was expanded from 3,358 kilometers in 2000 to 6,211 kilometers in 2008. Most of this expansion is in irrigated areas of the central and northern parts of the country, connecting big cities to rural areas, without extending to the vast rainfed production areas. The rural feeder roads are very poor, resulting in high transport costs, a small share of

FIGURE 10.6 Land cover and land use in Sudan, 2000

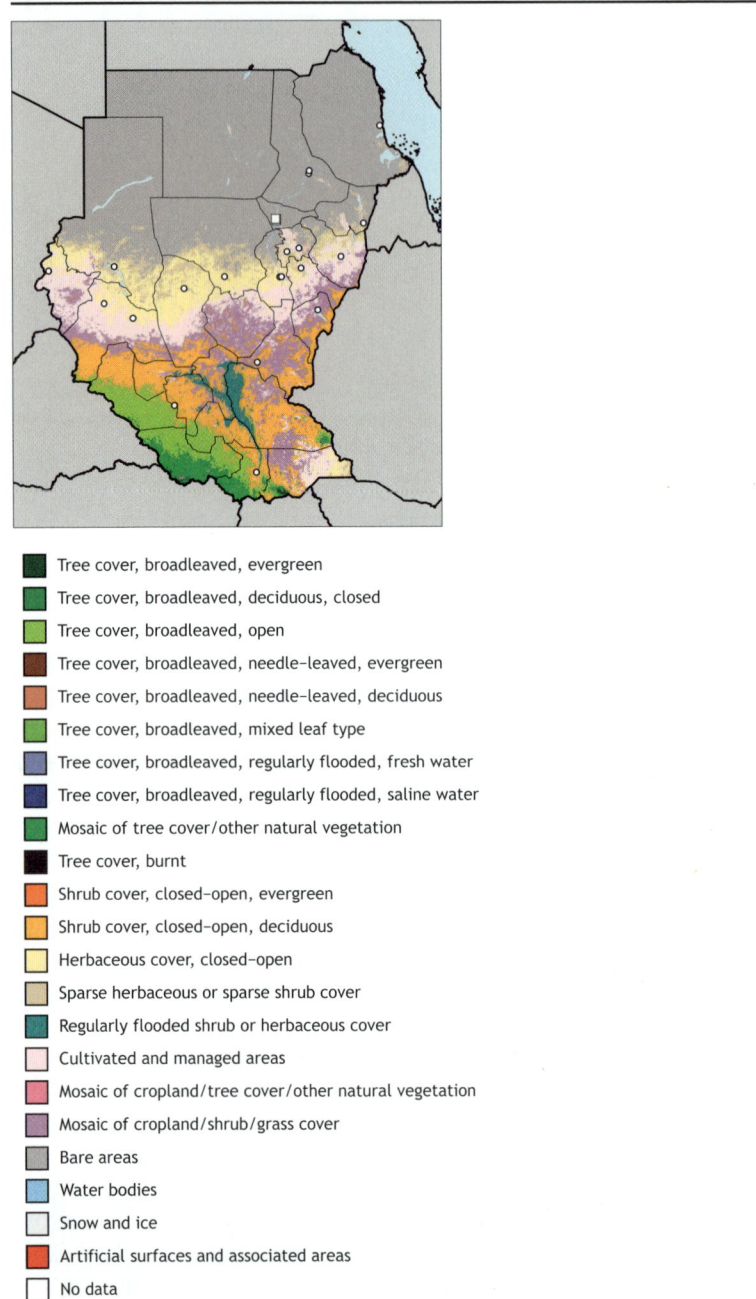

- Tree cover, broadleaved, evergreen
- Tree cover, broadleaved, deciduous, closed
- Tree cover, broadleaved, open
- Tree cover, broadleaved, needle-leaved, evergreen
- Tree cover, broadleaved, needle-leaved, deciduous
- Tree cover, broadleaved, mixed leaf type
- Tree cover, broadleaved, regularly flooded, fresh water
- Tree cover, broadleaved, regularly flooded, saline water
- Mosaic of tree cover/other natural vegetation
- Tree cover, burnt
- Shrub cover, closed–open, evergreen
- Shrub cover, closed–open, deciduous
- Herbaceous cover, closed–open
- Sparse herbaceous or sparse shrub cover
- Regularly flooded shrub or herbaceous cover
- Cultivated and managed areas
- Mosaic of cropland/tree cover/other natural vegetation
- Mosaic of cropland/shrub/grass cover
- Bare areas
- Water bodies
- Snow and ice
- Artificial surfaces and associated areas
- No data

Source: GLC2000 (Bartholome and Belward 2005).

FIGURE 10.7   Protected areas in Sudan, 2009

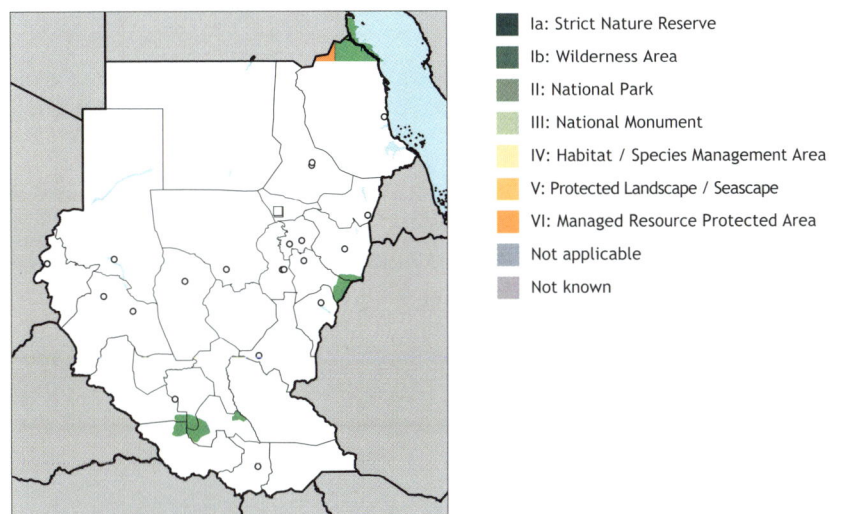

Ia: Strict Nature Reserve

Ib: Wilderness Area

II: National Park

III: National Monument

IV: Habitat / Species Management Area

V: Protected Landscape / Seascape

VI: Managed Resource Protected Area

Not applicable

Not known

Sources: Protected areas are from the World Database on Protected Areas (UNEP and IUCN 2009). Water bodies are from the World Wildlife Fund's Global Lakes and Wetlands Database (Lehner and Döll 2004).

product sales revenues for farmers, high input prices, and ultimately low incentives to producers.

Figures 10.9–10.11 show the estimated yield and growing areas for key cereal crops—sorghum, millet, and wheat. Sorghum is the main staple foodgrain in Sudan; it is produced under all production systems and occupies the largest cultivated area, at around 50 percent of field crop area (Table 10.3).

Sesame seeds and groundnuts are the main oil crops in terms of area and value of production as well as export earnings (Tables 10.3–10.5). In cultivated areas, the two oilseed crops are among the top five crops, at 12 percent and 6 percent for sesame and groundnuts, respectively.

Wheat is one of the main foodgrains; demand is increasing in urban areas due to changes in consumption patterns (see Table 10.5). Consumption has increased from less than 500,000 tons in the 1970s to around 2 million tons in 2009.

Millet is also an important foodcrop, particularly in western Sudan. It is second to sorghum in cultivated area, occupying more than 18 percent of the total area of field crops (see Table 10.3). It is produced under both traditional and mechanized rainfed systems, predominantly on small farms.

FIGURE 10.8   Travel time to urban areas of various sizes in Sudan, circa 2000

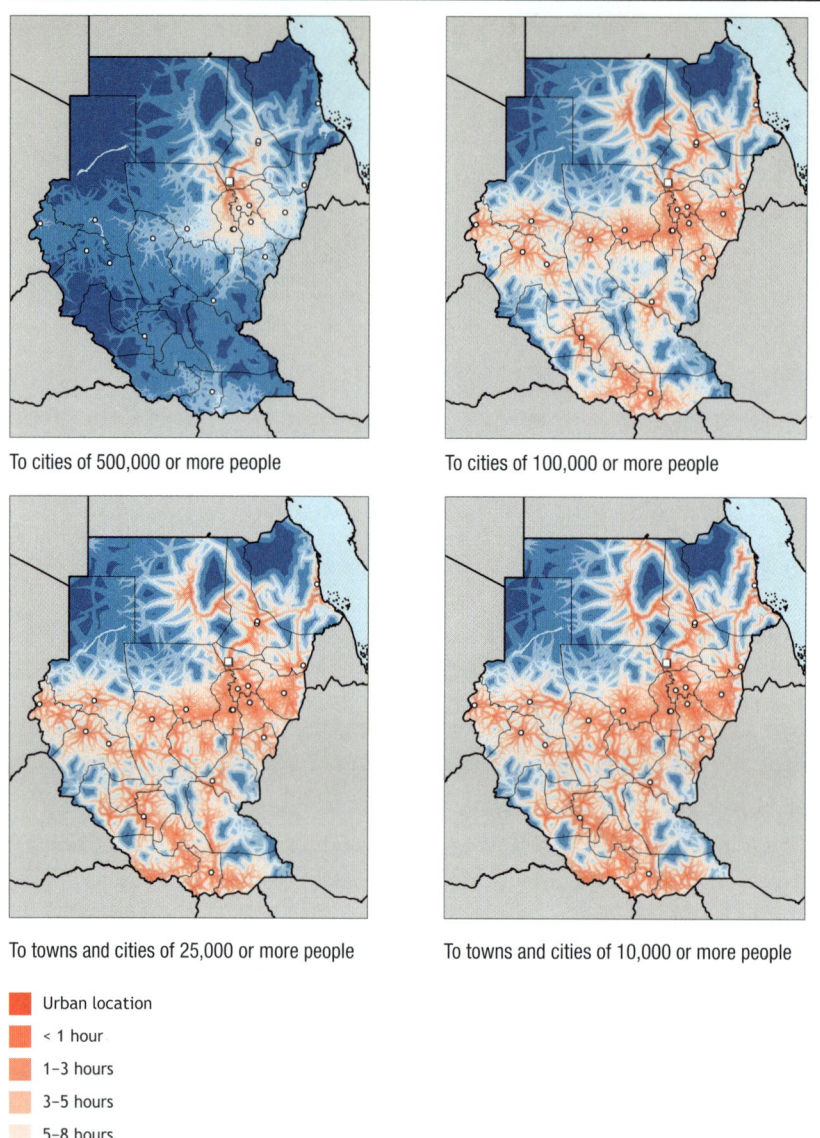

To cities of 500,000 or more people

To cities of 100,000 or more people

To towns and cities of 25,000 or more people

To towns and cities of 10,000 or more people

Urban location
< 1 hour
1–3 hours
3–5 hours
5–8 hours
8–11 hours
11–16 hours
16–26 hours
> 26 hours

Source: Authors' calculations.

FIGURE 10.9   Yield (metric tons per hectare) and harvest area density (hectares) for rainfed sorghum in Sudan, 2000

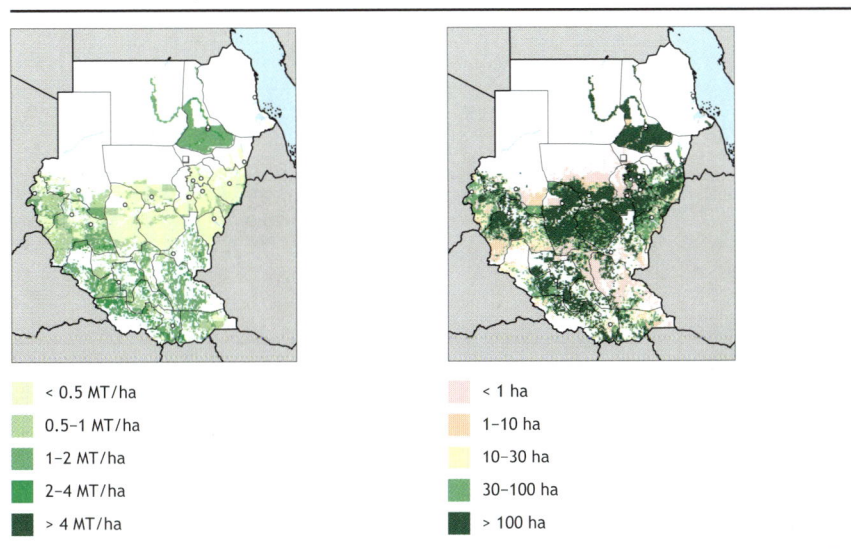

< 0.5 MT/ha	< 1 ha
0.5–1 MT/ha	1–10 ha
1–2 MT/ha	10–30 ha
2–4 MT/ha	30–100 ha
> 4 MT/ha	> 100 ha

Source: SPAM (Spatial Production Allocation Model) (You and Wood 2006; You, Wood, and Wood-Sichra 2006, 2009).
Note: ha = hectare; MT/ha = metric tons per hectare.

FIGURE 10.10   Yield (metric tons per hectare) and harvest area density (hectares) for irrigated wheat in Sudan, 2000

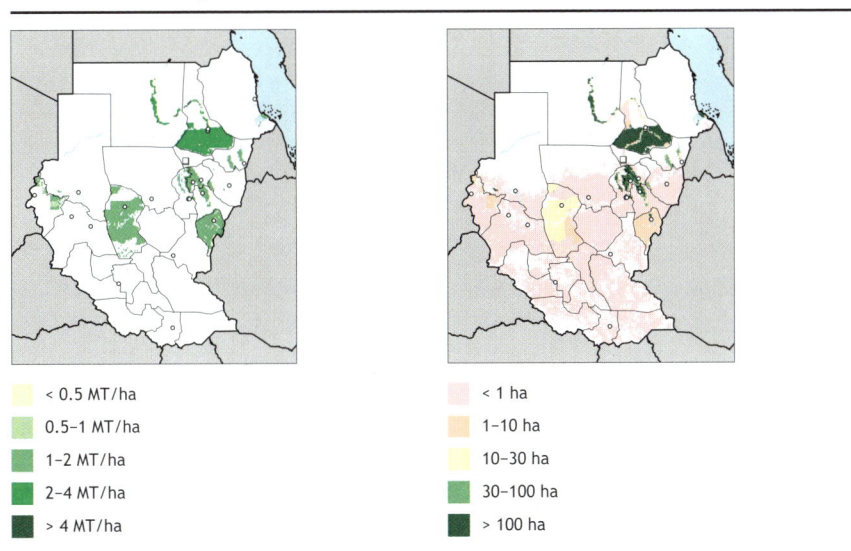

< 0.5 MT/ha	< 1 ha
0.5–1 MT/ha	1–10 ha
1–2 MT/ha	10–30 ha
2–4 MT/ha	30–100 ha
> 4 MT/ha	> 100 ha

Source: SPAM (Spatial Production Allocation Model) (You and Wood 2006; You, Wood, and Wood-Sichra 2006, 2009).
Note: ha = hectare; MT/ha = metric tons per hectare.

FIGURE 10.11   Yield (metric tons per hectare) and harvest area density (hectares) for rainfed millet in Sudan, 2000

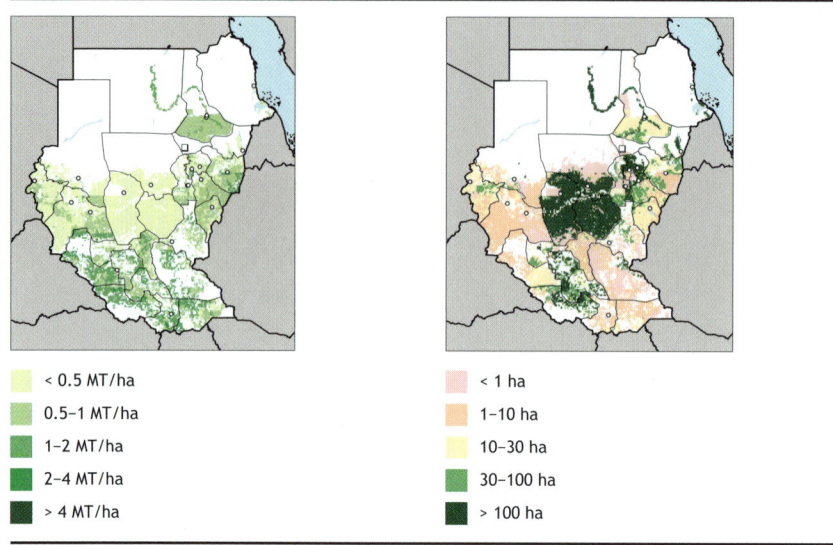

▢ < 0.5 MT/ha		▢ < 1 ha	
▢ 0.5–1 MT/ha		▢ 1–10 ha	
▢ 1–2 MT/ha		▢ 10–30 ha	
▢ 2–4 MT/ha		▢ 30–100 ha	
▢ > 4 MT/ha		▢ > 100 ha	

Source: SPAM (Spatial Production Allocation Model) (You and Wood 2006; You, Wood, and Wood-Sichra 2006, 2009).
Note: ha = hectare; MT/ha = metric tons per hectare.

TABLE 10.3   Harvest area of leading agricultural commodities in Sudan, 2006–2008 (thousands of hectares)

Rank	Crop	Percent of total	Harvest area
	Total	100.0	12,561
1	Sorghum	52.1	6,543
2	Millet	18.3	2,296
3	Sesame seeds	11.7	1,470
4	Groundnuts	5.7	715
5	Wheat	2.0	253
6	Other fresh vegetables	1.9	240
7	Seed cotton	1.2	151
8	Cowpeas	0.9	110
9	Other pulses	0.7	90
10	Melon seeds	0.6	81

Source: FAOSTAT (FAO 2010).
Note: All values are based on the three-year average for 2006–2008.

TABLE 10.4   Value of production for leading agricultural commodities in Sudan, 2005–2007 (millions of constant 2000 US$)

Rank	Crop	Percent of total	Value of production
	Total	100.0	4,539.0
1	Sorghum	15.0	682.6
2	Other fresh fruit	8.8	399.8
3	Potatoes	7.0	318.5
4	Tomatoes	6.7	302.3
5	Groundnuts	5.9	269.4
6	Dates	5.9	267.7
7	Other fresh vegetables	5.8	262.0
8	Mangoes, mangosteens, and guavas	5.4	246.5
9	Sesame seeds	5.2	235.7
10	Okra	4.4	200.1

Source: FAOSTAT (FAO 2010).
Notes: All values are based on the three-year average for 2005–2007. US$ = US dollars.

TABLE 10.5   Consumption of leading food commodities in Sudan, 2003–2005 (thousands of metric tons)

Rank	Crop	Percent of total	Food consumption
	Total	100.0	10,940
1	Sorghum	23.3	2,553
2	Wheat	13.5	1,474
3	Other vegetables	8.8	959
4	Sugar	7.3	799
5	Fermented beverages	6.2	678
6	Other fruits	4.7	514
7	Tomatoes	4.4	483
8	Millet	4.3	467
9	Sugarcane	3.1	344
10	Beef	3.1	343

Source: FAOSTAT (FAO 2010).
Note: All values are based on the three-year average for 2003–2005.

Sudan is also endowed with a wealth of livestock resources, estimated at more than 140 million cattle, sheep, goats, and camels. Various livestock production systems are used in almost all agroecological zones, predominantly rain-fed. Milk and meat are the leading agricultural commodities in terms of value of production. For the period 2006–2008, cow's milk and beef accounted for 26 percent and 13 percent, respectively, of the total value of the top 20 agricultural commodities.

Scenarios for the Future

Economic and Demographic Indicators

Population projections by the UN Population Division through 2050 show annual population growth rates falling below current rates but population continuing to increase (Figure 10.12). Challenges associated with such population growth include increased demand for food, intensive pressure on land and other natural resources, and a probable increase of rural–urban migration, with implications for urban water, health, and education services. In the low-variant scenario, the population increases by only 50 percent between 2010 and 2050, but in the high-variant scenario the population doubles in that time period.

FIGURE 10.12   Population projections for Sudan, 2010–2050

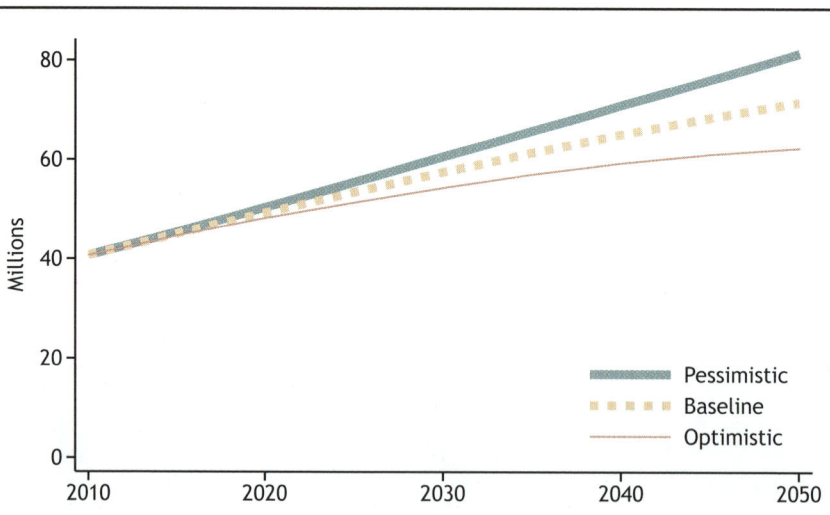

Source: UNPOP (2009).

FIGURE 10.13   Gross domestic product (GDP) per capita in Sudan, future scenarios, 2010–2050

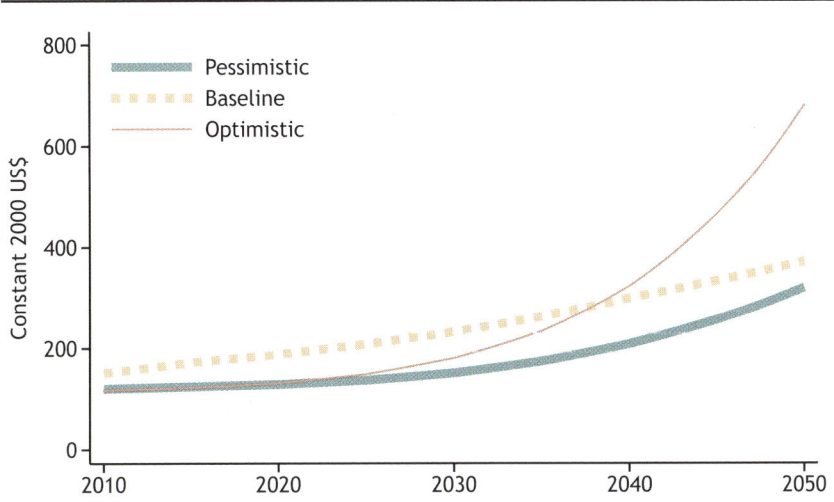

Sources: Computed from GDP data from the World Bank Economic Adaptation to Climate Change project (World Bank 2010), from the Millennium Ecosystem Assessment (2005) reports, and from population data from the United Nations (UNPOP 2009).
Note: US$ = US dollars.

Projections of future per capita GDP also show an increasing trend (Figure 10.13). However, unless there is improvement in the performance of agriculture to enhance its contribution to the national GDP, the disparities between urban and rural areas and between regions will continue.

Biophysical Analysis

Climate Models

In Figure 10.14 we see the changes in annual precipitation projected by the four general circulation models (GCMs) used in our analysis with the A1B scenario.[1] Two of the models, ECHAM 5 and MIROC 3.2, generally show most of the southern part of Sudan getting wetter, a very favorable outcome, particularly for the semiarid regions. The CNRM-CM3 model shows the

---

1   CNRM-CM3 is National Meteorological Research Center–Climate Model 3. MIROC 3.2 is the Model for Interdisciplinary Research on Climate, developed at the University of Tokyo Center for Climate System Research. CSIRO Mark 3 is a climate model developed at the Australia Commonwealth Scientific and Industrial Research Organisation. ECHAM 5 is a fifth-generation climate model developed at the Max Planck Institute for Meteorology in Hamburg. The A1B scenario is a greenhouse gas emissions scenario that assumes fast economic growth, a population that peaks midcentury, and the development of new and efficient technologies, along with a balanced use of energy sources.

western part of Sudan getting wetter. There are a few patches of lower annual precipitation in these maps, but they are small.

Although changes in annual precipitation are predicted to be mostly non-existent or positive, the story for temperature change is different. All four GCMs show Sudan getting warmer, by 0.5°C to as much as 3°C for two of the models in the northern reach of the country (Figure 10.15). The ECHAM 5 model does not show increases of 3°C, but it does project increases of 2°–2.5°C in the northern three-quarters of the country. Higher temperatures would increase evaporation and reduce soil moisture, increasing plants' water requirements—an unfavorable trend, particularly if associated with a lower level of precipitation and insufficient irrigation water. The CSIRO Mark 3 model shows only moderate increases in temperature, with all but very small patches in the range of 1°–1.5°C. The MIROC 3.2 model shows only modest temperature increases in the south, unlike in the north.

Crop Models

The Decision Support Software for Agrotechnology Transfer crop simulation software was used to compute yields under the current climate and compare those to computations based on the future climate. All four models show a yield loss of 5–25 percent of baseline (2000 climate) over most of the country's sorghum harvest area (Figure 10.16). In the marginal cultivated areas of the semidry zone, three of the four models show some loss of the baseline area. These results would have serious implications for Sudan's food security, because sorghum is the main staple cereal grain supporting the rural population. Those same three models show some limited gain in baseline area that could potentially offset part of the loss in yield in most of the rest of the cropped land. The CNRM-CM3 GCM is different from the other three in that it predicts an increase in cultivable area and large yield declines (greater than 25 percent) in much of the eastern portion of the country.

For wheat (Figure 10.17), all four models show negative impacts, ranging from a complete loss of the baseline area to a yield loss of between 5 percent and more than 25 percent of baseline. The areas most affected will be central Sudan (in the area of the Gezira Scheme irrigation project and along the White and Blue Niles) and part of the River Nile state. Although these areas, particularly central Sudan, produce 75 percent of the country's wheat, they are considered marginal areas for wheat production, with the current temperatures relatively warm for wheat growing. The research challenge will be to develop appropriate wheat production technologies that will mitigate the effects of climate change.

FIGURE 10.14 Changes in mean annual precipitation in Sudan, 2000–2050, A1B scenario (millimeters)

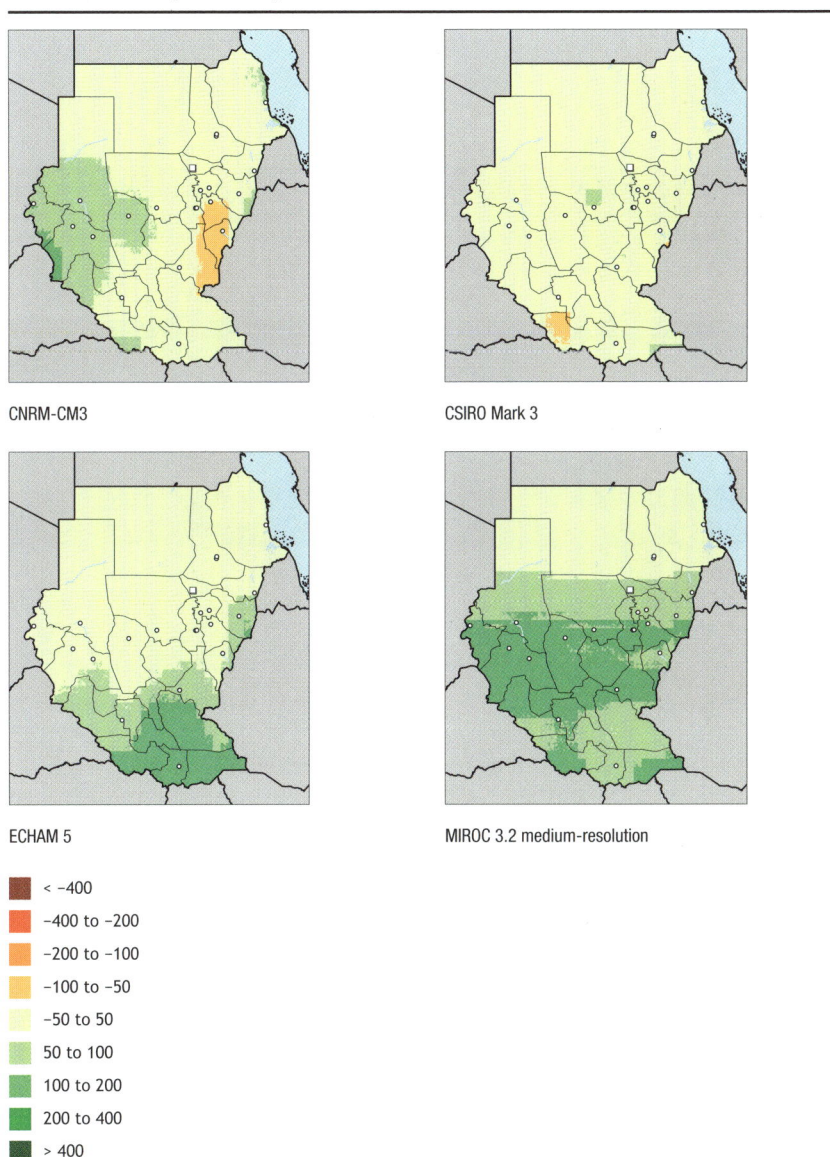

CNRM-CM3

CSIRO Mark 3

ECHAM 5

MIROC 3.2 medium-resolution

■ < –400
■ –400 to –200
■ –200 to –100
■ –100 to –50
■ –50 to 50
■ 50 to 100
■ 100 to 200
■ 200 to 400
■ > 400

Source: Authors' calculations based on Jones, Thornton, and Heinke (2009).

Notes: A1B = greenhouse gas emissions scenario that assumes fast economic growth, a population that peaks midcentury, and the development of new and efficient technologies, along with a balanced use of energy sources; CNRM-CM3 = National Meteorological Research Center–Climate Model 3; CSIRO = climate model developed at the Australia Commonwealth Scientific and Industrial Research Organisation; ECHAM 5 = fifth-generation climate model developed at the Max Planck Institute for Meteorology (Hamburg); GCM = general circulation model; MIROC = Model for Interdisciplinary Research on Climate, developed by the University of Tokyo Center for Climate System Research.

FIGURE 10.15 Changes in monthly mean maximum daily temperature in Sudan for the warmest month, 2000–2050, A1B scenario (°C)

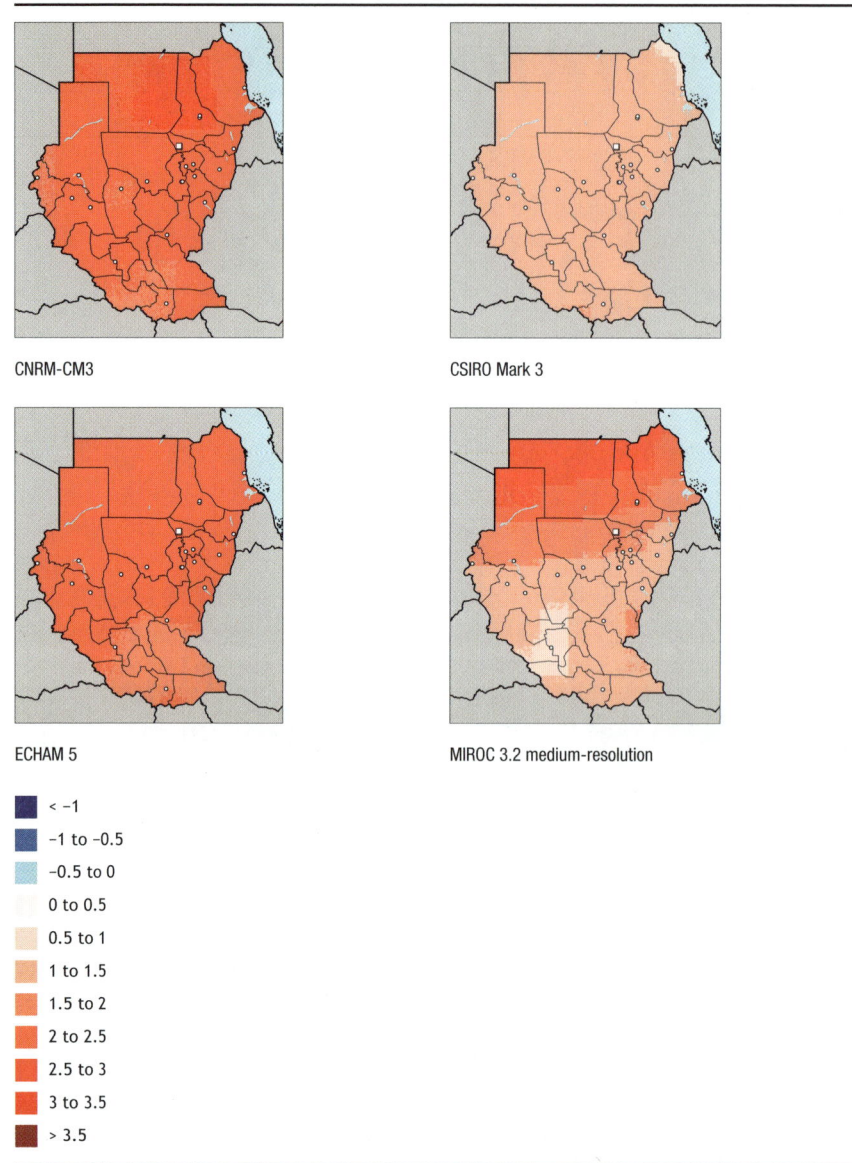

CNRM-CM3

CSIRO Mark 3

ECHAM 5

MIROC 3.2 medium-resolution

■	< −1
■	−1 to −0.5
■	−0.5 to 0
□	0 to 0.5
■	0.5 to 1
■	1 to 1.5
■	1.5 to 2
■	2 to 2.5
■	2.5 to 3
■	3 to 3.5
■	> 3.5

Source: Authors' calculations based on Jones, Thornton, and Heinke (2009).

Notes: A1B = greenhouse gas emissions scenario that assumes fast economic growth, a population that peaks midcentury, and the development of new and efficient technologies, along with a balanced use of energy sources; CNRM-CM3 = National Meteorological Research Center–Climate Model 3; CSIRO = climate model developed at the Australia Commonwealth Scientific and Industrial Research Organisation; ECHAM 5 = fifth-generation climate model developed at the Max Planck Institute for Meteorology (Hamburg); GCM = general circulation model; MIROC = Model for Interdisciplinary Research on Climate, developed by the University of Tokyo Center for Climate System Research.

FIGURE 10.16 Yield change under climate change: Rainfed sorghum in Sudan, 2000–2050, A1B scenario

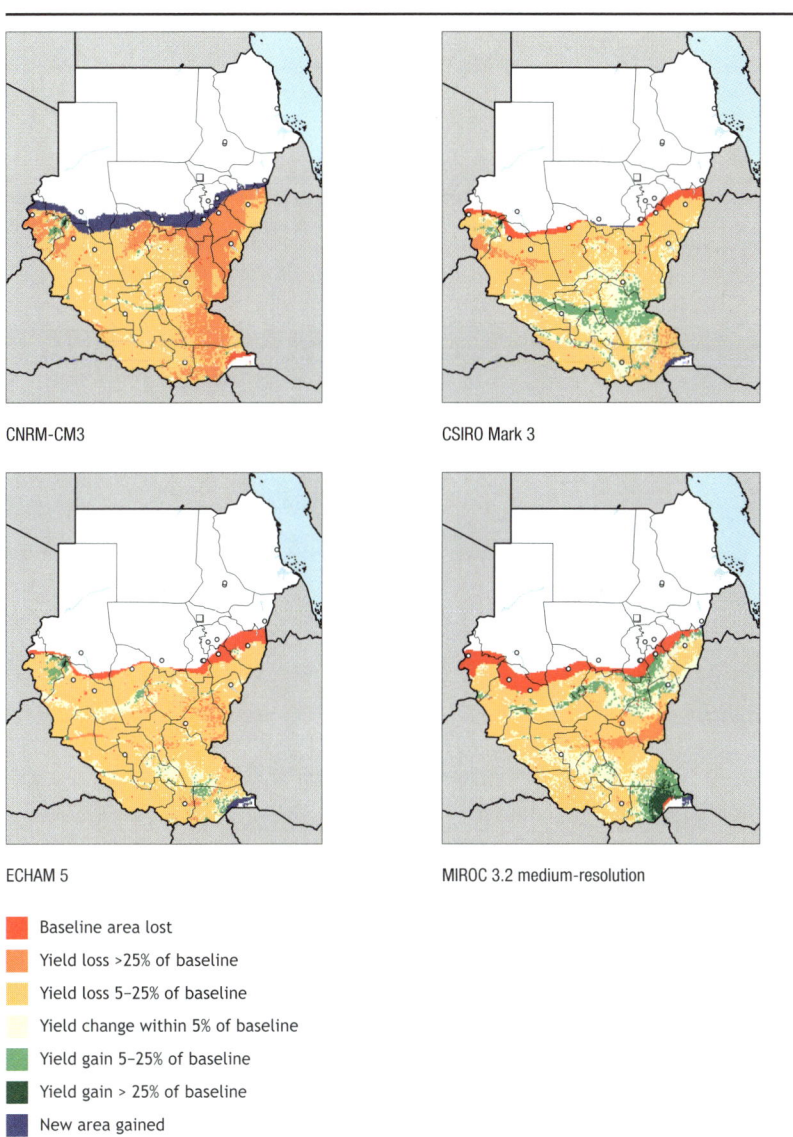

CNRM-CM3

CSIRO Mark 3

ECHAM 5

MIROC 3.2 medium-resolution

■ Baseline area lost
■ Yield loss >25% of baseline
■ Yield loss 5–25% of baseline
□ Yield change within 5% of baseline
■ Yield gain 5–25% of baseline
■ Yield gain > 25% of baseline
■ New area gained

Source: Authors' calculations.

Notes: A1B = greenhouse gas emissions scenario that assumes fast economic growth, a population that peaks midcentury, and the development of new and efficient technologies, along with a balanced use of energy sources; CNRM-CM3 = National Meteorological Research Center–Climate Model 3; CSIRO = climate model developed at the Australia Commonwealth Scientific and Industrial Research Organisation; ECHAM 5 = fifth-generation climate model developed at the Max Planck Institute for Meteorology (Hamburg); GCM = general circulation model; MIROC = Model for Interdisciplinary Research on Climate, developed by the University of Tokyo Center for Climate System Research.

FIGURE 10.17   Yield change under climate change: Irrigated wheat in Sudan, 2000–2050, A1B scenario

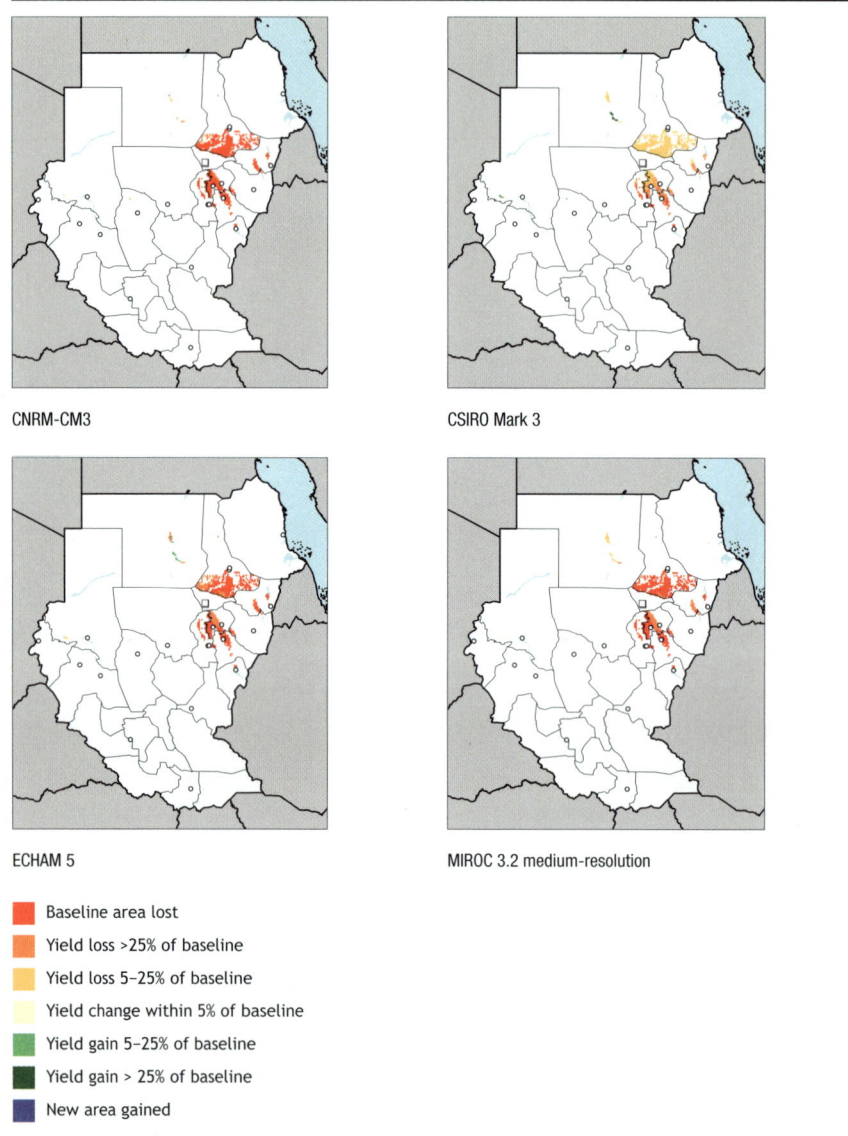

CNRM-CM3

CSIRO Mark 3

ECHAM 5

MIROC 3.2 medium-resolution

- Baseline area lost
- Yield loss >25% of baseline
- Yield loss 5–25% of baseline
- Yield change within 5% of baseline
- Yield gain 5–25% of baseline
- Yield gain > 25% of baseline
- New area gained

Source: Authors' calculations.

Notes: A1B = greenhouse gas emissions scenario that assumes fast economic growth, a population that peaks midcentury, and the development of new and efficient technologies, along with a balanced use of energy sources; CNRM-CM3 = National Meteorological Research Center–Climate Model 3; CSIRO = climate model developed at the Australia Commonwealth Scientific and Industrial Research Organisation; ECHAM 5 = fifth-generation climate model developed at the Max Planck Institute for Meteorology (Hamburg); GCM = general circulation model; MIROC = Model for Interdisciplinary Research on Climate, developed by the University of Tokyo Center for Climate System Research.

Vulnerability

Figure 10.18 shows the impact of future GDP and population scenarios on the number of malnourished children under age five in Sudan. Figure 10.19 shows the share of children who are malnourished. Figure 10.20 shows the average daily kilocalorie availability. In the optimistic scenario, lower demand for food and higher per capita GDP are favorable to consumers, and the optimistic trend accordingly shows increasing kilocalorie availability per capita and fewer malnourished children. Even the pessimistic scenario shows reasonably favorable outcomes: by 2050, there will be fewer malnourished children and greater kilocalorie consumption than in 2010. Although the number of malnourished children will rise slightly through 2030, the malnutrition rate will fall due to an increase in total population that will include an increase in the number of children under the age of five.

Agricultural Outcomes

The yield and area of sorghum will both increase, implying that total production must increase. The world price of sorghum will also rise (Figure 10.21). However, the demand for sorghum is likely to be driven mainly by population growth and

FIGURE 10.18 Number of malnourished children under five years of age in Sudan in multiple income and climate scenarios, 2010–2050

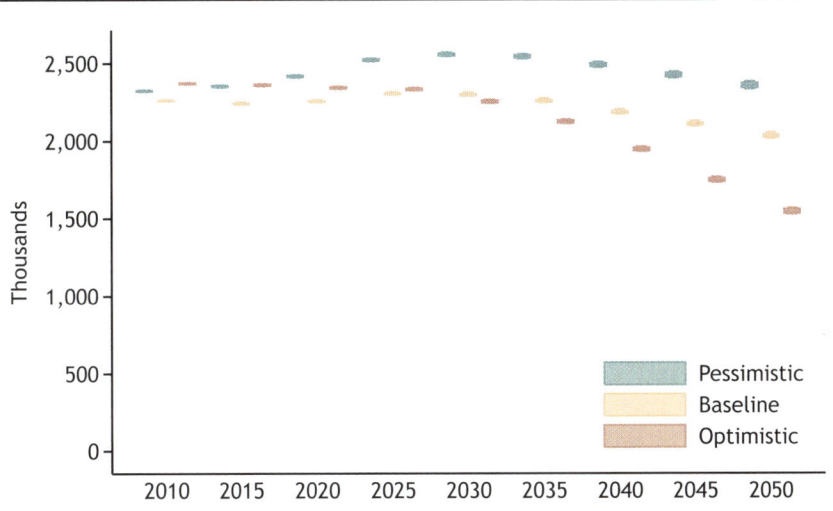

Source: Based on analysis conducted for Nelson et al. (2010).

Note: The box and whiskers plot for each socioeconomic scenario shows the range of effects from the four future climate scenarios.

FIGURE 10.19   Share of malnourished children under five years of age in Sudan in multiple income and climate scenarios, 2010–2050

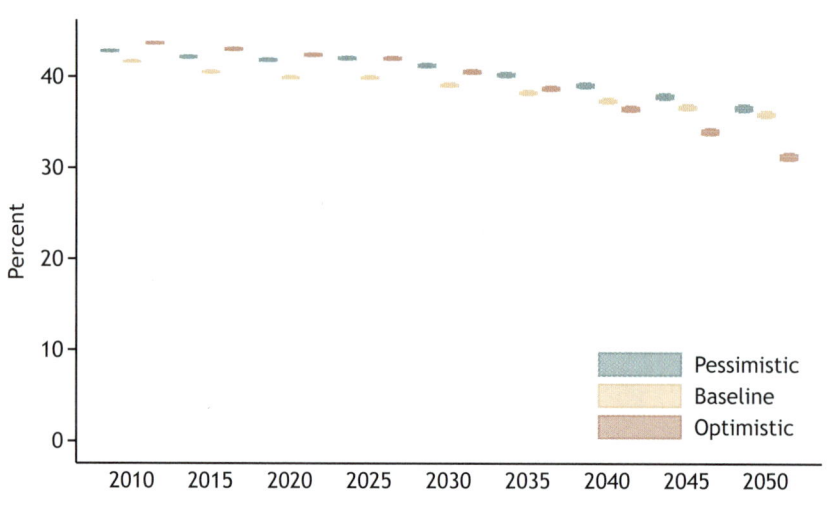

Source: Based on analysis conducted for Nelson et al. (2010).
Note: The box and whiskers plot for each socioeconomic scenario shows the range of effects from the four future climate scenarios.

FIGURE 10.20   Kilocalories per capita in Sudan in multiple income and climate scenarios, 2010–2050

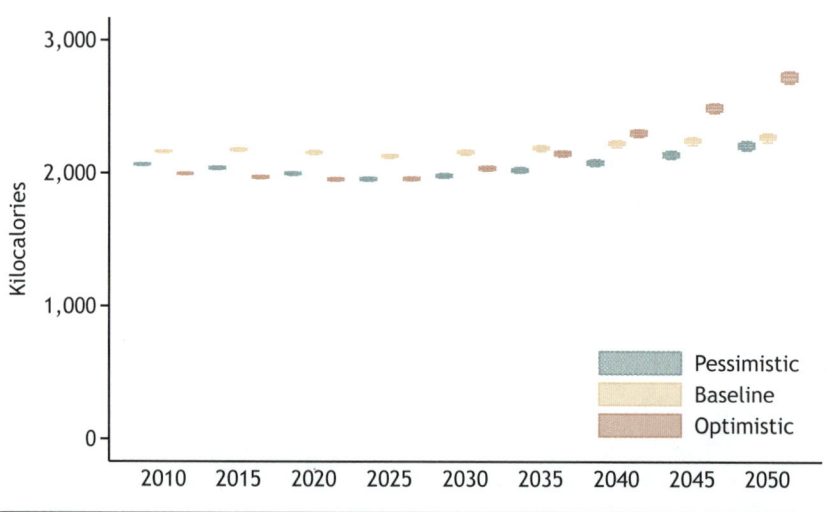

Source: Based on analysis conducted for Nelson et al. (2010).
Note: The box and whiskers plot for each socioeconomic scenario shows the range of effects from the four future climate scenarios.

FIGURE 10.21   Impact of changes in GDP and population on sorghum in Sudan, 2010–2050

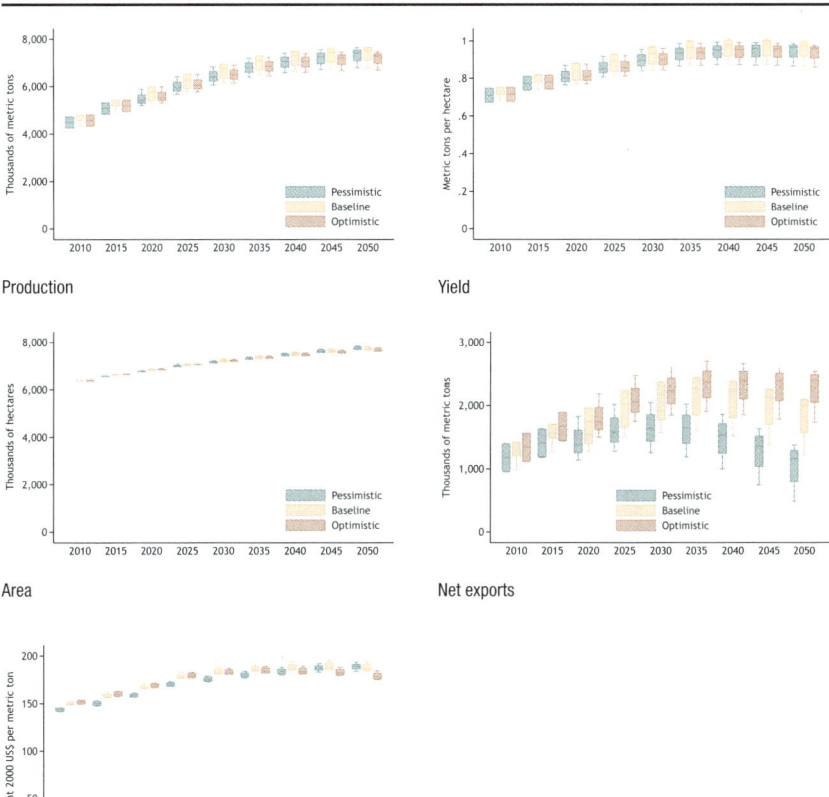

Production

Yield

Area

Net exports

Prices

Source: Based on analysis conducted for Nelson et al. (2010).

Notes: The box and whiskers plot for each socioeconomic scenario shows the range of effects from the four future climate scenarios. GDP = gross domestic product; US$ = US dollars.

internal market forces. The model shows more uncertainty about net exports of sorghum, as indicated by the lengthening of the whisker plots. This is because of variance in both production and consumption demand, which will lead to a higher variance in the difference between the two, which is net exports.

For wheat (Figure 10.22), the model has two distinct features: first, the modeled changes in production and yield show more uncertainty, and second, both the area harvested and production are shown decreasing, with area decreasing the more rapidly of the two, even though the world market price

FIGURE 10.22   Impact of changes in GDP and population on wheat in Sudan, 2010–2050

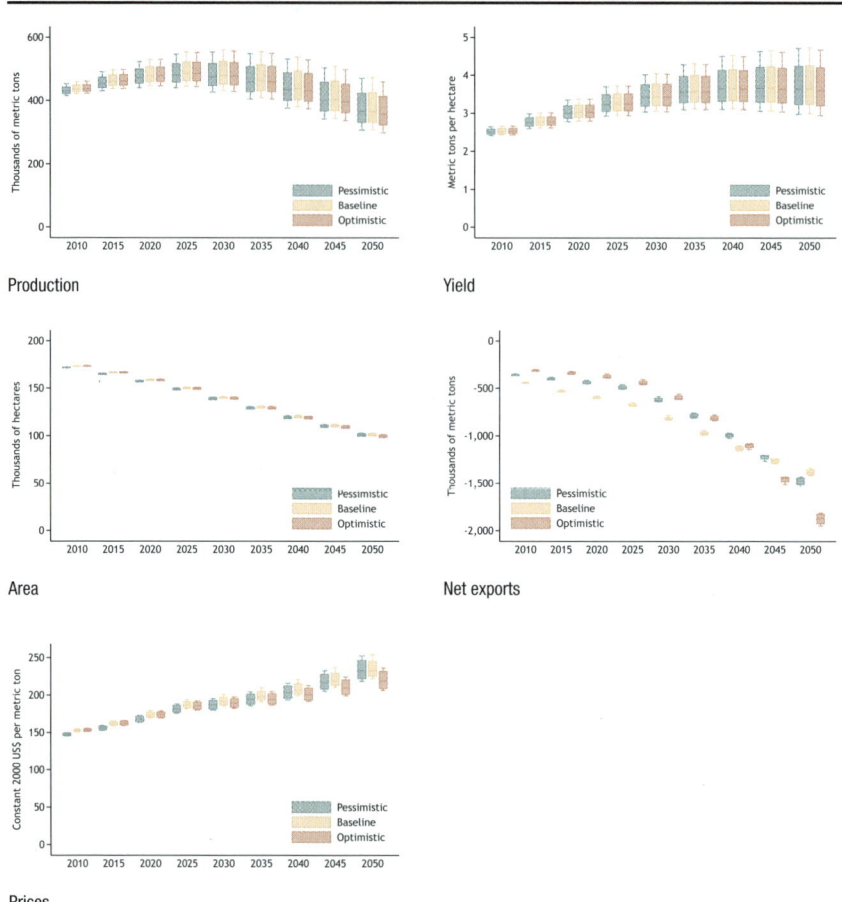

Production

Yield

Area

Net exports

Prices

Source: Based on analysis conducted for Nelson et al. (2010).

Notes: The box and whiskers plot for each socioeconomic scenario shows the range of effects from the four future climate scenarios. GDP = gross domestic product; US$ = US dollars.

will be increasing. This is a reflection of the climate model showing higher temperatures that will result in area and yield losses (as computed by the crop model) in the main wheat-producing areas of central Sudan. Sudan is likely to become more dependent on imports to meet its demand for wheat as a result of climate change over time.

FIGURE 10.23 Impact of changes in GDP and population on millet in Sudan, 2010–2050

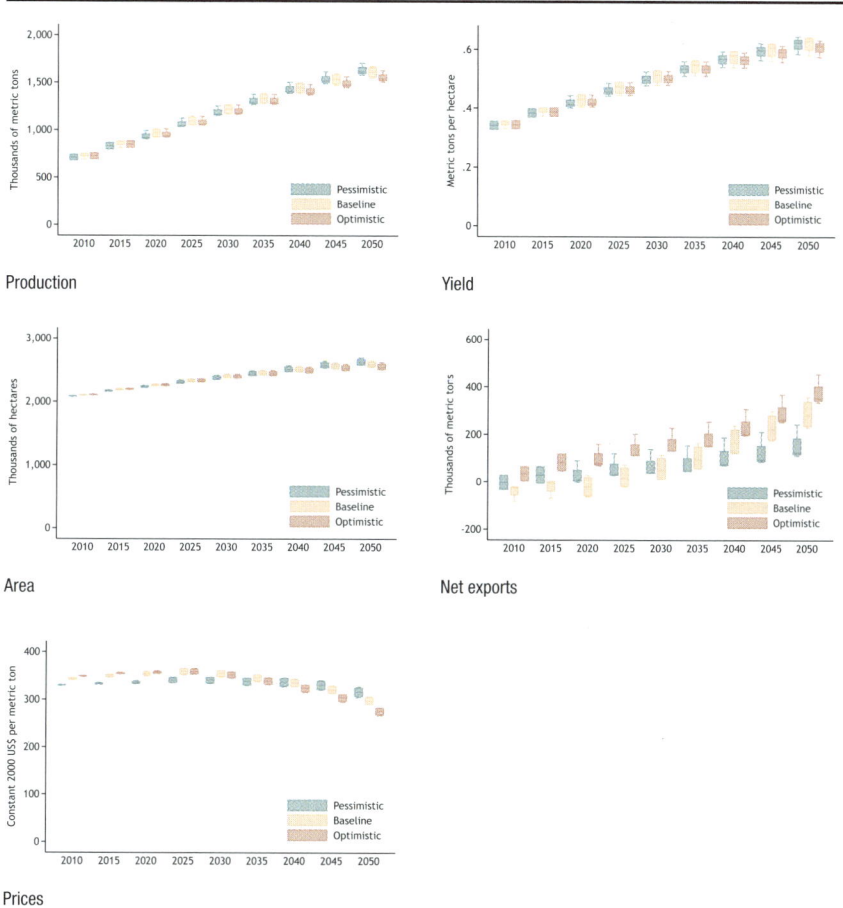

Production

Yield

Area

Net exports

Prices

Source: Based on analysis conducted for Nelson et al. (2010).
Notes: The box and whiskers plot for each socioeconomic scenario shows the range of effects from the four future climate scenarios. GDP = gross domestic product; US$ = US dollars.

In the model results for millet (Figure 10.23), yield increases by almost 70 percent, and with area increasing by more than 20 percent, production necessarily more than doubles. The production boost outpaces any increase in consumer demand, resulting in an increase in exports. The world millet price holds mostly constant through 2030, then falls off gradually.

Conclusions and Policy Recommendations

The majority of Sudan is prone to risk of climate change and indeed may already be seeing its effects through decreased and erratic rainfall patterns, extended droughts, and desertification. These events, coupled with such actions of farmers as nutrient mining of the soils and poor erosion defenses in a bid to meet their basic needs, have led to land degradation and natural destruction. Given that the country is heavily reliant on climate-dependent agriculture, appropriate coping strategies are vital, especially for the vulnerable groups who depend on agriculture for their livelihood. Suitable strategies come from understanding the form in which and the extent to which the country is exposed to climate risks. This chapter has analyzed the vulnerability of Sudan to climate change based on various economic, demographic, agricultural, and land use indicators. Several GCMs were used to assess the country's susceptibility under different scenarios.

The crop models generally indicated negative outcomes, at least for rain-fed sorghum and irrigated wheat. Although the yield losses and area losses projected for sorghum are significant, the losses for wheat appear devastating. The crop models do not allow for technological change, but the global International Model for Policy Analysis of Agricultural Commodities and Trade (IMPACT) does. With technological change, the situation for sorghum does not appear nearly as bad as it does in the crop model, and the production of sorghum appears likely to keep pace with domestic demand. This is not the case for wheat. Unless a major scientific breakthrough for wheat is forthcoming, Sudan will end up importing most of its wheat, because the temperatures from climate change are likely to become too high for successful wheat cultivation. IMPACT suggests that millet would be a very successful crop in Sudan and that exports of millet would likely rise.

Our analysis has also revealed an unclear pattern of rainfall but warmer conditions throughout the country, which could be detrimental to soil moisture and crop production. Our findings also predict future loss of yields for both irrigated wheat and sorghum from previously cultivated land. But in the case of sorghum, some of the losses may be offset by gains in yield stemming from both formerly and newly cultivated lands. Although increased production is predicted for all key crops with the exception of wheat, such gains will not be accompanied by exports to the world market, suggesting that the internal demand for food will be more than what can be supplied by the domestic markets.

Policies that promote the development and testing of new agricultural technologies (such as high-yielding varieties and hybrids that are resistant to drought and heat stress) and improved water harvesting and management techniques should be developed to enable the enhanced productivity that matches the food needs of a rising population. But development and testing of technologies is only part of the solution: an active agricultural extension service, integrated with the agricultural research institutions, will be important for communicating these developments to farmers. Establishing early warning systems responsible for the collection, sharing, and distribution of weather data in a timely manner will also moderate the impacts of climate change.

Institutional partnerships, coordination, and collaboration are required to integrate and coordinate approaches to ensure the effective long-term adaptation of strategies. It is important that programs work with vulnerable communities at the local level, applying a bottom-up approach to project planning. Such an approach can generate valuable lessons for adaptation success and also for sharing with vulnerable communities outside Sudan.

References

Balgis-Osman, E., N. G. Elhassan, H. Ahmed, and S. Zakieldin. 2005. *Sustainable Livelihood Approach for Assessing Community Resilience to Climate Change: Case Studies from Sudan.* AIACC (Assessments of Impacts and Adaptations of Climate Change) Working Paper 17. Accessed September 22, 2010. www.aiaccproject.org/working_papers/Working%20Papers/AIACC_WP_No017.pdf.

Bartholome, E., and A. S. Belward. 2005. "GLC2000: A New Approach to Global Land Cover Mapping from Earth Observation Data." *International Journal of Remote Sensing* 26 (9–10): 1959–1977.

CIESIN (Center for International Earth Science Information Network), Columbia University, IFPRI (International Food Policy Research Institute), World Bank, and CIAT (Centro Internacional de Agricultura Tropical). 2004. *Global Rural–Urban Mapping Project (GRUMP), Alpha Version: Population Density Grids.* Palisades, NY, US: Socioeconomic Data and Applications Center (SEDAC), Columbia University. http://sedac.ciesin.columbia.edu/gpw.

Cohen, M. J., C. Tirado, N.-L. Aberman, and B. Thompson. *Impact of Climate Change and Bi-energy on Nutrition.* Rome: Food and Agriculture Organization of the United Nations and Washington, DC: International Food Policy Research Institute.

Faki, H., E. M. Nur, and A. Abdelfattah. 2009. *Poverty Assessment and Mapping in the Sudan.* Wad Medani, Sudan: Agricultural Economics and Policy Research Centre, Agricultural Research Corporation, and the International Centre for Agricultural Research in the Dry Areas.

FAO (Food and Agriculture Organization of the United Nations). 2010. FAOSTAT. Rome. http://faostat.fao.org.

————. 2011. "Sudan: Country Information." Accessed August 21, 2011. www.fao.org/emergencies/country-information/list/africa/sudan/en/.

IFAD (International Fund for Agricultural Development). 2011. *Rural Poverty Report 2011*: *New Realities, New Challenges; New Opportunities for Tomorrow's Generation.* Accessed July 11, 2012. www.ifad.org/rpr2011/report/e/rpr2011.pdf.

Jones, P. G., P. K. Thornton, and J. Heinke. 2009. *Generating Characteristic Daily Weather Data Using Downscaled Climate Model Data from the IPCC's Fourth Assessment.* Project report for the International Livestock Research Institute. Geneva: International Panel on Climate Change.

Lehner, B., and P. Döll. 2004. "Development and Validation of a Global Database of Lakes, Reservoirs, and Wetlands." *Journal of Hydrology* 296 (1–4): 1–22.

Library of Congress Federal Research Division. 2004. "Country Profile: Sudan, December 2004." Accessed August 21, 2010. http://lcweb2.loc.gov/frd/cs/profiles/Sudan.pdf.

Metz, H. C., ed. 1991. *Sudan: A Country Study.* Washington, DC: Government Printing Office for the Library of Congress.

Millennium Ecosystem Assessment. 2005. *Ecosystems and Human Well-being: Synthesis.* Washington, DC: Island Press. www.maweb.org/en/Global.aspx.

Nelson, G. C., M. W. Rosegrant, A. Palazzo, I. Gray, C. Ingersoll, R. Robertson, S. Tokgoz, et al. 2010. *Food Security, Farming, and Climate Change to 2050: Scenarios, Results, Policy Options.* Washington, DC: International Food Policy Research Institute.

Republic of Sudan, CBS (Central Bureau of Statistics). 2010. *Statistical Year Book for 2009.* Khartoum. www.cbs.gov.sd/sites/default/files/Publications/Stat_book%20_2009.pdf.

Republic of Sudan, MEPD (Ministry of Environment and Physical Development), Higher Council for Environment and Natural Resources. 2003. *Sudan's First National Communications under the United Nations Framework Convention on Climate Change*, vol. 1: *Main Communication.* Khartoum.

————. 2007. *National Adaptation Programme of Action.* Khartoum.

Republic of Sudan, Ministry of Welfare and Social Security, NPC/GS (National Population Council General Secretariat). 2010. *Sudan Millennium Development Goals Progress Report.* Khartoum. www.sd.undp.org/doc/Sudan%20MDGs%20Report%202010.pdf. Accessed May 30, 2013.

SIFSIA–N/FAO (Sudan Institutional Capacity Programme: Food Security Information for Action–North Sudan/Food and Agriculture Organization of the United Nations). 2008. "Determinants of Current Food Price Hikes and Their Implications in the Northern States of Sudan." National consultancy report submitted to SIFSIA-N/FAO, Khartoum. www.fao.org/fileadmin/user_upload/sifsia/docs/Food%20Price%20Hikes%20and%20Impacts%20in%20NS%20-%20Jan%202009.pdf.

UNDP (United Nations Development Programme). 2008. *The UN Millennium Development Goals (MGDs): Status of MDGs in Sudan in 2008.* New York.

———. 2011. "Sudan Millennium Development Goals Progress Report 2010." Accessed August 20. www.sd.undp.org/doc/Sudan percent20MDGs percent20Report percent202010.pdf.

UNEP (United Nations Environment Programme). 2007. *Sudan Post-Conflict Environmental Assessment.* Accessed May 30, 2013. http://postconflict.unep.ch/publications/UNEP_Sudan .pdf.

UNEP (United Nations Environment Programme) and IUCN (International Union for the Conservation of Nature). 2009. World Database on Protected Areas (WDPA): Annual Release. No longer available online.

UNPOP (United Nations Secretariat, Department of Economic and Social Affairs, Population Division). 2009. *World Population Prospects: The 2008 Revision.* Accessed April 06, 2010. http://esa.un.org/unpp.

World Bank. 2009. *World Development Indicators.* Accessed May 2011. http://data.worldbank.org/ data-catalog/world-development-indicators.

———. 2010. *Economics of Adaptation to Climate Change: Synthesis Report.* Washington, DC. http://climatechange.worldbank.org/content/economics-adaptation-climate-change-study -homepage. Accessed July 17, 2012.

Wood, S., G. Hyman, U. Deichmann, E. Barona, R. Tenorio, Z. Guo, S. Castano, O. Rivera, E. Diaz, and J. Marin. 2010. "Sub-national Poverty Maps for the Developing World Using International Poverty Lines: Preliminary Data Release." Accessed May 6, 2010. http://povertymap.info.

You, L., and S. Wood. 2006. "An Entropy Approach to Spatial Disaggregation of Agricultural Production." *Agricultural Systems* 90 (1–3): 329–347.

You, L., S. Wood, and U. Wood-Sichra. 2006. "Generating Global Crop Distribution Maps: From Census to Grid." Paper presented at the International Association of Agricultural Economists Conference, Brisbane, Australia, August 11–18.

———. 2009. "Generating Plausible Crop Distribution and Performance Maps for Sub-Saharan Africa Using a Spatially Disaggregated Data Fusion and Optimization Approach." *Agricultural Systems* 99 (2–3): 126–140.

Zaki-Eldeen, S. A. 2009. "Adaptation to Climate Change: A Vulnerability Assessment for Sudan." *Gatekeeper* 142 (November).

Chapter 11

TANZANIA

Caroline Kilembe, Timothy S. Thomas, Michael Waithaka, Miriam Kyotalimye, and Siza Tumbo

The Government of Tanzania has formulated a number of policies, strategies, and programs guided by its *Tanzania Development Vision 2025* (URT, Planning Commission 2005), which sets long-term development goals to turn Tanzania into a middle-income country by 2025 and make it competitive in the globalized world economy. The vision articulates five attributes that Tanzania should possess by 2025: (1) high-quality livelihoods; (2) peace, stability, and unity; (3) good governance; (4) a well-educated and learning society; and (5) a competitive economy capable of producing sustainable growth and shared benefits. *Vision 2025* is meant to inspire efforts toward sustainable development in conformity with the United Nations Millennium Development Goals.

Because the Tanzanian economy depends on agriculture, all strategies for sustainable development have identified agriculture and food security as critical pillars to poverty alleviation. The following policy documents have been informed by the aspirations of *Vision 2025:* the National Adaptation Programme of Action (NAPA), the Agriculture and Livestock Policy of 1997 (currently under review), the National Food Security Policy (final draft), the Agricultural Sector Development Strategy, the Agricultural Sector Development Program, and the National Strategy for Growth and Poverty Alleviation. These policy documents recognize agriculture as the mainstay of the country's economy and acknowledge that sustainable development will require short- and long-term strategic actions to address the potential impacts of climate change on agriculture and food security. All these development strategies conform to the commitment of the Government of Tanzania to reduce the vulnerability of the agricultural and food security sectors to the adverse impacts of climate change.

NAPA was formulated in 2006. It calls for the identification of immediate and urgent climate change adaptation actions that are geared toward long-term sustainable development. This chapter complements the government's efforts in addressing climate change challenges and is intended to help researchers and policymakers to better understand and anticipate the likely impacts of climate change on agriculture and food security. Our study reviews current data on agriculture and economic development, models anticipated changes in climate between 2000 and 2050, uses crop models to assess the impact of climate changes on agricultural production, and models global supply and demand for food commodities in order to predict food price trends.

Review of the Current Situation and Trends

Economic and Demographic Indicators

Population
Figure 11.1 shows the total and rural population (left axis) and the share of the population that is urban (right axis). Table 11.1 provides additional information concerning rates of population growth. The table indicates that the total, rural, and urban populations of Tanzania are increasing but at a decreasing rate. The reasons for this general decline in growth are a reduced fertility rate (although it is still high enough to cause increases in population), mortality due to HIV/AIDS, and increased use of family planning methods among

FIGURE 11.1   Population trends in Tanzania: Total population, rural population, and percent urban, 1960–2008

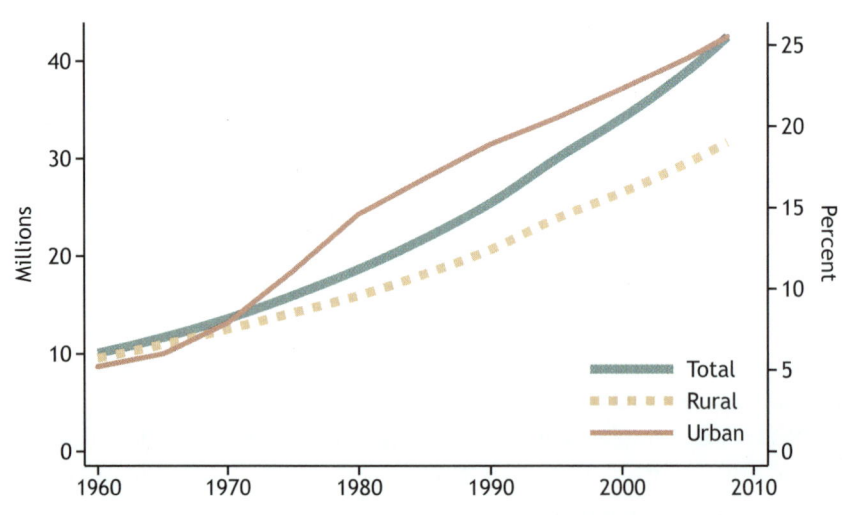

Source: *World Development Indicators* (World Bank 2009).

TABLE 11.1   Population growth rates in Tanzania, 1960–2008 (percent)

Decade	Total growth rate	Rural growth rate	Urban growth rate
1960–1969	3.0	2.7	7.1
1970–1979	3.2	2.4	9.4
1980–1989	3.1	2.6	5.7
1990–1999	3.0	2.6	4.6
2000–2008	2.7	2.2	4.4

Source: Authors' calculations based on *World Development Indicators* (World Bank 2009).

the population. The growth of the urban population is progressing at a much faster rate than is rural population growth or that of the entire country. One reason for this is weaknesses in the agricultural system, which has caused rural populations to migrate to urban centers in search of alternative means of livelihood.

Figure 11.2 shows the geographic distribution of the population in Tanzania based on census data and other sources. Populations are larger in the northern and southwestern parts of the country close to the lake regions. The elevated populations in the north are due to the fishing and mining activities as well as the influx of refugees from neighboring countries. The high distribution of people in the southwest is linked to increased opportunities for

FIGURE 11.2   Population distribution in Tanzania, 2000 (persons per square kilometer)

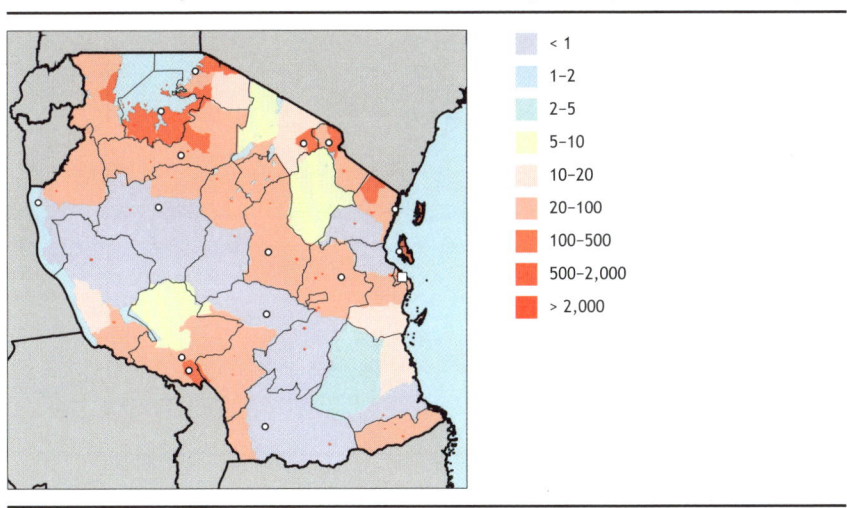

- < 1
- 1–2
- 2–5
- 5–10
- 10–20
- 20–100
- 100–500
- 500–2,000
- > 2,000

Source: CIESIN et al. (2004).

practicing pastoralism and agropastoralism (Madulu n.d.). The presence of a large urban or rural population can lead to reduced resources and the enhancement of unfavorable environmental and climatic impacts such as flooding, deforestation, and water and air pollution (Madulu 2004). Given that a majority of the Tanzanian populace is involved in agriculture, efforts to boost the rural economy amid the challenges of climate change can be helpful. Such efforts should be geared toward increasing agricultural production and productivity and improving the processing of agricultural produce to add value and thus improve the incomes of producers. Further efforts should be made to ensure market access to allow farmers to more easily engage in profitable agriculture trading.

Income

Figure 11.3 shows trends in gross domestic product (GDP) per capita and proportion of GDP from agriculture from 1988 to 2010. Agriculture is included as an indicator of its importance, as a sector that is vulnerable to climate change impacts. Per capita income has increased since 1994, while agriculture's share in GDP is decreasing, though ever so slightly.

FIGURE 11.3   Per capita GDP in Tanzania (constant 2000 US$) and share of GDP from agriculture (percent), 1988–2008

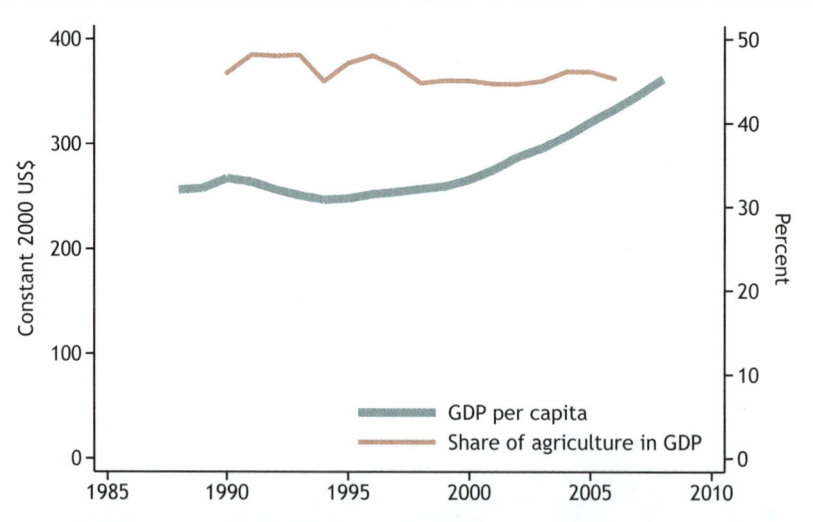

Source: *World Development Indicators* (World Bank 2009).
Note: GDP = gross domestic product; US$ = US dollars.

Vulnerability to Climate Change

Table 11.2 provides some data on indicators of a population's vulnerability and resiliency to economic shocks: education level, literacy, and concentration of labor in poorer or less dynamic sectors. The table shows that in 2007 about 70 percent of the population was engaged in agriculture activities; in 2006, almost 90 percent were engaged in vulnerable employment. An adult literacy rate of just over 70 percent suggests that the population has some skills that will help it adapt to the emerging challenges in the sector. If efforts are made to ensure that the young generation is well educated, there will likely be increased capacity to deal with the challenges of climate change to the agricultural sector. The rate of malnutrition of children under age five was about 16 percent in 2010. Low levels of adaptation to climate change would likely heighten these levels if on-farm production were adversely affected. Table 11.2 shows that the enrollment rate for primary education is significantly higher than for secondary education.

Figure 11.4 shows two noneconomic correlates of poverty, life expectancy and under-five mortality. Both under-five mortality and life expectancy at birth improved significantly in Tanzania from 1960 to 2010. However, gains in these indicators are threatened by potential changes in climate. Warmer temperatures and high levels of rainfall have been linked to the increased spread of malaria parasites and cholera (Yanda, Kangalawe, and Sigalla 2005; Traerup, Ortiz, and Markandya 2010). Malaria, for example, is a leading cause of illness and deaths in children under age five and in all persons over age five in the country (URT, Ministry of Health and Social Welfare 2008) and accounts for more than one-third of the health expenditures (Jowett and Miller 2005). Further improvements in the health indicators will contribute to growth in the country's agricultural production. This suggests a potential synergy from the integration of health and agricultural adaptations to climate change.

TABLE 11.2 Education and labor statistics for Tanzania, 1990s and 2000s

Indicator	Year	Percent
Primary school enrollment: Percent gross (three-year average)	2010	112.4
Secondary school enrollment: Percent gross (three-year average)	1999	6.1
Adult literacy rate	2007	72.3
Percent employed in agriculture	2007	72.4
Percent with vulnerable employment (in agriculture on own farm or as a day laborer)	2006	87.7
Under-five malnutrition (weight for age)	2010	15.8

Source: *World Development Indicators* (World Bank 2009).

FIGURE 11.4 Well-being indicators in Tanzania, 1960–2008

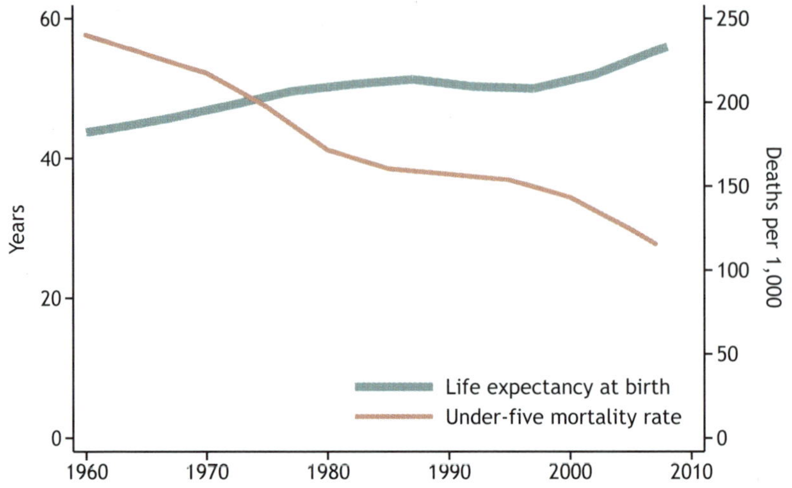

Source: *World Development Indicators* (World Bank 2009).

Figure 11.5 shows the proportion of the population living on less than $2 per day. The map seems to suggest that in many parts of the country that are primarily rural and reliant on agriculture, 95 percent of the population is living in poverty, suggesting that there are very limited household resources that could be used to adapt to climate change.

Review of Land Use, Potential, and Limitations

Land Use Overview

Figure 11.6 shows land cover and land use in Tanzania as of 2000. Most of the tree-covered areas are attractive for the expansion of agriculture, exposing them to land degradation and deforestation. This is because the agricultural system most often practices expansion through tree cutting and forest burning. Farmers need to be educated on how to protect the environment amid climate change to preserve the potential of the land for agriculture and livestock keeping. It is in fact possible that the shrub-cover areas (in northern and central Tanzania) were once covered by trees but have been degraded by agricultural and livestock-keeping practices.

Figure 11.7 shows the locations of protected areas, including parks and reserves. These locations provide important protection for fragile environments,

FIGURE 11.5   Poverty in Tanzania, circa 2005 (percentage of population below US$2 per day)

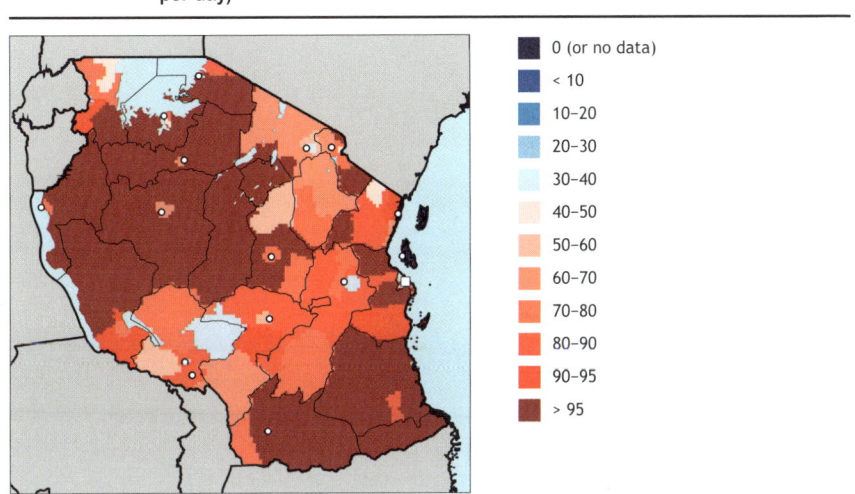

■	0 (or no data)
■	< 10
■	10–20
■	20–30
■	30–40
■	40–50
■	50–60
■	60–70
■	70–80
■	80–90
■	90–95
■	> 95

Source: Wood et al. (2010).
Note: Based on 2005 US$ (US dollars) and on purchasing power parity value.

which may also be important for the tourism industry. Incursions into reserved and protected areas expose these areas to the adverse effects of climate change.

Figure 11.8 shows travel times to urban areas of various sizes, which are potential markets for agricultural products and also serve as sources of agricultural inputs and rural consumer goods. It takes more than 26 hours for farmers in most places in Tanzania to access large markets such as Dar es Salaam, Mwanza, and Arusha, with 500,000 people or more. There are a few areas, such as Morogoro, Kilimanjaro, Tanga, Dodoma, and Kagera, whose farmers can easily access these large markets, with travel times, on average, of less than three hours due to improved trunk roads.

Regarding travel times to cities of 100,000, farmers in more areas can access regional markets in less than three hours. However, it is still difficult for those in remote areas and locations near protected areas to access the regional markets, requiring more than 26 hours due to poor infrastructure.

For towns of 25,000, more people from remote areas can access district markets in less than three hours than can get to regional markets. However, it still takes more than 26 hours from places around protected areas and from locations along swamps and wetlands to access the district markets.

For towns of 10,000 people, the figure indicates that people in almost every area can access ward markets in less than three hours, except for those in

FIGURE 11.6   Land cover and land use in Tanzania, 2000

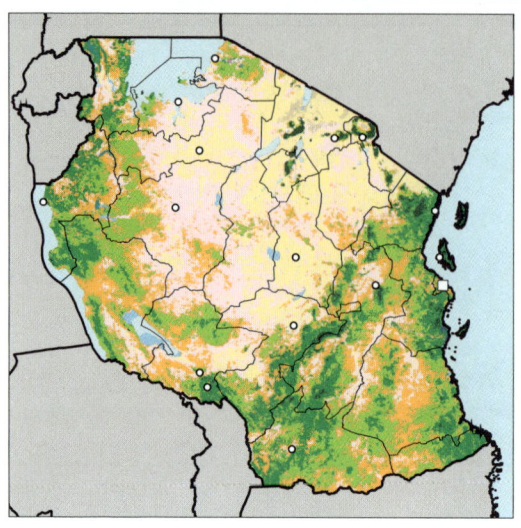

<!-- legend -->
■ Tree cover, broadleaved, evergreen

■ Tree cover, broadleaved, deciduous, closed

■ Tree cover, broadleaved, open

■ Tree cover, broadleaved, needle-leaved, evergreen

■ Tree cover, broadleaved, needle-leaved, deciduous

■ Tree cover, broadleaved, mixed leaf type

■ Tree cover, broadleaved, regularly flooded, fresh water

■ Tree cover, broadleaved, regularly flooded, saline water

■ Mosaic of tree cover/other natural vegetation

■ Tree cover, burnt

■ Shrub cover, closed-open, evergreen

■ Shrub cover, closed-open, deciduous

■ Herbaceous cover, closed-open

■ Sparse herbaceous or sparse shrub cover

■ Regularly flooded shrub or herbaceous cover

■ Cultivated and managed areas

■ Mosaic of cropland/tree cover/other natural vegetation

■ Mosaic of cropland/shrub/grass cover

■ Bare areas

■ Water bodies

■ Snow and ice

■ Artificial surfaces and associated areas

■ No data

Source: GLC2000 (Bartholome and Belward 2005).

FIGURE 11.7   Protected areas in Tanzania, 2009

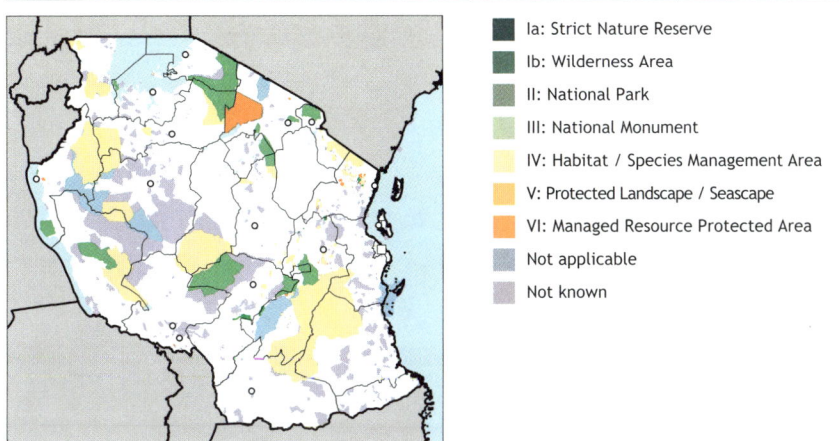

Ia: Strict Nature Reserve

Ib: Wilderness Area

II: National Park

III: National Monument

IV: Habitat / Species Management Area

V: Protected Landscape / Seascape

VI: Managed Resource Protected Area

Not applicable

Not known

Sources: Protected areas are from the World Database on Protected Areas (UNEP and IUCN 2009). Water bodies are from the World Wildlife Fund's Global Lakes and Wetlands Database (Lehner and Döll 2004).

some locations near protected areas, who still have to travel 26 hours to reach such markets.

Generally, it is easier for most agricultural areas to access small markets (wards) than big markets. Poor infrastructure reduces the profit margin for those involved in the market chain. When producers are limited to small markets, they crowd each other and therefore sell at low prices with low profit margins.

Agriculture

Table 11.3 shows key agricultural commodities in Tanzania in terms of area harvested. Maize represents almost 30 percent of the cultivated land. The next four items—sorghum, rice, beans, and cassava—together make up almost another 30 percent. The agricultural commodities ranked 8–10 in the table are locally cultivated oil crops; though important for food security, they are not included on the national food balance sheet.

Table 11.4 shows consumption of leading food commodities ranked by weight. Those crops consumed more are assumed to be more important in the national food basket. We note that cassava leads the list, representing almost one-third of the diet of a typical Tanzanian, with maize coming in second, accounting for about one-sixth of the diet. Note that the national food balance sheets, which rank food commodities in relation to staple foods of preference, exclude some of these items (fermented beverages, other

FIGURE 11.8   Travel time to urban areas of various sizes in Tanzania, circa 2000

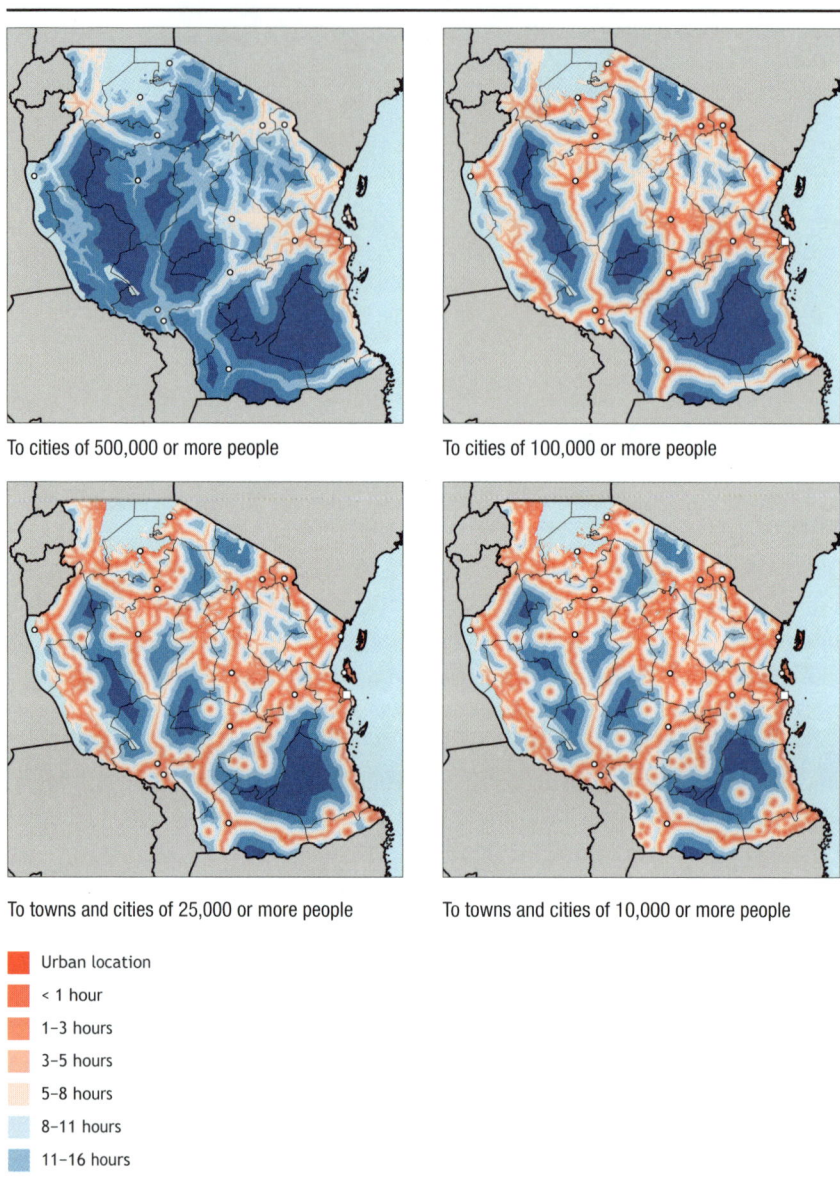

To cities of 500,000 or more people

To cities of 100,000 or more people

To towns and cities of 25,000 or more people

To towns and cities of 10,000 or more people

Urban location

< 1 hour

1–3 hours

3–5 hours

5–8 hours

8–11 hours

11–16 hours

16–26 hours

> 26 hours

Source: Authors' calculations.

TABLE 11.3   Harvest area of leading agricultural commodities in Tanzania, 2006–2008 (thousands of hectares)

Rank	Crop	Percent of total	Harvest area
	Total	100.0	10,308
1	Maize	29.8	3,067
2	Sorghum	8.7	897
3	Rice	7.0	723
4	Beans	7.0	717
5	Cassava	6.5	673
6	Bananas	4.7	480
7	Potatoes	4.9	505
8	Seed cotton	4.0	415
9	Groundnuts	4.0	413
10	Coconuts	3.0	310

Source: FAOSTAT (FAO 2010).
Note: All values are based on the three-year average for 2006–2008.

TABLE 11.4   Consumption of leading food commodities in Tanzania, 2003–2005 (thousands of metric tons)

Rank	Crop	Percent of total	Food consumption
	Total	100.0	17,283
1	Cassava	31.6	5,458
2	Maize	16.6	2,876
3	Fermented beverages	11.9	2,055
4	Potatoes	5.3	922
5	Other vegetables	5.2	894
6	Rice	4.2	725
7	Wheat	2.8	478
8	Other fruits	2.6	452
9	Plantains	2.5	427
10	Sorghum	1.9	323

Source: FAOSTAT (FAO 2010).
Note: All values are based on the three-year average for 2003–2005.

vegetables, and other fruits) (URT, Ministry of Agriculture, Food Security and Cooperatives 2006, 2007, 2008).

The maps in this section show the yield and area under cultivation of various crops of importance. Figure 11.9 shows rainfed maize production. The estimated yield for the year 2000 was on average between 0.5 and 2 tons per hectare on farms that ranged from 0.5 hectare to more than 100 hectares. Maize is widely grown throughout the country.

Figures 11.10–11.13 show the yield and cultivated area of cassava, sorghum, rice, and beans, respectively. Unlike maize, all of these four crops are limited in the geographic areas where they are cultivated, as shown by their distribution in the maps. Cassava is mostly cultivated in the southern and northwestern parts of the country. Sorghum is cultivated mostly along the corridor that stretches from Mwanza to Pwani Province. Rice and beans are cultivated in much smaller proportions in Mware, Mwanza, Kagera, and Tanga Provinces.

Production of major foodcrops (maize, sorghum, cassava, rice, potatoes, and bananas) increased from 10.6 million tons in 1991 to 14.6 million tons in 2009 (URT, Ministry of Agriculture, Food Security and Cooperatives

FIGURE 11.9   Yield (metric tons per hectare) and harvest area density (hectares) for rainfed maize in Tanzania, 2000

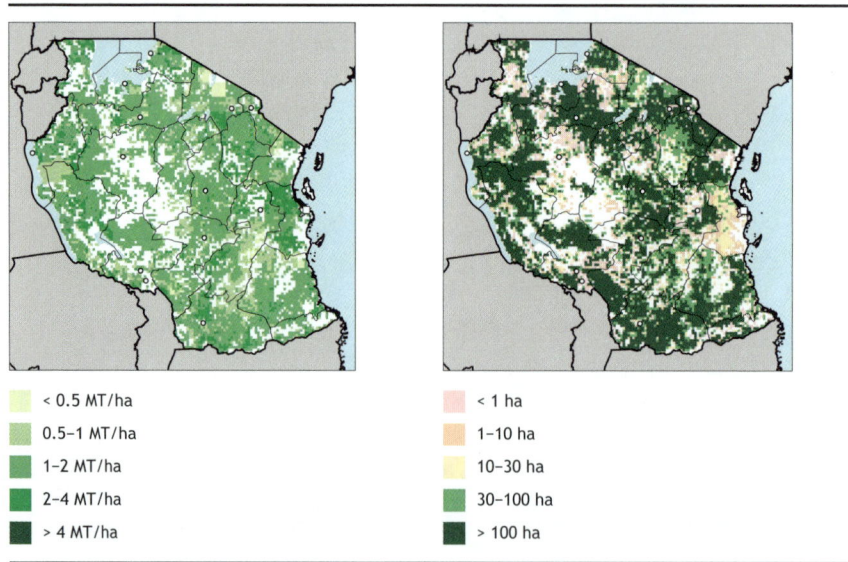

< 0.5 MT/ha	< 1 ha
0.5–1 MT/ha	1–10 ha
1–2 MT/ha	10–30 ha
2–4 MT/ha	30–100 ha
> 4 MT/ha	> 100 ha

Source: SPAM (Spatial Production Allocation Model) (You and Wood 2006; You, Wood, and Wood-Sichra 2006, 2009).
Note: ha = hectare; MT/ha = metric tons per hectare.

FIGURE 11.10   Yield (metric tons per hectare) and harvest area density (hectares) for rainfed cassava in Tanzania, 2000

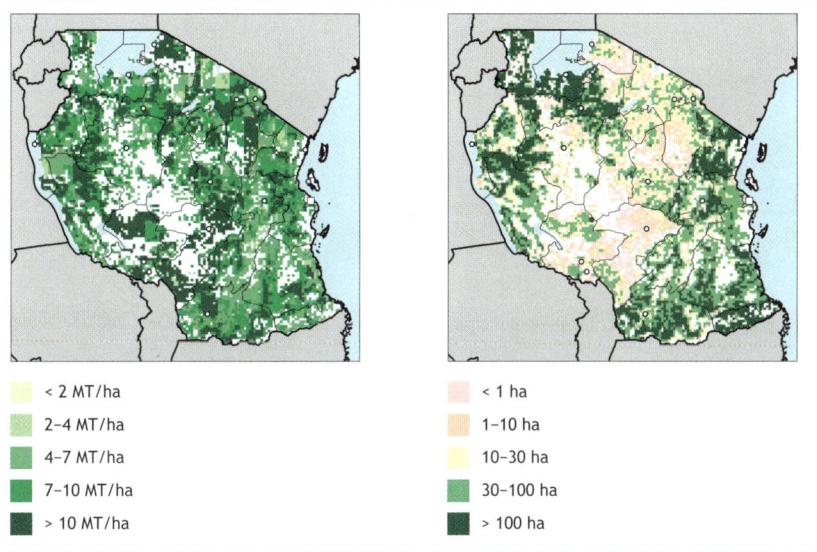

< 2 MT/ha	< 1 ha
2–4 MT/ha	1–10 ha
4–7 MT/ha	10–30 ha
7–10 MT/ha	30–100 ha
> 10 MT/ha	> 100 ha

Source: SPAM (Spatial Production Allocation Model) (You and Wood 2006; You, Wood, and Wood-Sichra 2006, 2009).
Note: ha = hectare; MT/ha = metric tons per hectare.

FIGURE 11.11   Yield (metric tons per hectare) and harvest area density (hectares) for rainfed sorghum in Tanzania, 2000

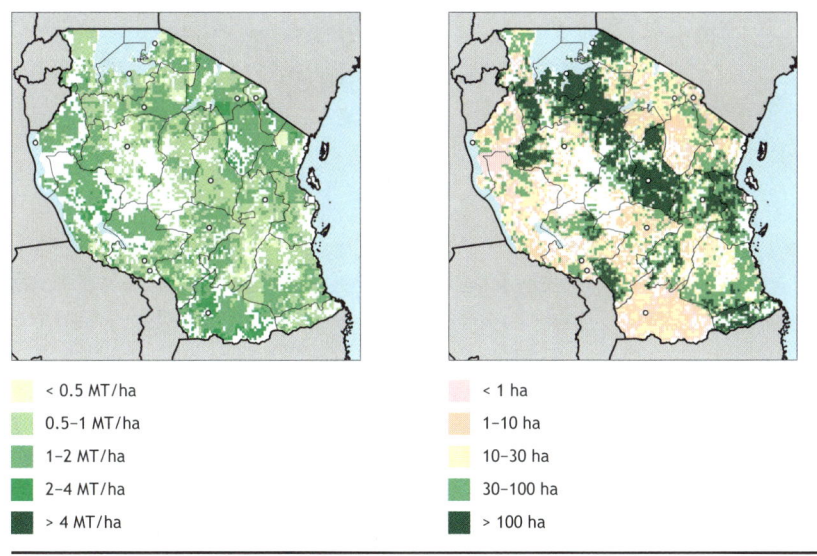

< 0.5 MT/ha	< 1 ha
0.5–1 MT/ha	1–10 ha
1–2 MT/ha	10–30 ha
2–4 MT/ha	30–100 ha
> 4 MT/ha	> 100 ha

Source: SPAM (Spatial Production Allocation Model) (You and Wood 2006; You, Wood, and Wood-Sichra 2006, 2009).
Note: ha = hectare; MT/ha = metric tons per hectare.

FIGURE 11.12   Yield (metric tons per hectare) and harvest area density (hectares) for rainfed rice in Tanzania, 2000

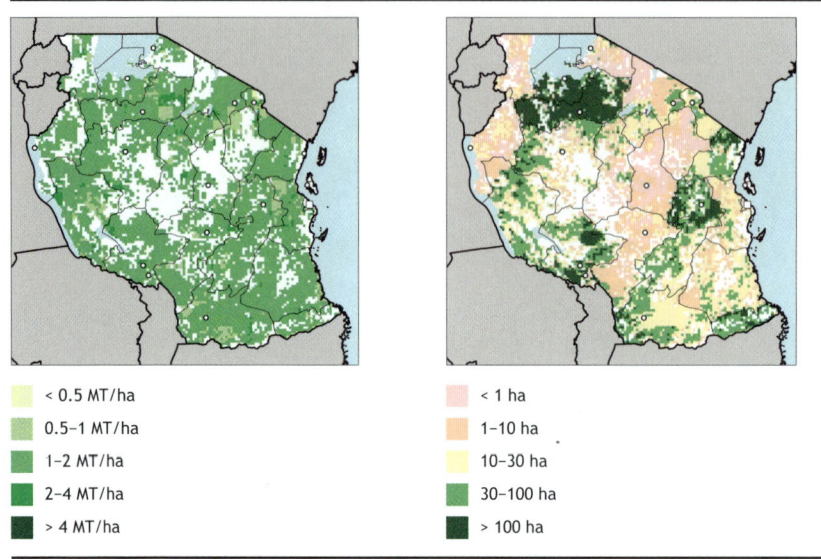

▢ < 0.5 MT/ha	▢ < 1 ha
▢ 0.5–1 MT/ha	▢ 1–10 ha
▢ 1–2 MT/ha	▢ 10–30 ha
▢ 2–4 MT/ha	▢ 30–100 ha
▢ > 4 MT/ha	▢ > 100 ha

Source: SPAM (Spatial Production Allocation Model) (You and Wood 2006; You, Wood, and Wood-Sichra 2006, 2009).
Note: ha = hectare; MT/ha = metric tons per hectare.

FIGURE 11.13   Yield (metric tons per hectare) and harvest area density (hectares) for rainfed beans in Tanzania, 2000

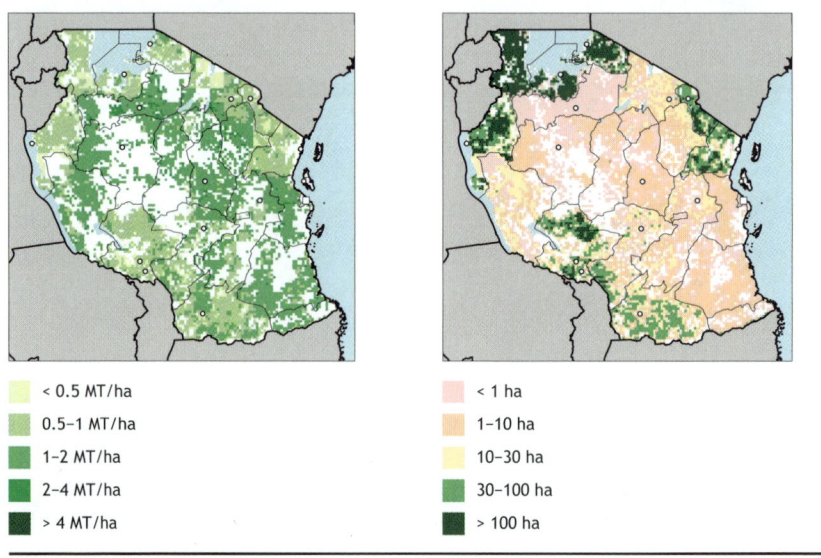

▢ < 0.5 MT/ha	▢ < 1 ha
▢ 0.5–1 MT/ha	▢ 1–10 ha
▢ 1–2 MT/ha	▢ 10–30 ha
▢ 2–4 MT/ha	▢ 30–100 ha
▢ > 4 MT/ha	▢ > 100 ha

Source: SPAM (Spatial Production Allocation Model) (You and Wood 2006; You, Wood, and Wood-Sichra 2006, 2009).
Note: ha = hectare; MT/ha = metric tons per hectare.

2010). The increase in production seems to have been driven by an increase in area cultivated rather than an increase in yield. The cultivated areas for major foodcrops increased from 3,300,759 hectares in 1991 to 6,526,531 hectares in 2009, but the productivity of the same crops *decreased* from 3.21 tons per hectare in 1991 to 2.24 tons per hectare in 2009. It was expected that both production and productivity might increase by 2010 due to government initiatives providing subsidies for inputs, such as fertilizers, seeds, and agrochemicals, to targeted smallholder farmers.

Scenarios for the Future

Economic and Demographic Indicators

Population
Figure 11.14 shows population projections for Tanzania from 2010 through 2050. Under the low variant the population level is expected to reach 95 million, more than double the population of 2010, while under the medium variant and high variant it is expected to reach 110 million and 125 million persons, respectively. A high rate of population growth can create challenges to food security by increasing population pressure on land and competition for resources.

FIGURE 11.14   Population projections for Tanzania, 2010–2050

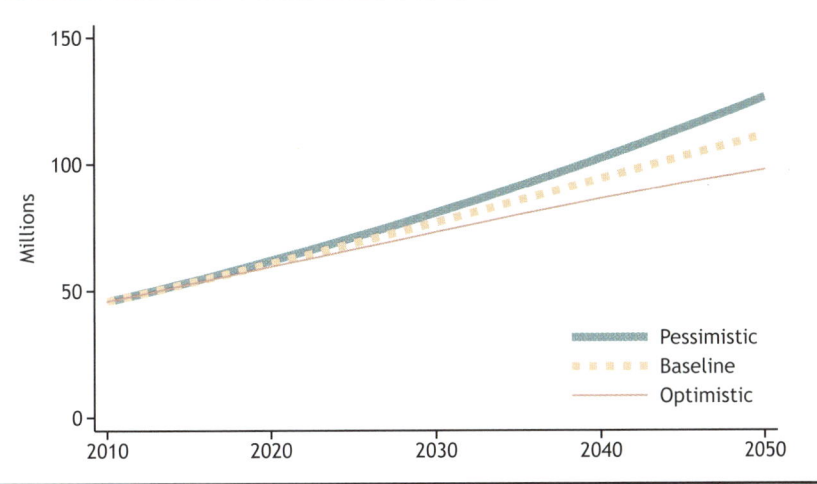

Source: UNPOP (2009).

FIGURE 11.15   Gross domestic product (GDP) per capita in Tanzania, future scenarios, 2010–2050

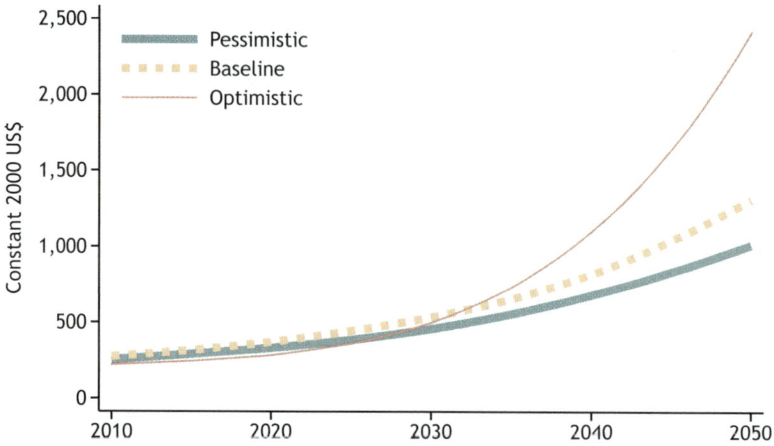

Sources: Computed from GDP data from the World Bank Economic Adaptation to Climate Change project (World Bank 2010), from the Millennium Ecosystem Assessment (2005) reports, and from population data from the United Nations (UNPOP 2009).
Note: US$ = US dollars.

Income

Figure 11.15 shows the three scenarios of GDP per capita used for this study. These are the results of combining three GDP projections with the three population projections of Figure 11.14, from the United Nations Population Division. The optimistic scenario combines high GDP with a low rate of population growth. The baseline scenario combines the medium GDP projection with the medium population projection. Finally, the pessimistic scenario combines the low GDP projection with the high population projection.

The graph shows an increase in per capita GDP in all scenarios; the best outcome is in the optimistic scenario, which predicts that per capita GDP will be $2,400 by 2050, though the pessimistic scenario projects a much lower $1,000 in 2050, which is still roughly four times larger than what it was in 2010. Overall, real GDP growth has averaged 6 percent in the past seven years as a result of economic reforms established to improve the country's performance. Agriculture (26.6 percent), industry (22 percent), and trade in exports ($3.8 billion) are the major contributors.[1] However, this contribution has not translated into improvement in the standard of living of ordinary Tanzanians. The country remains heavily dependent on foreign aid, with 30 percent of the budget financed

---

1   2010 estimate.

through donor aid (USDoS 2011). According to one group of analysts, although the per capita GDP at constant prices is predicted to rise (Trading Economics n.d.), such an increase will not be rapid given the country's economic situation and rising population. They suggest that future trends may be characterized by the pessimistic scenario. Improvements in per capita GDP would help increase the country's resilience to climate change shocks, because more income would enable families to access health services and food supplies.

Biophysical Analysis

Climate Models

Four general circulation models (GCMs) were used to model changes in average precipitation and temperatures in Tanzania between 2000 and 2050. Figure 11.16 shows projected precipitation changes under the four downscaled climate models used in this chapter with the A1B scenario.[2] There seem to be noteworthy differences among the models. Of the four, the MIROC 3.2 model is the wettest, with a median increase in precipitation of around 200 millimeters per year and some areas receiving up to 300 millimeters more per year. The ECHAM 5 model projects that most of Tanzania will not have any significant change in rainfall except on and around Lake Victoria, where rainfall is projected to increase by between 100 and 200 millimeters per year. The CRNM-CM3 model predicts the opposite for the area around Lake Victoria, with rainfall dropping by around 100 millimeters per year. The CNRM-CM3 model predicts an increase of 50–100 millimeters per year for the southern half of the country.

Figure 11.17 shows changes in normal mean daily maximum temperature for the month with the warmest mean daily maximum temperature. All maps show higher maximum temperatures, but there are key differences among the projections of the different models. The CNRM-CM3 and ECHAM 5 models have similar projections, with a median increase of 2.1°C. The CSIRO Mark 3 and MIROC 3.2 models also exhibit similarities to one another, with median temperature increases of around 1.0°C. The MIROC 3.2 model seems to exhibit the most spatial variability, with an increase of less than 0.5°C near Lake Victoria,

---

2 CNRM-CM3 is National Meteorological Research Center–Climate Model 3. MIROC 3.2 is the Model for Interdisciplinary Research on Climate, developed at the University of Tokyo Center for Climate System Research. CSIRO Mark 3 is a climate model developed at the Australia Commonwealth Scientific and Industrial Research Organisation. ECHAM 5 is a fifth-generation climate model developed at the Max Planck Institute for Meteorology in Hamburg. The A1B scenario is a greenhouse gas emissions scenario that assumes fast economic growth, a population that peaks midcentury, and the development of new and efficient technologies, along with a balanced use of energy sources.

FIGURE 11.16   Changes in mean annual precipitation in Tanzania, 2000–2050, A1B scenario (millimeters)

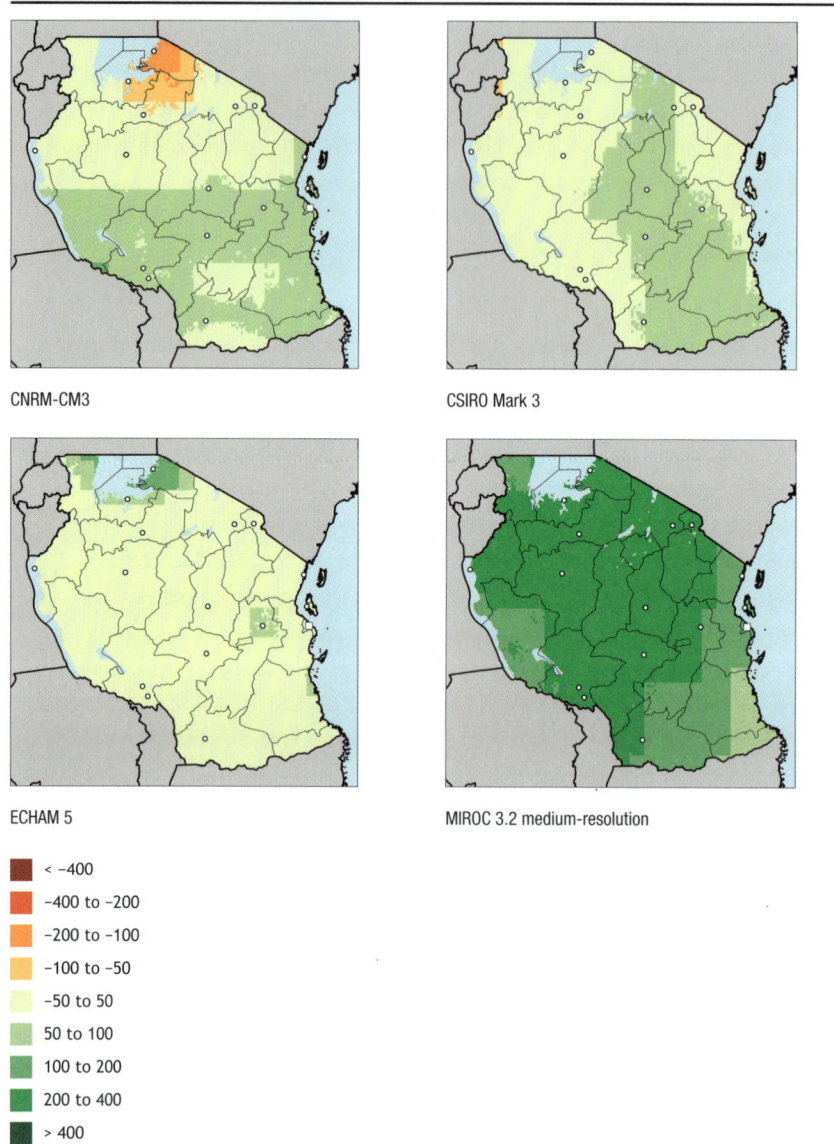

CNRM-CM3

CSIRO Mark 3

ECHAM 5

MIROC 3.2 medium-resolution

- < –400
- –400 to –200
- –200 to –100
- –100 to –50
- –50 to 50
- 50 to 100
- 100 to 200
- 200 to 400
- > 400

Source: Authors' calculations based on Jones, Thornton, and Heinke (2009).

Notes: A1B = greenhouse gas emissions scenario that assumes fast economic growth, a population that peaks midcentury, and the development of new and efficient technologies, along with a balanced use of energy sources; CNRM-CM3 = National Meteorological Research Center–Climate Model 3; CSIRO = climate model developed at the Australia Commonwealth Scientific and Industrial Research Organisation; ECHAM 5 = fifth-generation climate model developed at the Max Planck Institute for Meteorology (Hamburg); GCM = general circulation model; MIROC = Model for Interdisciplinary Research on Climate, developed by the University of Tokyo Center for Climate System Research.

FIGURE 11.17   Changes in monthly mean maximum daily temperature in Tanzania for the warmest month, 2000–2050, A1B scenario (°C)

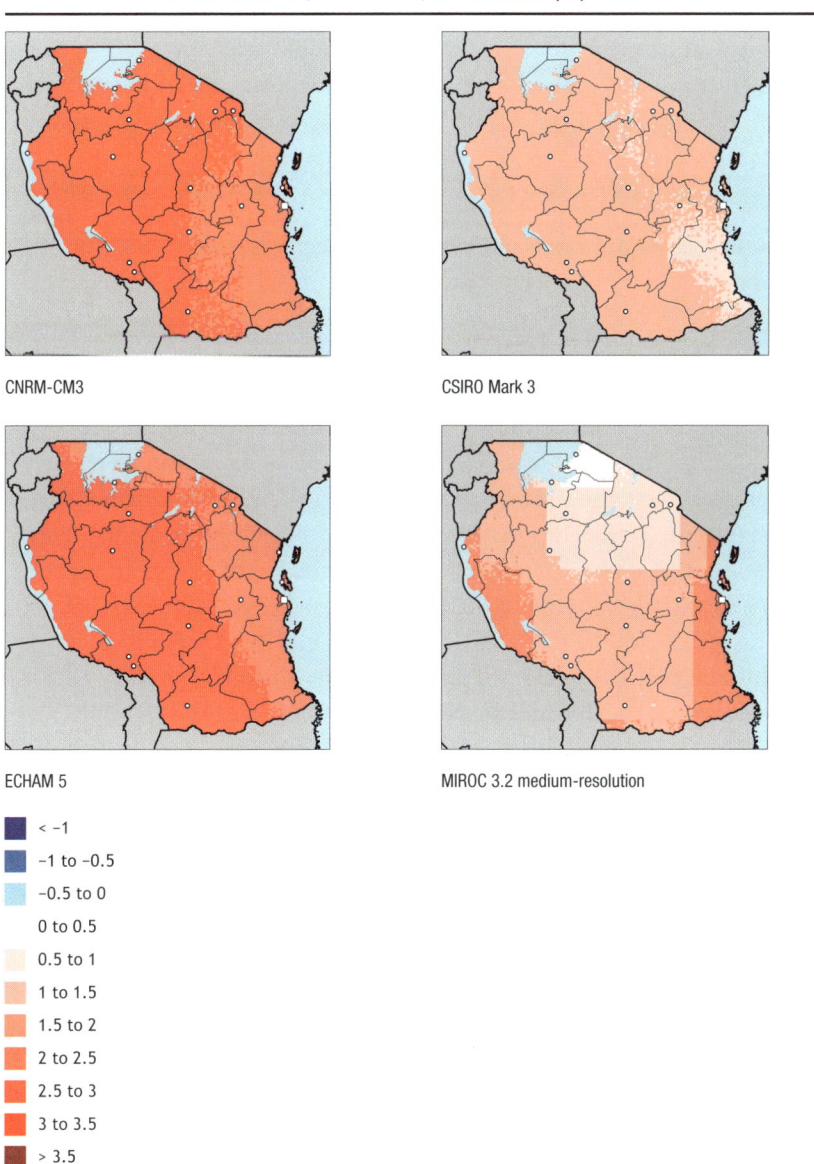

CNRM-CM3

CSIRO Mark 3

ECHAM 5

MIROC 3.2 medium-resolution

- < –1
- –1 to –0.5
- –0.5 to 0
- 0 to 0.5
- 0.5 to 1
- 1 to 1.5
- 1.5 to 2
- 2 to 2.5
- 2.5 to 3
- 3 to 3.5
- > 3.5

Source: Authors' calculations based on Jones, Thornton, and Heinke (2009).

Notes: A1B = greenhouse gas emissions scenario that assumes fast economic growth, a population that peaks midcentury, and the development of new and efficient technologies, along with a balanced use of energy sources; CNRM-CM3 = National Meteorological Research Center–Climate Model 3; CSIRO = climate model developed at the Australia Commonwealth Scientific and Industrial Research Organisation; ECHAM 5 = fifth-generation climate model developed at the Max Planck Institute for Meteorology (Hamburg); GCM = general circulation model; MIROC = Model for Interdisciplinary Research on Climate, developed by the University of Tokyo Center for Climate System Research.

which increases as one travels southward, reaching as high as a 1.5°C increase. Then, as one moves to either the eastern or the western border of the country, the temperature increase rises even more, to 2°C. Significantly higher temperatures could have consequences for agricultural productivity due to an increased spread of crop pests and diseases and also human diseases given that warmer temperatures favor their multiplication and transmission.

Crop Models

The Decision Support Software for Agrotechnology Transfer (DSSAT) software was used to compute yields for the climates of 2010 and 2050, and then the yields for each were compared. Yield changes for key crops from the DSSAT model are mapped in Figures 11.18–11.20.

The results in Figure 11.18, for maize, show vast differences in results depending on the GCM and the location. In the CNRM-CM3 and the MIROC 3.2 results, many areas show gains in yields of more than 25 percent, though both of them show areas in which yields decline by more than 25 percent. By contrast, the CSIRO Mark 3 results show mostly more moderate results of less than a 25 percent increase or decrease. For the southern part of Tanzania, the ECHAM 5 results predict widespread losses, though the other models disagree, with the CNRM-CM3 and the CSIRO Mark 3 predicting that yields in most of the areas will actually gain in southern Tanzania. All of the models have particular areas in which major yield reductions take place. The CNRM-CM3 model predicts losses in the northern part of the country.

Without significant geographic agreement across models, it is not possible to focus on particular strategies at this point. The results do suggest that it would be helpful to prepare a number of different responses from which farmers could choose as appropriate. It is also helpful to note that although there can be negative consequences of climate change in some places, there may be positive outcomes in other places, and it would be wise to capitalize on potential improvements as well as to prepare for potential declines.

The results in Figure 11.19 show a much more dire picture for sorghum, with much of the country experiencing losses in yields of from 5 percent to more than 25 percent of the baseline; this prediction includes the main sorghum-growing areas in the country. However, all four models show at least a hint of yield increase in provinces along the border with Kenya. Indeed, the MIROC 3.2 and ECHAM 5 results show much more than a hint; the MIROC 3.2 results show a large area with gains of more than 25 percent. In the MIROC 3.2 results there is also an area of significant loss along the southern coast of the Indian Ocean and inland to Morogoro Region.

FIGURE 11.18   Yield change under climate change: Rainfed maize in Tanzania, 2000–2050, A1B scenario

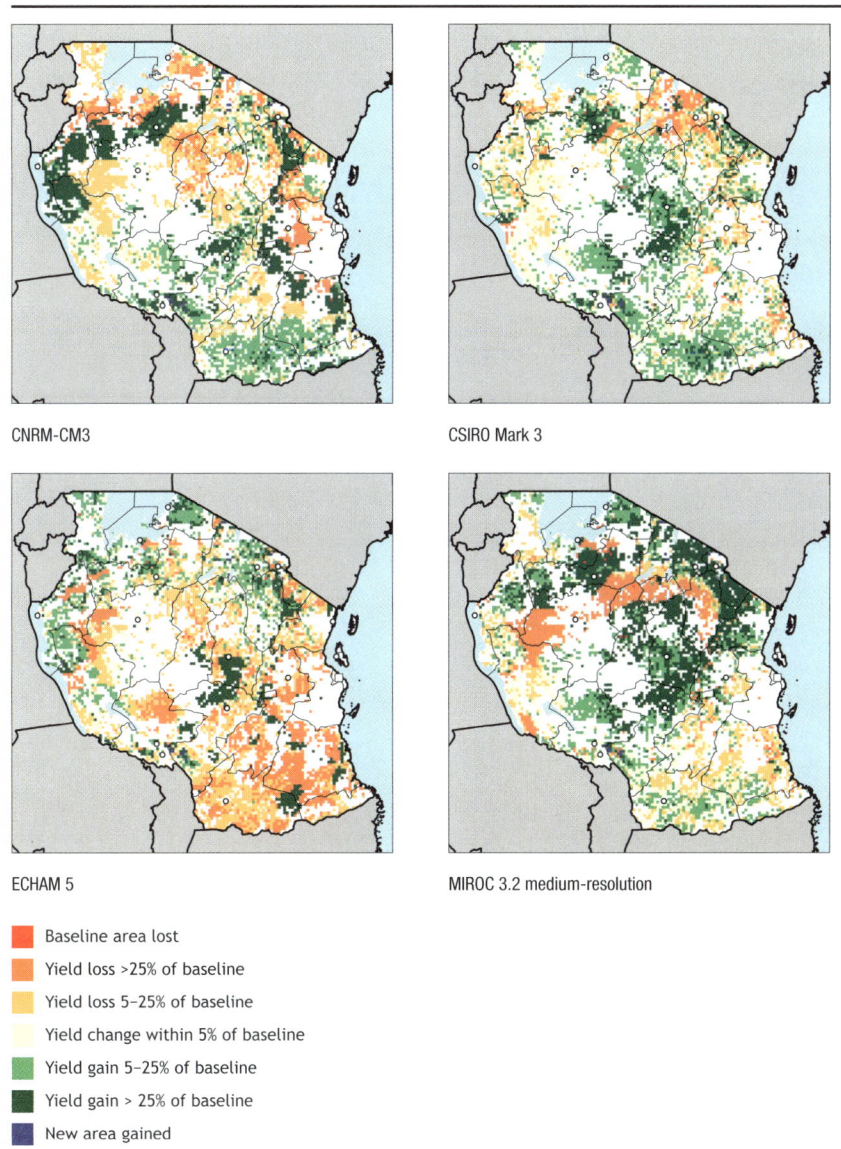

CNRM-CM3

CSIRO Mark 3

ECHAM 5

MIROC 3.2 medium-resolution

■ Baseline area lost
■ Yield loss >25% of baseline
■ Yield loss 5–25% of baseline
□ Yield change within 5% of baseline
■ Yield gain 5–25% of baseline
■ Yield gain > 25% of baseline
■ New area gained

Source: Authors' calculations.

Notes: A1B = greenhouse gas emissions scenario that assumes fast economic growth, a population that peaks midcentury, and the development of new and efficient technologies, along with a balanced use of energy sources; CNRM-CM3 = National Meteorological Research Center–Climate Model 3; CSIRO = climate model developed at the Australia Commonwealth Scientific and Industrial Research Organisation; ECHAM 5 = fifth-generation climate model developed at the Max Planck Institute for Meteorology (Hamburg); GCM = general circulation model; MIROC = Model for Interdisciplinary Research on Climate, developed by the University of Tokyo Center for Climate System Research.

FIGURE 11.19   Yield change under climate change: Rainfed sorghum in Tanzania, 2000–2050, A1B scenario

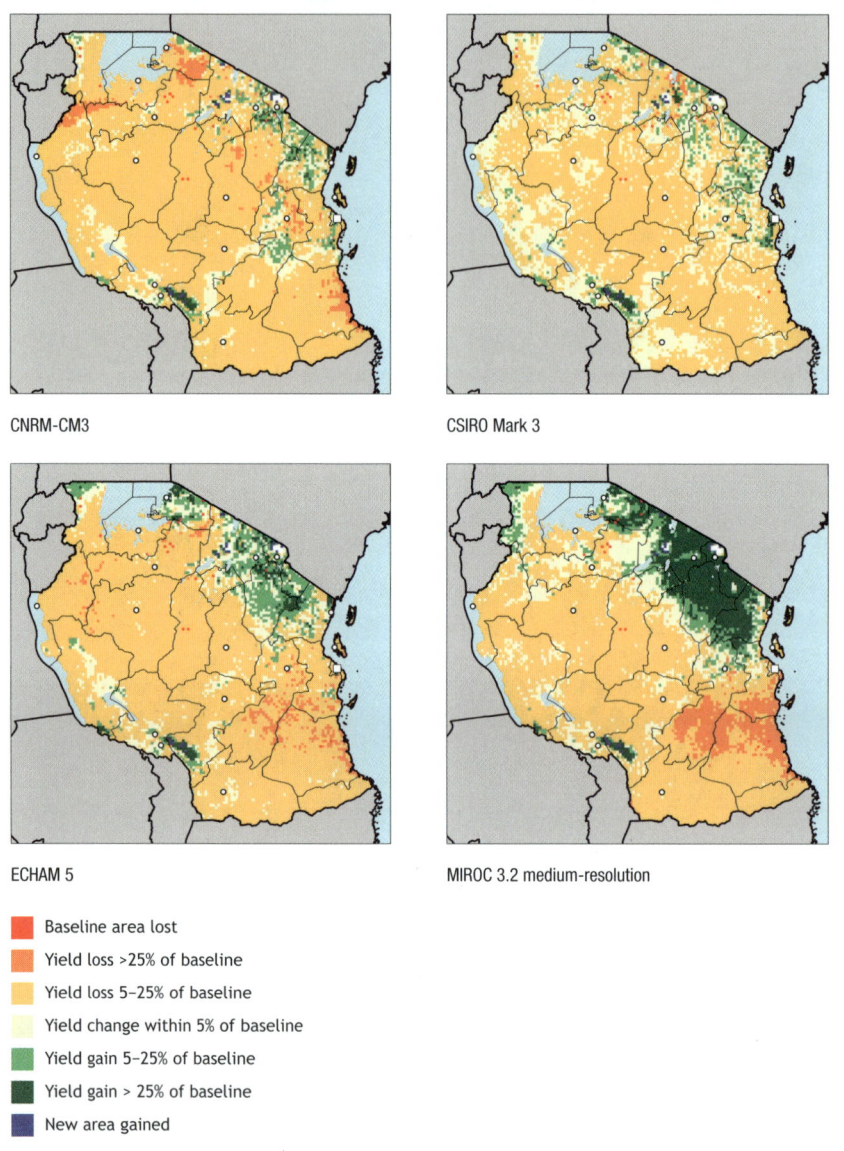

CNRM-CM3

CSIRO Mark 3

ECHAM 5

MIROC 3.2 medium-resolution

■ Baseline area lost
■ Yield loss >25% of baseline
■ Yield loss 5–25% of baseline
□ Yield change within 5% of baseline
■ Yield gain 5–25% of baseline
■ Yield gain > 25% of baseline
■ New area gained

Source: Authors' calculations.

Notes: A1B = greenhouse gas emissions scenario that assumes fast economic growth, a population that peaks midcentury, and the development of new and efficient technologies, along with a balanced use of energy sources; CNRM-CM3 = National Meteorological Research Center–Climate Model 3; CSIRO = climate model developed at the Australia Commonwealth Scientific and Industrial Research Organisation; ECHAM 5 = fifth-generation climate model developed at the Max Planck Institute for Meteorology (Hamburg); GCM = general circulation model; MIROC = Model for Interdisciplinary Research on Climate, developed by the University of Tokyo Center for Climate System Research.

FIGURE 11.20   Yield change under climate change: Rainfed rice in Tanzania, 2000–2050, A1B scenario

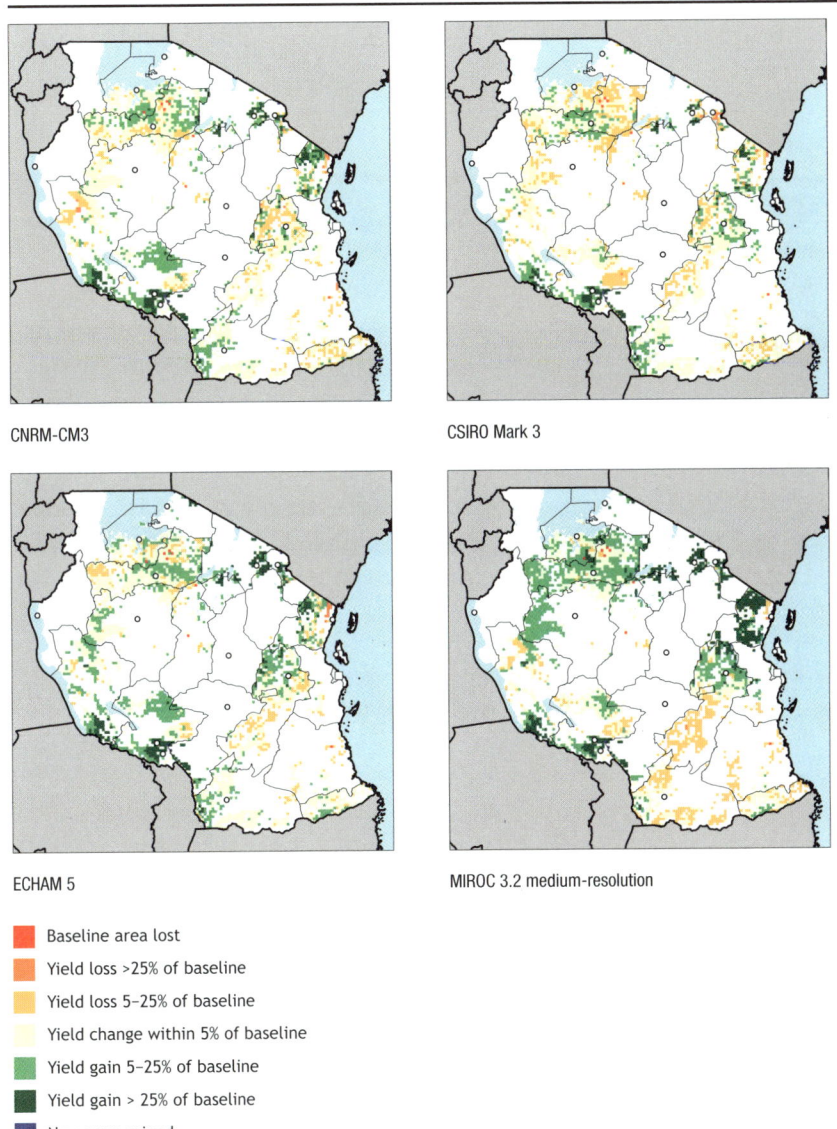

CNRM-CM3

CSIRO Mark 3

ECHAM 5

MIROC 3.2 medium-resolution

- Baseline area lost
- Yield loss >25% of baseline
- Yield loss 5–25% of baseline
- Yield change within 5% of baseline
- Yield gain 5–25% of baseline
- Yield gain > 25% of baseline
- New area gained

Source: Authors' calculations.

Notes: A1B = greenhouse gas emissions scenario that assumes fast economic growth, a population that peaks midcentury, and the development of new and efficient technologies, along with a balanced use of energy sources; CNRM-CM3 = National Meteorological Research Center–Climate Model 3; CSIRO = climate model developed at the Australia Commonwealth Scientific and Industrial Research Organisation; ECHAM 5 = fifth-generation climate model developed at the Max Planck Institute for Meteorology (Hamburg); GCM = general circulation model; MIROC = Model for Interdisciplinary Research on Climate, developed by the University of Tokyo Center for Climate System Research.

The findings presented in Figure 11.20 show both gains and losses in rice yields depending on location, though the response is far from uniform geographically. The general observation that was made for maize—that there will be areas of potential gains as well as losses—pertains to the case of rainfed rice as well.

Vulnerability

Figure 11.21 shows the impact of future GDP and population scenarios on the number of malnourished children under age five. Figure 11.22 shows the share of children who are malnourished.

All scenarios show an increase in the number of malnourished children until the year 2025. The number in the optimistic scenario falls before that in the baseline, which in turn falls before that in the pessimistic scenario. With the population rising, these gradual increases in the number of malnourished children will result in a gradual decline in the malnutrition rate. The trends for malnourished children are inversely related to caloric availability, as depicted in Figure 11.23. That is, malnutrition falls with a rise in per capita kilocalories. A rise in GDP per capita will enable families to access better healthcare and food supplies. Possibly a better contributor to farm households' well-being will

FIGURE 11.21 Number of malnourished children under five years of age in Tanzania in multiple income and climate scenarios, 2010–2050

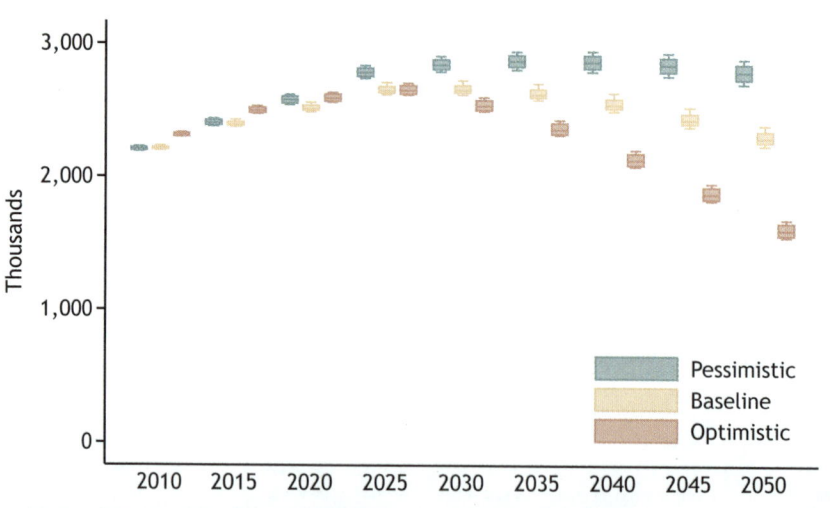

Source: Based on analysis conducted for Nelson et al. (2010).

Note: The box and whiskers plot for each socioeconomic scenario shows the range of effects from the four future climate scenarios.

FIGURE 11.22   Share of malnourished children under five years of age in Tanzania in multiple income and climate scenarios, 2010–2050

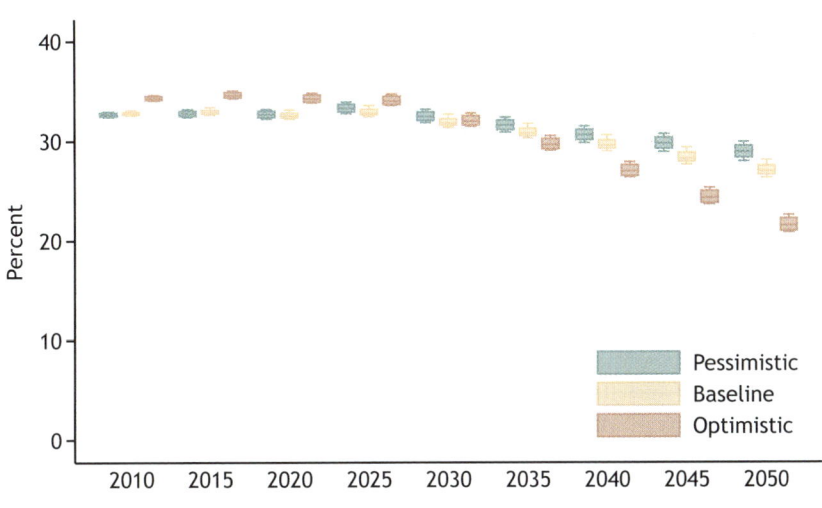

Source: Based on analysis conducted for Nelson et al. (2010).
Note: The box and whiskers plot for each socioeconomic scenario shows the range of effects from the four future climate scenarios.

FIGURE 11.23   Kilocalories per capita in Tanzania in multiple income and climate scenarios, 2010–2050

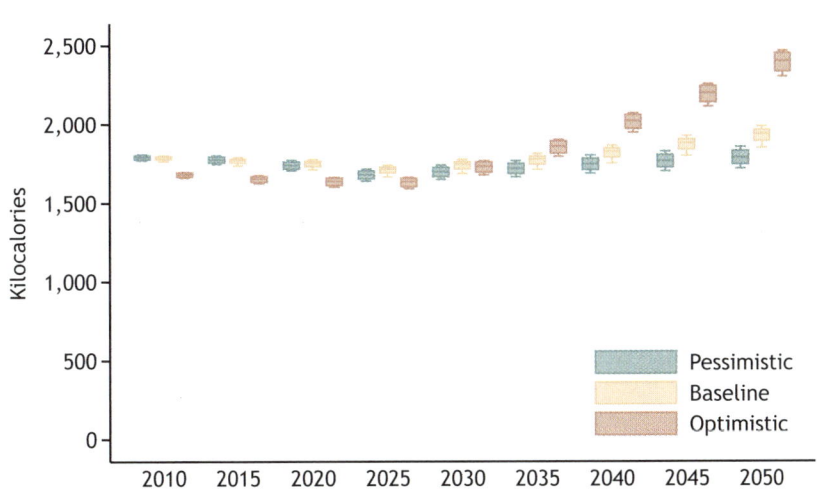

Source: Based on analysis conducted for Nelson et al. (2010).
Note: The box and whiskers plot for each socioeconomic scenario shows the range of effects from the four future climate scenarios.

be the increased food supplies from increased production in 2050, which will increase the kilocalories available to families. Given the relatively large increase in per capita income between 2010 and 2050 in the pessimistic scenario, we would expect a positive and fairly large response in terms of calorie consumption, even taking into consideration the large projected increase in prices for staple foods. The lack of a greater increase reflects the fact that the International Model for Policy Analysis of Agricultural Commodities and Trade (IMPACT) used own-price elasticities that were too large, amplifying the price effect more than it should have.

Agricultural Outcomes

Figures 11.24–11.27 show simulation results from the IMPACT model associated with key agricultural crops in Tanzania. Each featured crop has five graphs: production, yield, harvested area, net exports, and world price.

The findings show a slightly higher maize yield that, when coupled with the very small increase in projected crop area, indicates a small increase in overall production. With such a high rate of population growth and very little expansion of production, the outcome can only be to increase imports of maize. With maize prices almost doubling, the amount of money spent on maize imports will be quite a bit higher than at present.

With the relatively high variance for cassava yields between GCMs, it is difficult to tell in which direction the trend will go, but it would probably be reasonably accurate to say that yield will remain largely unchanged between 2010 and 2050. There will be a slight increase in area, but ultimately production is not projected to change much. With at least two times the population by 2050, Tanzania's demand for cassava will far outstrip supply, and the imports of cassava will increase dramatically, as they will for maize. We also note that, unlike in the case of maize, it appears that the world prices for cassava will fall slightly between 2010 and 2050 (see Figure 11.25).

The story for sorghum is completely different than what we have seen for maize and cassava. Yields will more than triple (see Figure 11.26). This seems to contradict the findings of the DSSAT crop model, but that crop model did not allow for adaptation, keeping the same seed variety, the same amount of fertilizer, and the same location for planting. But if these things were varied and there were other types of technological innovation, the IMPACT models suggests that sorghum would do well with climate change in Tanzania. We see that the area producing sorghum will also expand by around 40 percent, leading production to more than quadruple. With that kind of production

FIGURE 11.24 Impact of changes in GDP and population on maize in Tanzania, 2010–2050

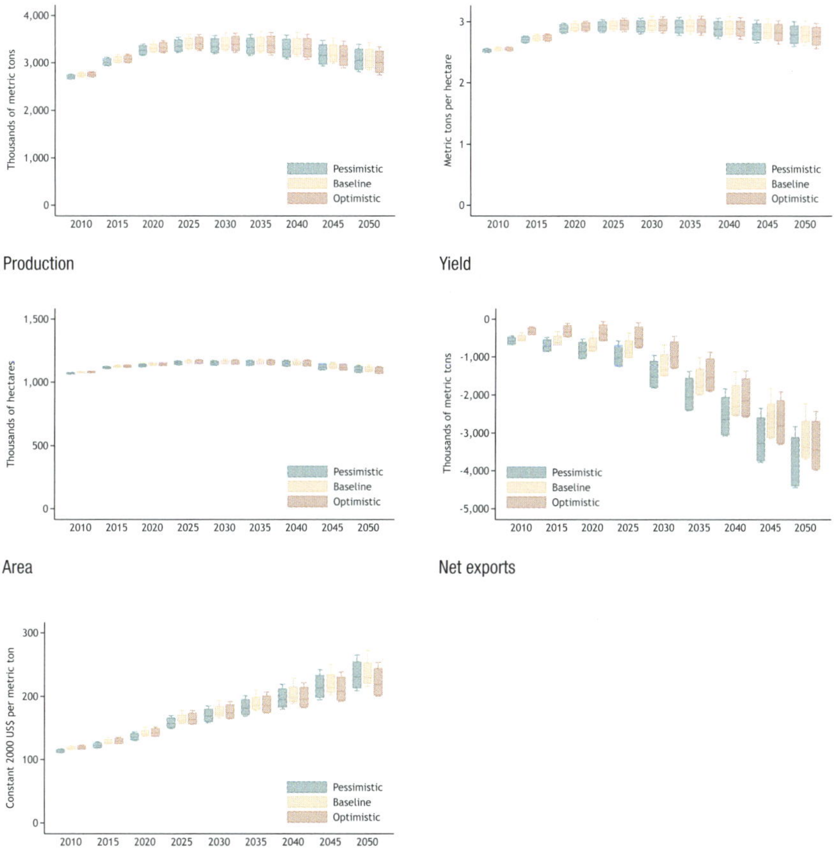

Production

Yield

Area

Net exports

Prices

Source: Based on analysis conducted for Nelson et al. (2010).

Note: The box and whiskers plot for each socioeconomic scenario shows the range of effects from the four future climate scenarios. GDP = gross domestic product; US$ = US dollars.

increase, Tanzania will be able to export around 70 percent of the sorghum it produces in 2050 if the projections are reasonably accurate.

The IMPACT model projects that under climate change the rice yield, with technological improvement, will approximately double between 2010 and 2050 (see Figure 11.27). With a small increase in area, the production of rice will slightly more than double. However, with such a rapidly growing population, production will not keep up with demand, and imports of rice will increase, much as will those of maize and cassava.

FIGURE 11.25   Impact of changes in GDP and population on cassava in Tanzania, 2010–2050

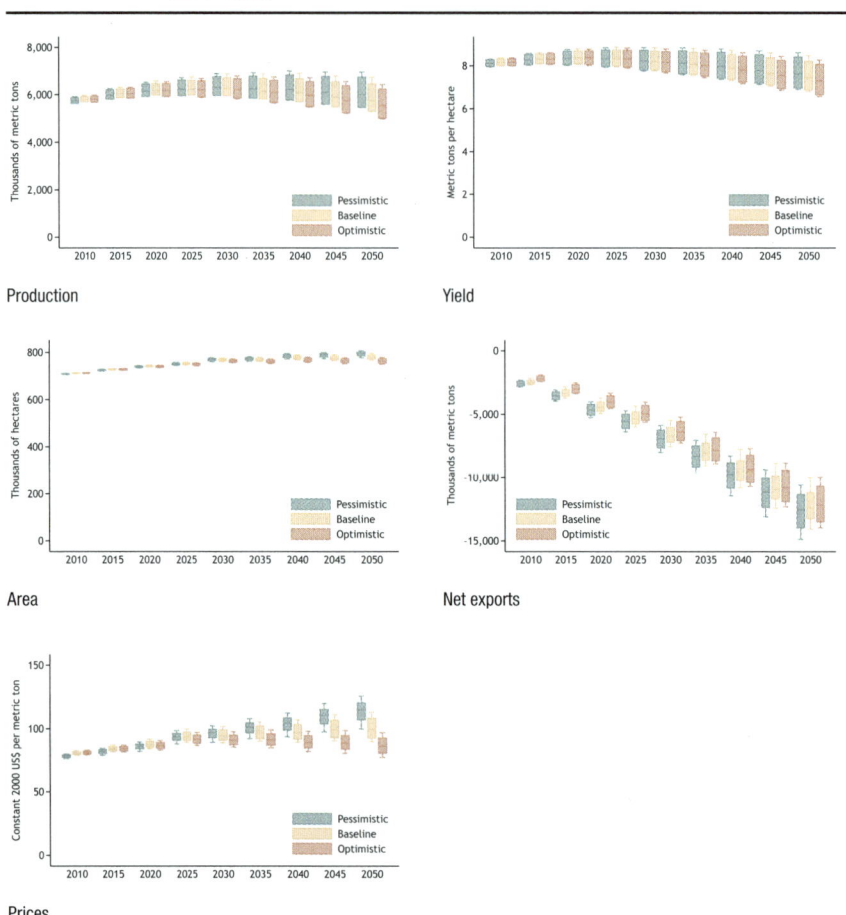

Production

Yield

Area

Net exports

Prices

Source: Based on analysis conducted for Nelson et al. (2010).

Note: The box and whiskers plot for each socioeconomic scenario shows the range of effects from the four future climate scenarios. GDP = gross domestic product; US$ = US dollars.

Conclusions and Policy Recommendations

Tanzania's population is expected to more than double over the next four decades, which could increase the competition for agricultural land and other resources, exacerbating the effects of climate change. But projections from reputable sources for GDP per capita seem to suggest a generally brighter future for Tanzanians, enabling average people to expand their calorie consumption in all but the pessimistic scenario, in which it will likely stay near where it is now.

FIGURE 11.26 Impact of changes in GDP and population on sorghum in Tanzania, 2010–2050

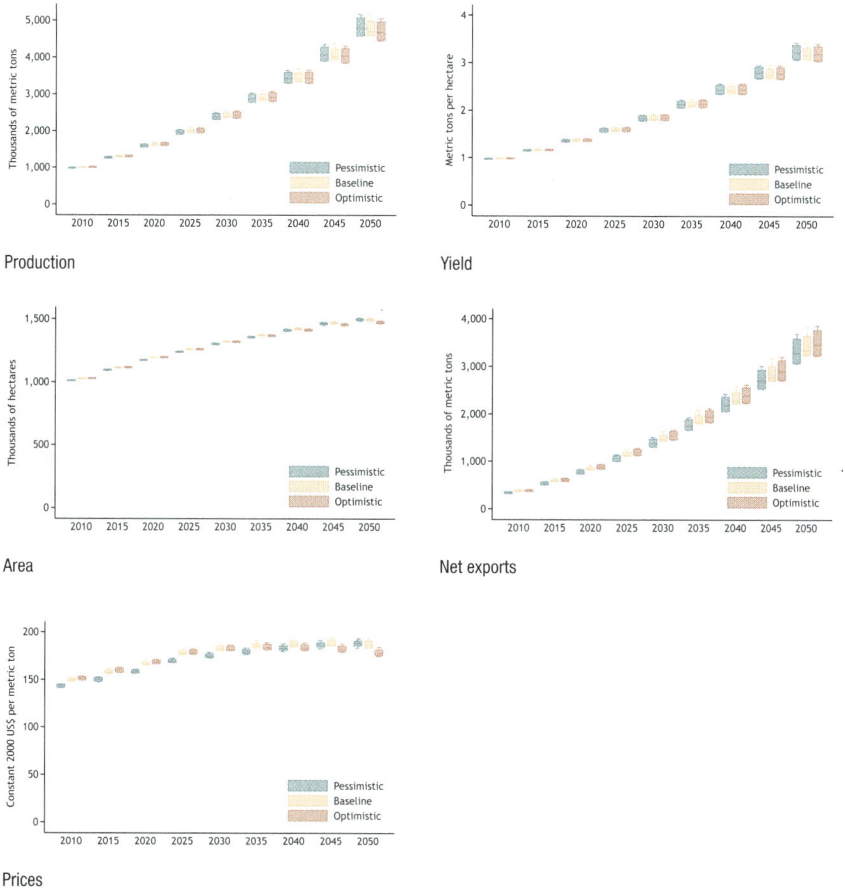

Production

Yield

Area

Net exports

Prices

Source: Based on analysis conducted for Nelson et al. (2010).
Note: The box and whiskers plot for each socioeconomic scenario shows the range of effects from the four future climate scenarios. GDP = gross domestic product; US$ = US dollars.

The climate models used in this chapter suggest that there will be either no change in rainfall or some increase in annual rainfall, which is good news given that many nations are expecting to receive less rainfall when they already have too little. The climate models differ over how much warmer it will become by 2050: two models suggest around 1°C, and the other two models suggest around 2°C. The hotter it becomes, the more problems are anticipated: plants will suffer heat stress, and some crop diseases and pests will probably become more difficult to combat in warmer temperatures because they flourish in heat.

FIGURE 11.27   Impact of changes in GDP and population on rice in Tanzania, 2010–2050

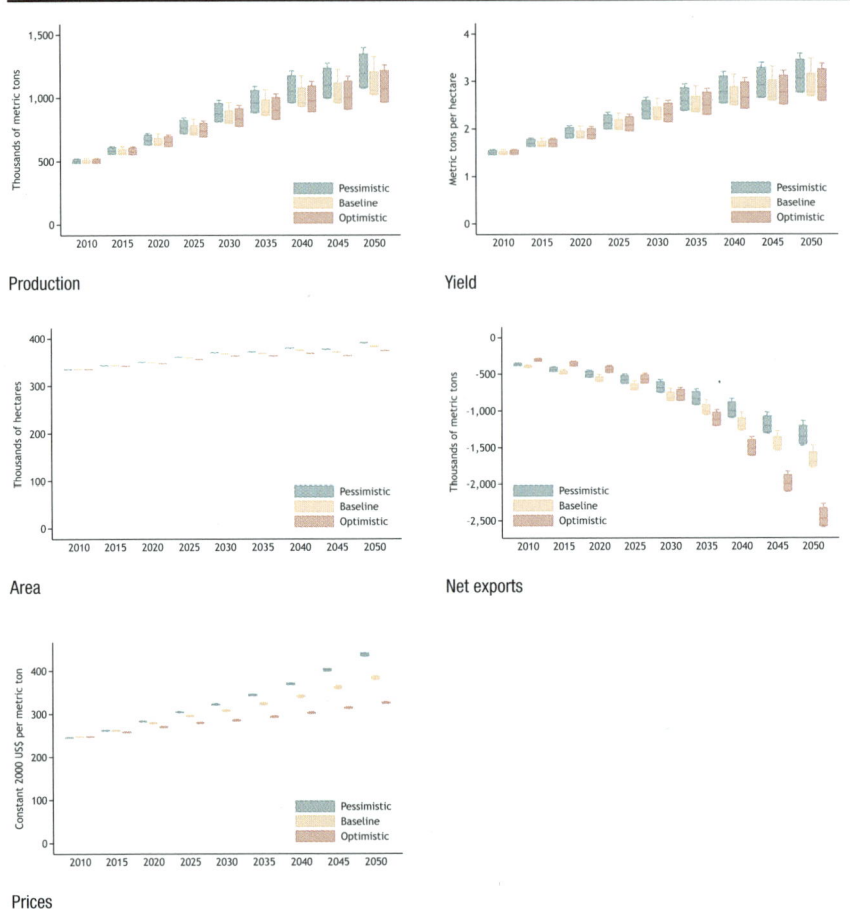

Source: Based on analysis conducted for Nelson et al. (2010).

Note: The box and whiskers plot for each socioeconomic scenario shows the range of effects from the four future climate scenarios. GDP = gross domestic product; US$ = US dollars.

The crop model results presented in this chapter are mixed: some parts of the country would likely see yield increases from climate change, and other parts would have yield decreases. Sorghum seems likely to suffer the most of the three presented, but nonetheless there is a region in which it may experience a significant increase in yields as well. These crop models do not allow for adaptation, so we would necessarily expect that farmers will do better than the models predict.

Nonetheless, the models suggest some useful points for policymakers. First, with some locations gaining and other locations losing yield, the farmers in the losing areas are going to struggle and will need help in the form of field-tested advice on what new varieties to try or what crops to switch to or how to manage new techniques to counteract climate change. The farmers in losing areas will also tend to migrate to urban areas, to other rural areas that are doing better, or to plots in unsettled areas. Establishing a legal framework either to protect unsettled areas or to assist farmers in migrating might be very important.

Helping farmers to adapt is a function of both the national agricultural research institutes and the national agricultural extension services. In order for these groups to succeed, they will need more funding and will need to be able to focus more on getting workable innovations out to farmers.

The IMPACT model, which assumes technological change and adaptation, suggests that the projected increases in population will create a demand for maize, rice, and cassava that the farmers of Tanzania will be unable to meet, and imports of these foods will rise dramatically. On the other hand, the model suggests that sorghum yields will rise so steeply that Tanzania will be able to export most of its sorghum. Nonetheless, it seems that Tanzania will need to rely on its purchasing power abroad in order to satisfy its food security needs.

Implicit in the IMPACT model—though not modeled directly—is that at some point a demographic transition will occur and labor will shift out of agriculture and into services and manufacturing. This means that Tanzania should not neglect these sectors, because ultimately for GDP to grow by as much as some of the experts suggest it will, Tanzania will have to shift people out of agriculture—not all people, because farming will need to continue—because farming will not be able to absorb the projected population.

If it is true that climate change will increase the incidence and severity of weather-related shocks such as droughts and floods, it will also be important for policymakers to shore up the social safety net to prevent people from dying or falling into deep poverty and to get farmers back on their feet so they can plant for the next season.

References

Bartholome, E., and A. S. Belward. 2005. "GLC2000: A New Approach to Global Land Cover Mapping from Earth Observation Data." *International Journal of Remote Sensing* 26 (9–10): 1959–1977.

CIESIN (Center for International Earth Science Information Network), Columbia University, IFPRI (International Food Policy Research Institute), World Bank, and CIAT (Centro Internacional de Agricultura Tropical). 2004. *Global Rural–Urban Mapping Project (GRUMP), Alpha Version: Population Density Grids.* Palisades, NY, US: Socioeconomic Data and Applications Center (SEDAC), Columbia University. http://sedac.ciesin.columbia.edu/gpw.

FAO (Food and Agriculture Organization of the United Nations). 2010. FAOSTAT. Rome. http://faostat.fao.org.

Jones, P. G., P. K. Thornton, and J. Heinke. 2009. *Generating Characteristic Daily Weather Data Using Downscaled Climate Model Data from the IPCC's Fourth Assessment.* Project report for the International Livestock Research Institute. Geneva: International Panel on Climate Change.

Jowett, M., and Miller, J. N. 2005. "The Financial Barrier of Malaria in Tanzania: Implications for Future Government Policy." *International Journal of Health Planning and Management* 20 (1): 67–84.

Lehner, B., and P. Döll. 2004. "Development and Validation of a Global Database of Lakes, Reservoirs, and Wetlands." *Journal of Hydrology* 296 (1–4): 1–22.

Madulu, N. F. 2004. "Assessment of Linkages between Population Dynamics and Environmental Change in Tanzania." *African Journal of Environmental Assessment and Management* 9: 88–102.

———. n.d. "Population Distribution and Density in Tanzania: Experiences from 2002 Population and Housing Census." Unpublished report. Accessed September 23, 2010. http://ccs.ukzn.ac.za/files/madulu.pdf.

Millennium Ecosystem Assessment. 2005. *Ecosystems and Human Well-being: Synthesis.* Washington, DC: Island Press. www.maweb.org/en/Global.aspx.

Nelson, G. C., M. W. Rosegrant, A. Palazzo, I. Gray, C. Ingersoll, R. Robertson, S. Tokgoz, et al. 2010. *Food Security, Farming, and Climate Change to 2050: Scenarios, Results, Policy Options.* Washington, DC: International Food Policy Research Institute.

Trading Economics. n.d. "Tanzania GDP per Capita at Constant Prices." Accessed August 20, 2011. www.tradingeconomics.com/tanzania/gdp-per-capita-at-constant-prices-imf-data.html.

Traerup, S.L.M., R. A. Ortiz, and A. Markandya. 2010. *The Health Impacts of Climate Change: A Study of Cholera in Tanzania.* BC3 (Basque Centre for Climate Change) Working Paper 2010–01. Bilbao, Spain: BC3.

UNEP (United Nations Environment Programme) and IUCN (International Union for the Conservation of Nature). 2009. World Database on Protected Areas (WDPA) Annual Release. No longer available online.

UNPOP (United Nations Secretariat, Department of Economic and Social Affairs, Population Division). 2009. *World Population Prospects: The 2008 Revision.* Accessed April 6, 2010. http://esa.un.org/unpp.

URT (United Republic of Tanzania), Ministry of Agriculture, Food Security and Cooperatives. 2006. *AGSTAT for Food Security: Food Crop Production Forecast Reports.* Dodoma.

———. 2007. *AGSTAT for Food Security: Food Crop Production Forecast Reports.* Dodoma.

———. 2008. *AGSTAT for Food Security: Food Crop Production Forecast Reports.* Dodoma.

———. 2010. *AGSTAT for Food Security: Food Crops Production Forecast Reports.* Dodoma.

URT (United Republic of Tanzania), Ministry of Health and Social Welfare. 2008. "Human Resource for Health Strategic Plan 2008–2013." Accessed August 11, 2011. www.unfpa.org/sowmy/resources/docs/library/R223_MOHTanzania_2008_HRH_Strategic_Plan_2008_2013.pdf.

URT (United Republic of Tanzania), Planning Commission. 2005. *Tanzania Development Vision 2025.* Dar es Salaam.

USDoS (United States Department of State). 2011. "Background Note: Tanzania." Bureau of African Affairs. Accessed August 29. www.state.gov/r/pa/ei/bgn/2843.htm.

Wood, S., G. Hyman, U. Deichmann, E. Barona, R. Tenorio, Z. Guo, S. Castano, O. Rivera, E. Diaz, and J. Marin. 2010. "Sub-national Poverty Maps for the Developing World Using International Poverty Lines: Preliminary Data Release." Accessed May 6. http://povertymap.info.

World Bank. 2009. *World Development Indicators.* Accessed May 2011. http://data.worldbank.org/data-catalog/world-development-indicators.

———. 2010. *Economics of Adaptation to Climate Change: Synthesis Report.* Washington, DC. http://climatechange.worldbank.org/content/economics-adaptation-climate-change-study-homepage.

Yanda, P. Z., R.Y.M. Kangalawe, and R. J. Sigalla. 2005. *Climatic and Socio-Economic Influences on Malaria and Cholera Risks in the Lake Victoria Region of Tanzania.* AIACC (Assessments of Impacts and Adaptations to Climate Change) Working Paper 12. Accessed August 11, 2011. www.aiaccproject.org/working_papers/Working %20Papers/AIACC_WP_No012.pdf.

You, L., and S. Wood. 2006. "An Entropy Approach to Spatial Disaggregation of Agricultural Production." *Agricultural Systems* 90 (1–3): 329–347.

You, L., S. Wood, and U. Wood-Sichra. 2006. "Generating Global Crop Distribution Maps: From Census to Grid." Paper presented at the International Association of Agricultural Economists Conference, Brisbane, Australia, August 11–18.

———. 2009. "Generating Plausible Crop Distribution and Performance Maps for Sub-Saharan Africa Using a Spatially Disaggregated Data Fusion and Optimization Approach." *Agricultural Systems* 99 (2–3): 126–140.

Chapter 12

UGANDA

Bernard Bashaasha, Timothy S. Thomas, Michael Waithaka,
and Miriam Kyotalimye

The climate of Uganda is regarded as its most valuable natural resource, and a major determinant of other natural resources, such as water, forests, and wildlife, as well as human activities based on these resources, such as agriculture and ecotourism (Republic of Uganda, MWE 2007). Together these resources provide the means of livelihood for many Ugandans and enhance economic growth, which is predominantly agriculture based. However, the past few decades have been marked by climate variability, resulting in increased frequency of extreme weather events such as droughts, floods, and landslides, causing damage to natural resources and slowing social and economic development.

Prolonged droughts can have serious impacts on agricultural production. Even long dry spells during the rainy season are sufficient to reduce agricultural production, thus seriously affecting the livelihoods of the rural communities. They lead to reduced incomes and therefore to decreased health status and standards of living. Thus assessing the vulnerability of agriculture to climate change is critical to understanding the impact of climate change on the welfare of the population and the country's preparedness to deal with climate change in general.

Review of the Current Situation and Trends

Economic and Demographic Indicators

Population
Figure 12.1 shows the total and rural populations of Uganda (left axis) as well as the share of the population that is urban (right axis). In the 48 years displayed in the graph, we see that the population has quadrupled. We also note that although the proportion of the population living in urban areas has steadily grown, 87 percent of the population still lives in rural areas. Table 12.1 provides additional information concerning rates of population

FIGURE 12.1   Population trends in Uganda: Total population, rural population, and percent urban, 1960–2008

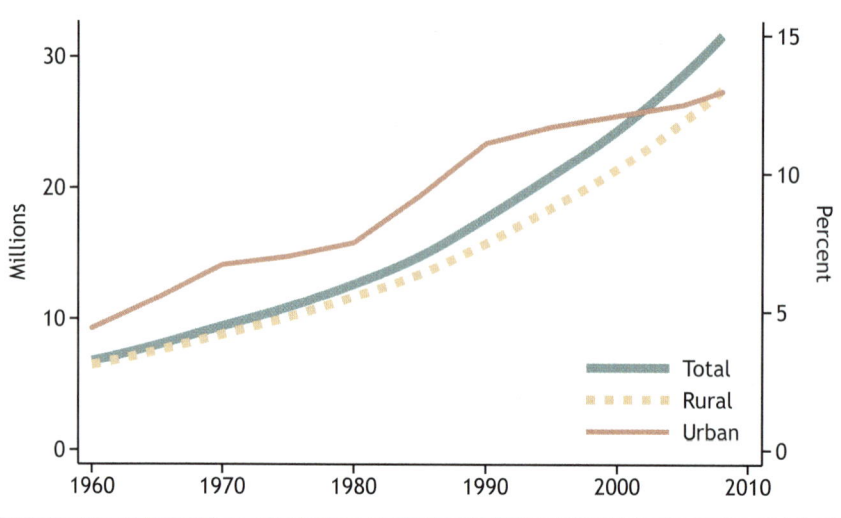

Source: *World Development Indicators* (World Bank 2009).

growth. It confirms the high growth rate of the population, with no decade showing growth of less than 2.9 percent per year.

Figure 12.2 shows the geographic distribution of the population in Uganda, estimated from census data and other sources. Recent data show that the county's population continues to grow at a very fast rate. According to the Ministry of Finance, Planning and Economic Development (Republic of Uganda, MFPED 2010), Uganda's total population grew by 4.8 percent between 2007 and 2008, by 3.7 percent between 2008 and 2009, and by

TABLE 12.1   Population growth rates in Uganda, 1960–2008 (percent)

Decade	Total growth rate	Rural growth rate	Urban growth rate
1960–1969	3.3	3.1	7.6
1970–1979	2.9	2.8	4.0
1980–1989	3.4	3.0	7.3
1990–1999	3.2	3.1	4.1
2000–2008	3.2	3.1	4.1

Source: Authors' calculations based on *World Development Indicators* (World Bank 2009).

FIGURE 12.2 Population distribution in Uganda, 2000 (persons per square kilometer)

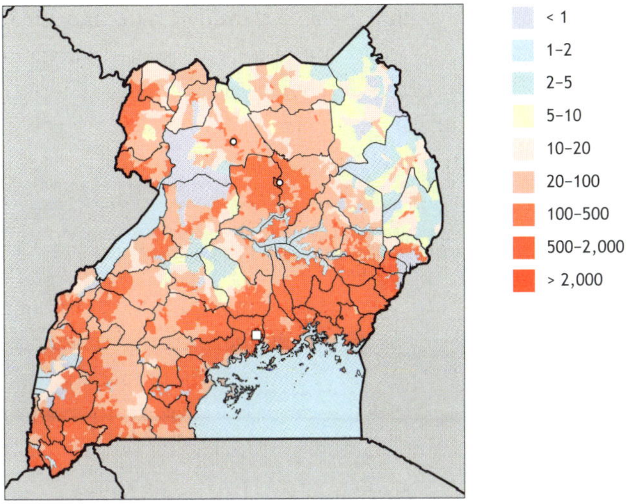

Legend:
- < 1
- 1–2
- 2–5
- 5–10
- 10–20
- 20–100
- 100–500
- 500–2,000
- > 2,000

Source: CIESIN et al. (2004).

3.7 percent between 2009 and 2010. The total population was estimated at 31.8 million in 2010. According to the Uganda Bureau of Statistics (UBOS 2010), Uganda's population remains predominantly rural (85 percent in 2010). This has not changed since the Uganda National Household Survey (UNHS) of 2005–2006; if anything, there has been a slight increase in the proportion of the rural population (Table 12.2).

Though generally stable, the population distribution shows a slight increase in the eastern region and a decline in the central region, while the

TABLE 12.2 Distribution of population in Uganda by residence and region, 2005/2006 and 2009/2010 (percent)

Residence	2005/2006	2009/2010
Rural	84.6	85.0
Urban	15.4	15.0
Region		
Central	29.2	26.5
Eastern	25.2	29.6
Northern	19.7	20.0
Western	25.9	24.0

Source: UBOS (2010).

northern region remains sparsely populated. The evidence from the UNHS (UBOS 2010) seems to be consistent with Figure 12.2, even though the map represents the year 2000. Although the proportion of the rural population remains the same, Uganda's population growth rate means that the rural population is also increasing rapidly. Agriculture thus remains very important as the main activity of the rural population.

Income

Figure 12.3 shows trends in gross domestic product (GDP) per capita and the proportion of GDP from agriculture. The figure shows that the proportion of agricultural GDP declines with economic transformation, because the bulk of economic output comes increasingly from the industry and service sectors. Per capita GDP (at constant 2002 prices) has increased from 448,860 Uganda shillings (Ush) in 2000 to Ush 650,773 in 2009—an increase of 45 percent over the nine-year period. However, as shown in Table 12.3, the growth rate of per capita GDP has not been uniform and actually dropped dramatically, to 1.9 percent, between 2008 and 2009, from a high of 6.9 percent between 2007 and 2008, largely on account of unfavorable weather conditions that negatively affected agriculture. This may reflect the perennially low

FIGURE 12.3   Per capita GDP in Uganda (constant 2000 US$) and share of GDP from agriculture (percent), 1960–2008

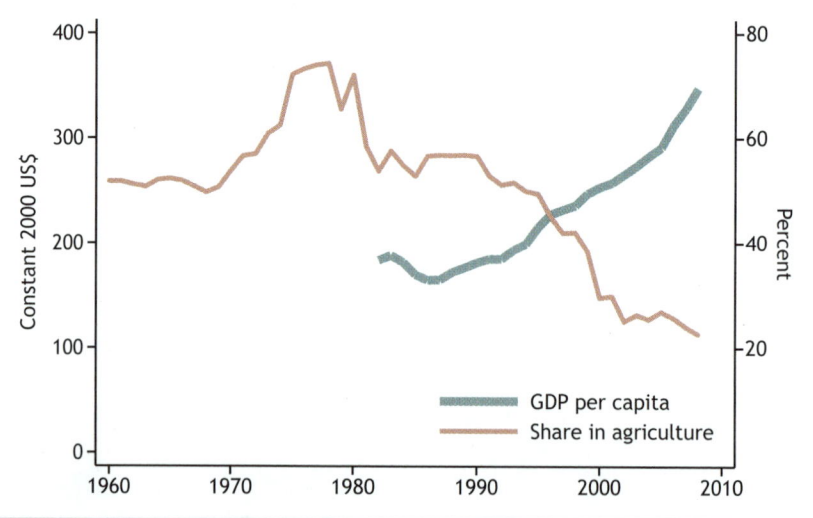

Source: *World Development Indicators* (World Bank 2009).
Note: GDP = gross domestic product; US$ = US dollars.

TABLE 12.3   Per capita gross domestic product (GDP) (Uganda shillings) and growth rate (percent) in Uganda at constant 2002 prices, 2000–2009

Year	Per capita GDP	Growth rate
2000	448,860	n.d.
2001	472,816	5.3
2002	490,190	3.7
2003	503,980	2.8
2004	516,420	2.5
2005	550,193	6.5
2006	570,410	3.7
2007	596,979	4.7
2008	638,443	6.9
2009	650,773	1.9

Source: Republic of Uganda, MFPED (2010).
Note: n.d. = no data because 2000 was the first year.

growth rate of the agricultural sector, averaging 1.3 percent per year between 2005/2006 and 2009/2010 (Republic of Uganda, MFPED 2010) in the context of a more sluggish manufacturing and service economy. Ugandans are, on average, increasingly enjoying a higher standard of living.

Vulnerability to Climate Change

Table 12.4 provides some data on indicators of a population's vulnerability and resiliency to economic shocks: level of education, literacy, and concentration of labor in poorer or less dynamic sectors. The reasonably high levels of education and literacy rates depicted in the table augur well for Uganda's response to

TABLE 12.4   Education and labor statistics for Uganda, 2000s

Indicator	Year	Percent
Primary school enrollment: Percent gross (three-year average)	2007	116.2
Secondary school enrollment: Percent gross (three-year average)	2007	22.5
Adult literacy rate	2007	73.6
Percent employed in agriculture	2003	68.7
Percent with vulnerable employment (in agriculture on own farm or as a day laborer)	2003	85.2
Under-five malnutrition (weight for age)	2001	19.0

Source: *World Development Indicators* (World Bank 2009).

climate change. Improved education establishes a platform for mass education and sensitization to climate change and enhances the probability of adopting climate change adaptation and mitigation measures. Research has consistently demonstrated (Deressa and Hassan 2009) that the probability of adopting a technology increases with the level of education. However, the education and literacy differential between rural and urban areas will need to be taken into account in program design.

A large proportion of Uganda's population still derives a living from the primary sector, including agriculture, hunting, mining, and quarrying. Primary-sector employment was 72 percent in 2005/2006 but dropped by 6 percentage points, to 66 percent, in 2009/2010.

Figure 12.4 shows two noneconomic correlates of poverty: life expectancy and under-five mortality. According to the 2002 census, life expectancy in Uganda remains low, at 50.4 years (UBOS 2009), reflecting various diseases and inadequate medical services, but it has been rising since the late 1990s as the impact of HIV/AIDS has lessened. Malaria remains the most prevalent fatal illness (UBOS 2010). A high disease burden constrains a household's ability to respond to the impacts of climate change by adaptation or mitigation. The disease burden and associated costs in terms of time and money thus compound the country's vulnerability and undermine its response to climate change.

FIGURE 12.4   Well-being indicators in Uganda, 1960–2008

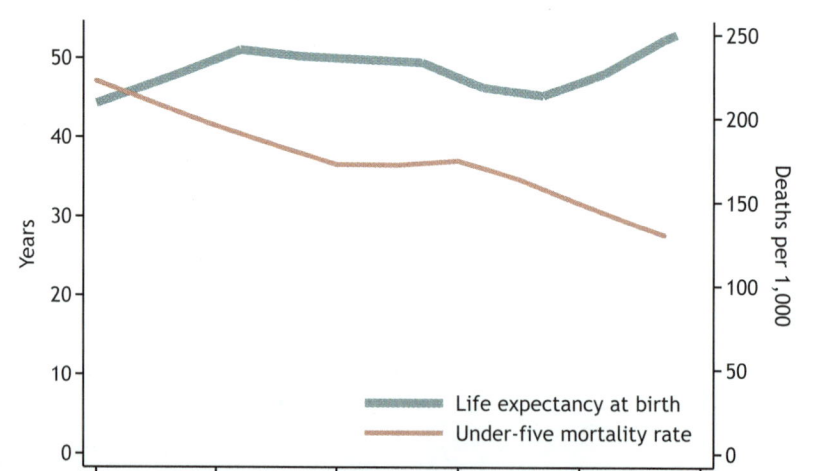

Source: *World Development Indicators* (World Bank 2009).

FIGURE 12.5 Poverty in Uganda, circa 2005 (percentage of population below US$2 per day)

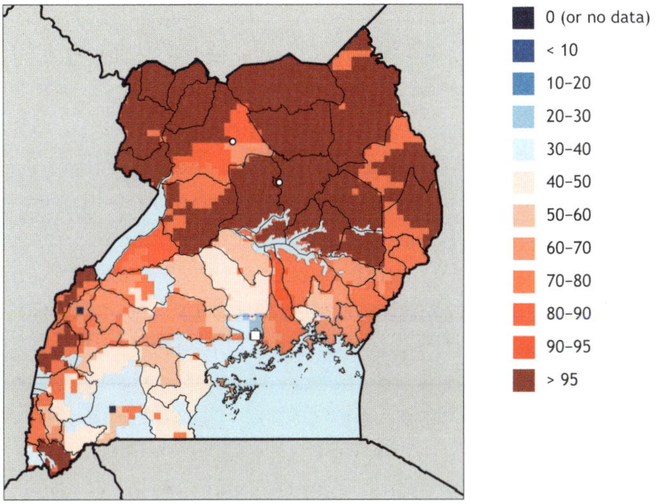

Source: Wood et al. (2010).
Note: US$ = US dollars.

Figure 12.5 shows the proportion of the population living on less than $2 per day. According to the Uganda Bureau of Statistics and the International Livestock Research Institute (UBOS and ILRI 2003–2004), the national poverty line in Uganda is Ush16,443 per month per adult equivalent (using 1989 prices). According to UBOS (2010), the poverty rate in Uganda is still high, at 24.5 percent, though it is down from 31 percent (as reported in the 2005/2006 UNHS). The climate-change consequences of extensive poverty are adverse, because households dependent on agriculture and constrained by poverty will have limited options for climate change adaptation, resilience, and mitigation. The incidence of poverty remains highest in northern Uganda and lowest in central Uganda.

Review of Land Use, Potential, and Limitations

Land Use Overview

As shown in Figure 12.6, the levels of broad-leaved evergreen and broad-leaved deciduous and closed tree cover are fairly low in Uganda. The dominant natural vegetation types appear to be open deciduous tree cover and a mosaic of trees and other natural vegetation. The figure also shows dominant patches

FIGURE 12.6   Land cover and land use in Uganda, 2000

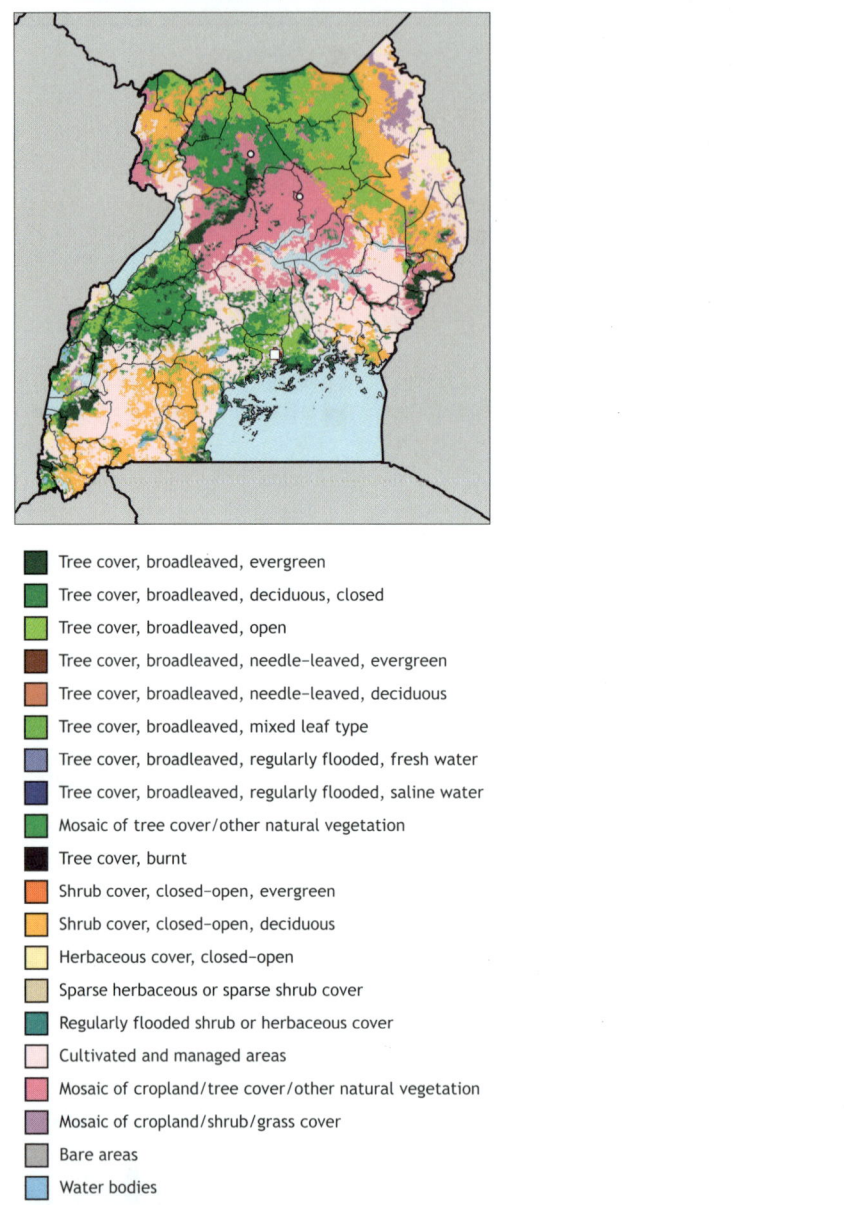

- ■ Tree cover, broadleaved, evergreen
- ■ Tree cover, broadleaved, deciduous, closed
- ■ Tree cover, broadleaved, open
- ■ Tree cover, broadleaved, needle–leaved, evergreen
- ■ Tree cover, broadleaved, needle–leaved, deciduous
- ■ Tree cover, broadleaved, mixed leaf type
- ■ Tree cover, broadleaved, regularly flooded, fresh water
- ■ Tree cover, broadleaved, regularly flooded, saline water
- ■ Mosaic of tree cover/other natural vegetation
- ■ Tree cover, burnt
- ■ Shrub cover, closed–open, evergreen
- ■ Shrub cover, closed–open, deciduous
- ■ Herbaceous cover, closed–open
- ■ Sparse herbaceous or sparse shrub cover
- ■ Regularly flooded shrub or herbaceous cover
- ■ Cultivated and managed areas
- ■ Mosaic of cropland/tree cover/other natural vegetation
- ■ Mosaic of cropland/shrub/grass cover
- ■ Bare areas
- ■ Water bodies
- ■ Snow and ice
- ■ Artificial surfaces and associated areas
- ■ No data

Source: GLC2000 (Bartholome and Belward 2005).

of cultivated and cropland across the country. This scenario is consistent with the national biomass studies (Republic of Uganda, MWLE 2006), which show that the volume of total biomass declined by 20 percent, from 299.1 million tons in 2000 to 239.1 million tons in 2005. The same study shows that the amount of tropical high forest declined even faster, by 64 percent, from 44.2 million tons in 2000 to 15.8 million tons in 2005. The amount of woodland biomass is shown to have also declined by 31 percent over the same period. The scenario depicted is that of cultivated land or cropland continuously replacing forests, woodland, bushes, and even wetlands.

Figure 12.7 shows maps of travel times to urban areas of various sizes, illustrating the potential cost of transporting agricultural products to potential markets. Physical infrastructure, including roads, is critical to agricultural modernization (Republic of Uganda, MAAIF and MFPED 2000). Roads are important—particularly rural feeder roads and the associated network of bridges, footpaths, and bicycle paths—because they link rural areas to wider markets for their products and give them access to production inputs. The potential benefits include opportunities for competition, incentives for new innovations, reduced transaction costs, increased efficiency, and improved quality of services. These translate into low costs of production, increased productivity, higher farmgate prices, and increased profitability at the farm level. Travel time has been greatly reduced in much of Uganda, with the exception of such remote areas as the Karamoja subregion (the northeastern portion in Figure 12.7). Indeed, findings of the UNHS 2009/2010 (UBOS 2010) indicate that overall, 80 percent of the communities surveyed had easy access to all-season feeder roads.

Figure 12.8 shows the locations of protected areas, including parks and forest reserves. These locations provide important protection for fragile environments, which may also be important for tourism. The rate of deforestation in Uganda is estimated at 1.8 percent per year, largely through the expansion of subsistence agriculture (Republic of Uganda 2010, citing NEMA 2009). This has increased pressure on protected areas and undermined national climate change mitigation efforts (Republic of Uganda, MWE 2007).

Agriculture

Maize, beans, cassava, and cooking bananas (plantains) are among the most widely grown crops in Uganda in terms of hectares as well as proportion of households cultivating the crops (Table 12.5) (UBOS 2006a). Cassava and sweet potatoes are grown mainly in the central region, maize in the east, sorghum in the north, and finger millet, beans, and plantains in the west.

FIGURE 12.7 Travel time to urban areas of various sizes in Uganda, circa 2000

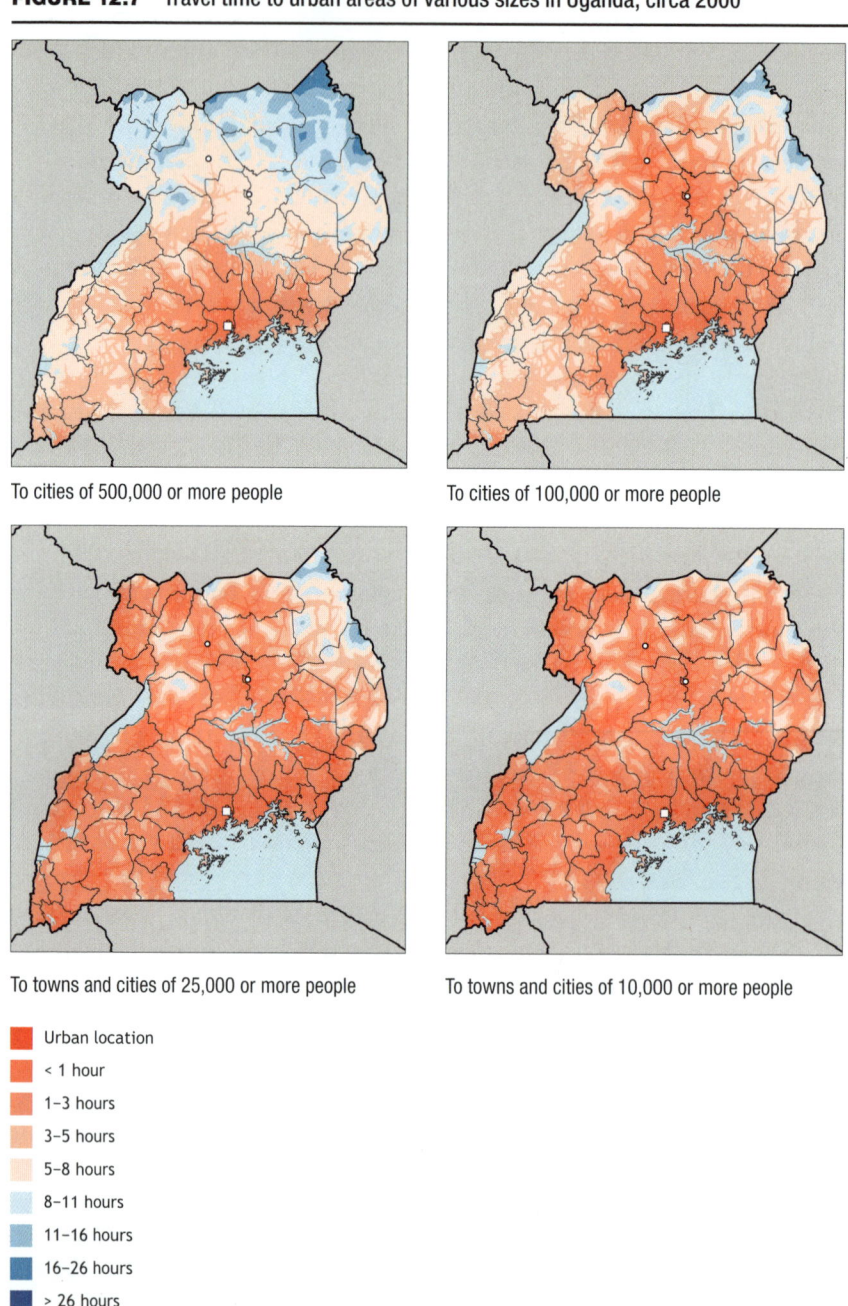

To cities of 500,000 or more people

To cities of 100,000 or more people

To towns and cities of 25,000 or more people

To towns and cities of 10,000 or more people

Urban location
< 1 hour
1–3 hours
3–5 hours
5–8 hours
8–11 hours
11–16 hours
16–26 hours
> 26 hours

Source: Authors' calculations.

FIGURE 12.8   Protected areas in Uganda, 2009

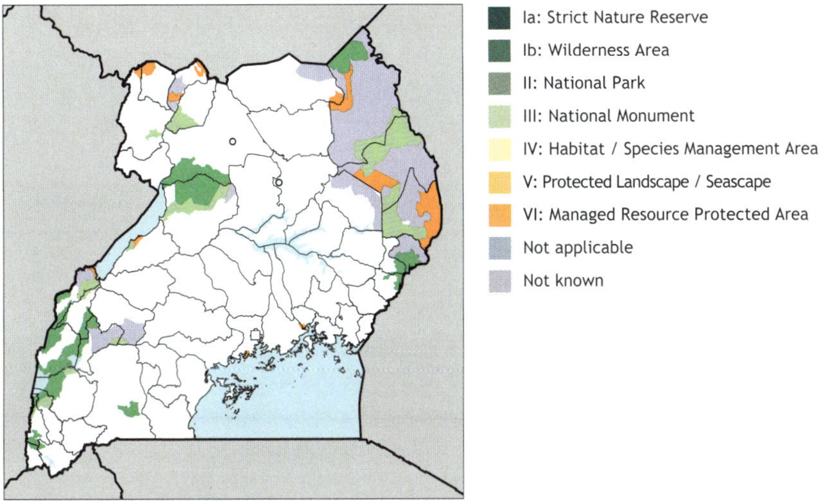

- Ia: Strict Nature Reserve
- Ib: Wilderness Area
- II: National Park
- III: National Monument
- IV: Habitat / Species Management Area
- V: Protected Landscape / Seascape
- VI: Managed Resource Protected Area
- Not applicable
- Not known

Sources: Protected areas are from the World Database on Protected Areas (UNEP and IUCN 2009). Water bodies are from the World Wildlife Fund's Global Lakes and Wetlands Database (Lehner and Döll 2004).

TABLE 12.5   Harvest area of leading agricultural commodities in Uganda, 2006–2008 (thousands of hectares)

Rank	Crop	Percent of total	Harvest area
	Total	100.0	7,054
1	Plantains	23.8	1,678
2	Beans	12.4	872
3	Maize	11.9	842
4	Sweet potatoes	8.3	587
5	Millet	6.2	438
6	Cassava	5.4	383
7	Sorghum	4.5	314
8	Sesame seeds	4.0	281
9	Coffee	3.5	250
10	Groundnuts	3.4	236

Source: FAOSTAT (FAO 2010).
Note: All values are based on the three-year average for 2006–2008.

The FAOSTAT data presented in Table 12.6 show that plantains constitute the most widely consumed food commodity, followed by fermented beverages, cassava, sweet potatoes, and maize, in that order. There are, however, important regional differences that are not captured in the FAO data. Figures 12.9 and 12.10 map out the estimated yields and growing areas for plantains and bananas and maize, respectively.

The harvested area and yield for plantains are highest in the western and central regions of the country, partly due to favorable weather and consumption preferences. Maize is more uniformly distributed across the country due to its ability to do well under diverse soil conditions, its marketability, and favorable consumption patterns. The maize yield is highest in the east, moderate in the central and western regions, and lowest in the north. Maize cultivation in the east benefits from hybrid varieties and increased fertilizer applications, both available from across the Kenyan border.

TABLE 12.6 Consumption of leading food commodities in Uganda, 2003–2005 (thousands of metric tons)

Rank	Crop	Percent of total	Food consumption
	Total	100.0	18,355
1	Plantains	23.8	4,362
2	Fermented beverages	18.1	3,327
3	Cassava	15.0	2,757
4	Sweet potatoes	12.1	2,228
5	Maize	4.1	757
6	Bananas	2.8	522
7	Millet	2.5	454
8	Beans	2.3	414
9	Potatoes	2.1	389
10	Other vegetables	1.9	353

Source: FAOSTAT (FAO 2010).
Note: All values are based on the three-year average for 2003–2005.

FIGURE 12.9 Yield (metric tons per hectare) and harvest area density (hectares) for rainfed plantains and bananas in Uganda, 2000

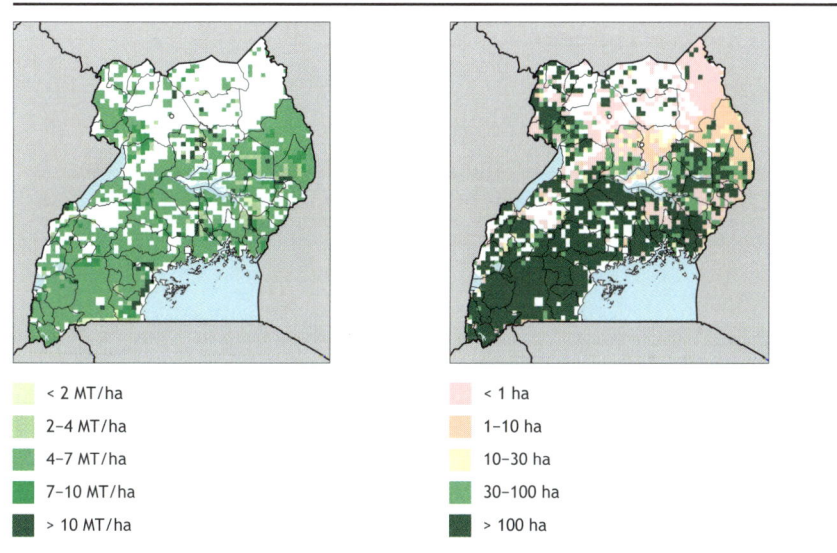

< 2 MT/ha	< 1 ha
2–4 MT/ha	1–10 ha
4–7 MT/ha	10–30 ha
7–10 MT/ha	30–100 ha
> 10 MT/ha	> 100 ha

Source: SPAM (Spatial Production Allocation Model) (You and Wood 2006; You, Wood, and Wood-Sichra 2006, 2009).
Note: ha = hectare; MT/ha = metric tons per hectare.

FIGURE 12.10 Yield (metric tons per hectare) and harvest area density (hectares) for rainfed maize in Uganda, 2000

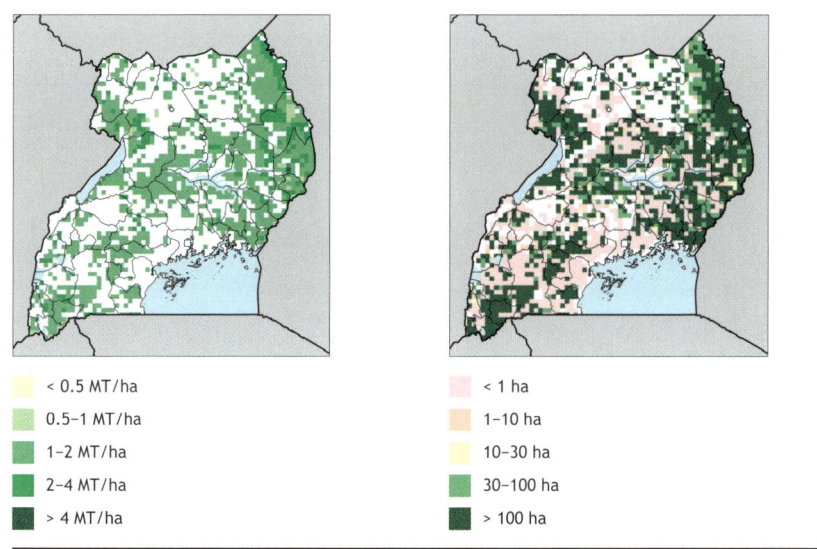

< 0.5 MT/ha	< 1 ha
0.5–1 MT/ha	1–10 ha
1–2 MT/ha	10–30 ha
2–4 MT/ha	30–100 ha
> 4 MT/ha	> 100 ha

Source: SPAM (Spatial Production Allocation Model) (You and Wood 2006; You, Wood, and Wood-Sichra 2006, 2009).
Note: ha = hectare; MT/ha = metric tons per hectare.

Scenarios for the Future

Economic and Demographic Indicators

Population

Figure 12.11 shows population projections by the UN Population Division through 2050. Uganda's population growth rate is the third highest in the world. It is higher than the average of 2.4 percent per year for Africa south of the Sahara, at 3.2 or possibly 3.4 percent (UNDP 2005). According to the high variant of the UN projections, Uganda's population is expected to reach 102.2 million in 2050, and this projection assumes that population growth will decline to 2.9 percent per year between 2040 and 2050. Even the low variant, with 80 million people, will likely present a major challenge.

The current rate of population growth has significant negative impacts on agricultural systems, the environment, labor markets, and the health and educational systems. Effective population policies will be critical to avoid straining the economy and to achieve further reductions in poverty. Two policies proposed to constrain Uganda's population growth include education (especially targeted at females) and promoting employment outside the primary sector to reduce pressure on the environment (NEMA 2008). In Figure 12.11, all three scenarios show a population close to or exceeding 100 million by 2050—and, judging

FIGURE 12.11   Population projections for Uganda, 2010–2050

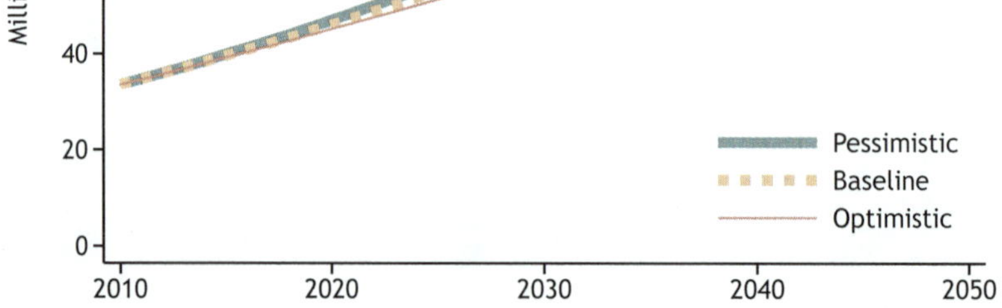

Source: UNPOP (2009).

from the pattern of the past three decades, the high-variant scenario appears the most likely, with negative implications for resilience to climate change.

Income

Figure 12.12 presents three overall scenarios for GDP per capita derived by combining three GDP scenarios with the three population scenarios of Figure 12.11 (based on UN population data). The optimistic scenario combines high GDP with low population scenarios for all countries, the baseline scenario combines the medium GDP projection with the medium population scenario, and the pessimistic scenario combines the low GDP scenario with the high population scenario. (The agricultural modeling in the next section uses these scenarios as well.)

Uganda's GDP per capita has been rising since the mid-1980s (see Figure 12.3) and was estimated at $504 as of June 2009. GDP growth has been driven by an economic transformation promoting the service and manufacturing sectors rather than primary-sector activities through a successful wave of economic policy reforms including liberalization and privatization. Uganda's economy also benefited from the worldwide boom in the telecommunications industry. Most experts now agree that Uganda's economy is running out of steam and will need a fresh wave of policy reforms.

FIGURE 12.12   Gross domestic product (GDP) per capita in Uganda, future scenarios, 2010–2050

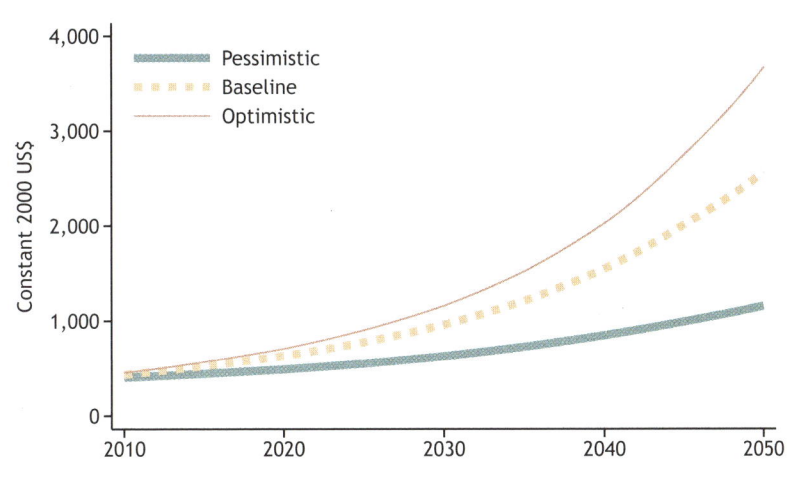

Sources: Computed from GDP data from the World Bank Economic Adaptation to Climate Change project (World Bank 2010), from the Millennium Ecosystem Assessment (2005) reports, and from population data from the United Nations (UNPOP 2009).

Note: US$ = US dollars.

Although Uganda's per capita GDP has consistently increased in absolute value, the growth rate recently dropped to 1.9 percent between 2008 and 2009 from a high of 6.9 percent between 2007 and 2008 (see Table 12.3).

Biophysical Analysis

Climate Models

Figure 12.13 shows projected annual precipitation changes under the four downscaled climate models (general circulation models, or GCMs) we use in this chapter with the A1B scenario.[1] The models seem quite diverse in their projections. The MIROC 3.2 model is the wettest model, predicting a median precipitation increase of 150 millimeters and a peak near 250 millimeters. The dry model is the CRNM-CM3 model, in which the southeastern quarter of Uganda is projected to receive more than 100 millimeters per year less rainfall, while much of the rest of the country will see no significant change. CSIRO Mark 3 suggests that rainfall will decline in the southwest rather than the southeast.

Climate change models developed for Uganda (Figure 12.14), as well as projections by the Intergovernmental Panel on Climate Change (IPCC 2001), point to an increase in temperature of between 0.5° and 1.5°C by the year 2020. The projections used in this chapter for 2050 show diversity (see Figure 12.14). Both the CNRM-CM3 and the ECHAM 5 models suggest that most of the country will experience a rise in high temperatures in the warmest month of between 2° and 2.5°C. Yet the MIROC 3.2 model suggests that the increase will be only around 1°C on average, and the CSIRO model predicts temperatures only slightly warmer. Official vulnerability assessments for Uganda identified precipitation as the most important variable related to climate change (NEMA 2006–2007).

The modeled climate changes are envisaged to have an impact on agriculture, forestry, and fisheries, which together presently account for 14.6 percent of Uganda's total GDP at constant 2002 prices (Republic of Uganda, MFPED 2010). Coffee production, accounting for 23.4 percent of total export earnings in 2008 (UBOS 2009), has been cited as the most vulnerable sector: the

---

1 CNRM-CM3 is National Meteorological Research Center–Climate Model 3. MIROC 3.2 is the Model for Interdisciplinary Research on Climate, developed at the University of Tokyo Center for Climate System Research. CSIRO Mark 3 is a climate model developed at the Australia Commonwealth Scientific and Industrial Research Organisation. ECHAM 5 is a fifth-generation climate model developed at the Max Planck Institute for Meteorology in Hamburg. The A1B scenario is a greenhouse gas emissions scenario that assumes fast economic growth, a population that peaks midcentury, and the development of new and efficient technologies, along with a balanced use of energy sources.

FIGURE 12.13 Changes in mean annual precipitation in Uganda, 2000–2050, A1B scenario (millimeters)

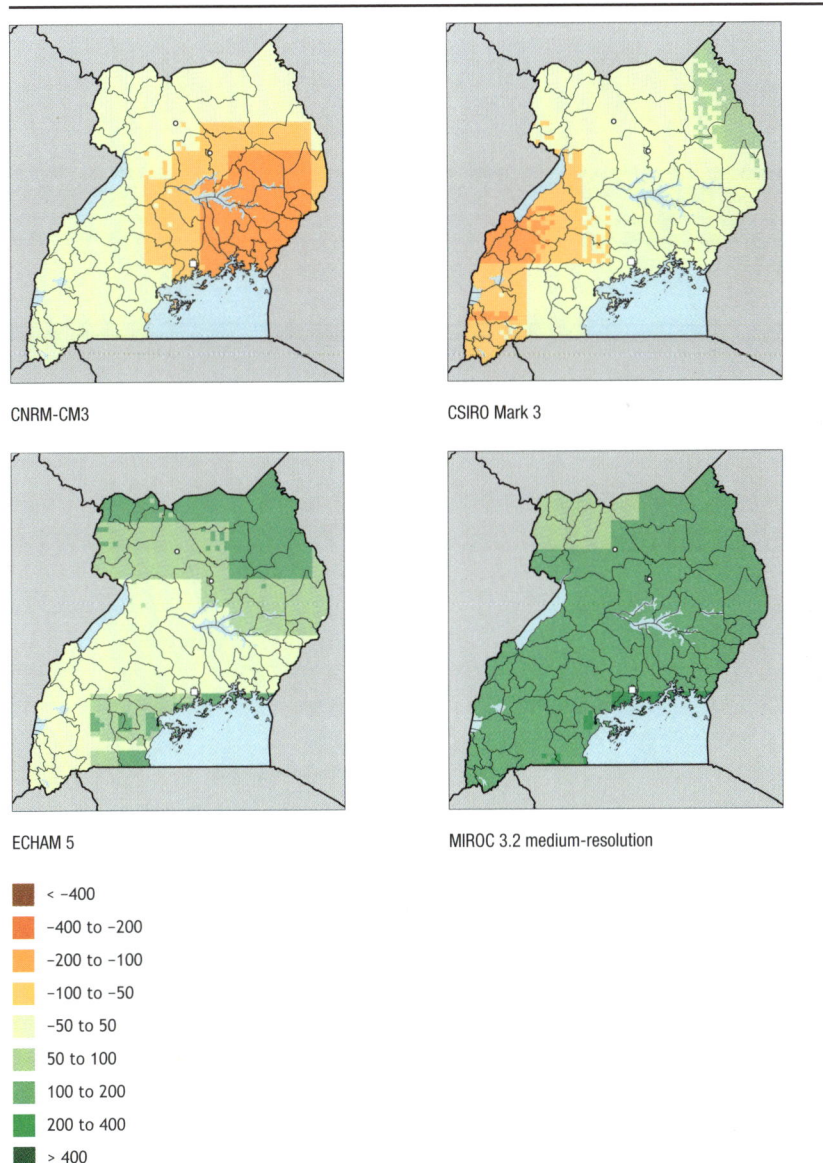

CNRM-CM3

CSIRO Mark 3

ECHAM 5

MIROC 3.2 medium-resolution

■ < –400
■ –400 to –200
■ –200 to –100
□ –100 to –50
□ –50 to 50
■ 50 to 100
■ 100 to 200
■ 200 to 400
■ > 400

Source: Authors' calculations based on Jones, Thornton, and Heinke (2009).

Notes: A1B = greenhouse gas emissions scenario that assumes fast economic growth, a population that peaks midcentury, and the development of new and efficient technologies, along with a balanced use of energy sources; CNRM-CM3 = National Meteorological Research Center–Climate Model 3; CSIRO = climate model developed at the Australia Commonwealth Scientific and Industrial Research Organisation; ECHAM 5 = fifth-generation climate model developed at the Max Planck Institute for Meteorology (Hamburg); GCM = general circulation model; MIROC = Model for Interdisciplinary Research on Climate, developed by the University of Tokyo Center for Climate System Research.

FIGURE 12.14   Changes in monthly mean maximum daily temperature in Uganda for the warmest month, 2000–2050, A1B scenario (°C)

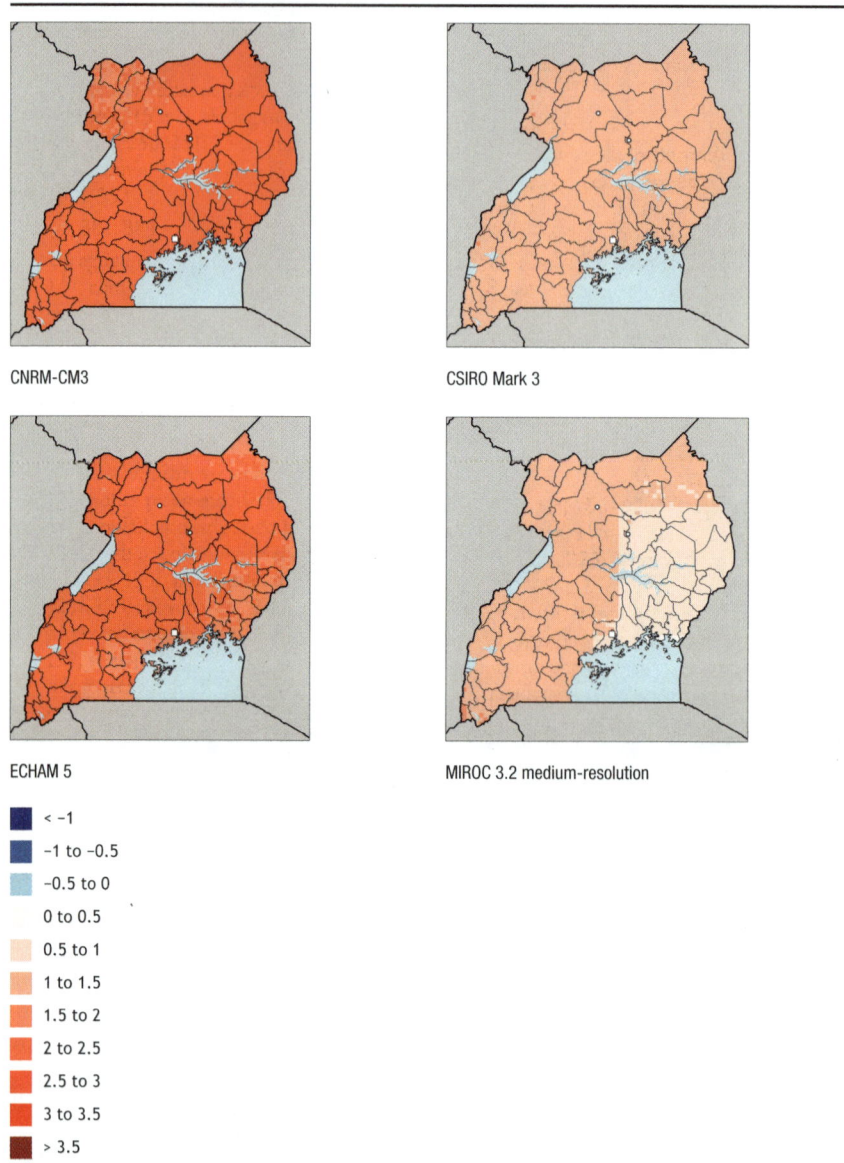

CNRM-CM3

CSIRO Mark 3

ECHAM 5

MIROC 3.2 medium-resolution

■ < −1
■ −1 to −0.5
■ −0.5 to 0
  0 to 0.5
  0.5 to 1
  1 to 1.5
  1.5 to 2
  2 to 2.5
  2.5 to 3
  3 to 3.5
■ > 3.5

Source: Authors' calculations based on Jones, Thornton, and Heinke (2009).

Notes: A1B = greenhouse gas emissions scenario that assumes fast economic growth, a population that peaks midcentury, and the development of new and efficient technologies, along with a balanced use of energy sources; CNRM-CM3 = National Meteorological Research Center–Climate Model 3; CSIRO = climate model developed at the Australia Commonwealth Scientific and Industrial Research Organisation; ECHAM 5 = fifth-generation climate model developed at the Max Planck Institute for Meteorology (Hamburg); GCM = general circulation model; MIROC = Model for Interdisciplinary Research on Climate, developed by the University of Tokyo Center for Climate System Research.

most pessimistic projections show Uganda's coffee production entirely wiped out in less than 100 years (Republic of Uganda, NPA 2010). The potential climate changes are also envisaged to have an impact on the lifespan and durability of infrastructure, including roads and the energy sector, especially because the largest portion of Uganda's clean energy requirements are met by hydroelectricity (Republic of Uganda, NPA 2010). Climate change may have negative effects on livelihood systems, leading to a decline in water rights, increased insecurity, rising unemployment, and increased spread of HIV/AIDS (NEMA 2006–2007). In 2007, with support from the Global Environmental Fund, Uganda launched a National Adaptation Programme of Action whose provisions have yet to be implemented.

According to the National Environment Management Authority (NEMA 2008), the average long-term annual rainfall (over the past decade) for Uganda has been about 1,318 millimeters; the main concern is the seasonal distribution of rain as well as the type of rain. The onset and cessation of rains are increasingly erratic, and the sporadic rainfall is heavier and more violent. According to Oxfam (2008), minimum temperatures have been rising faster than maximum temperatures; temperatures rose from about 1960 to 1982, then declined before rising again since the 1990s. The trend has been steadily upward, to an increase of about 1°C (Oxfam 2008). These observations conform to the mapped results from MIROC 3.2, which show increased amounts of rainfall and a moderate temperature increase as the most likely scenario by 2050.

Crop Models

Comparing future yield results to current or baseline yield results, both from the DSSAT software, produced the results for rainfed maize mapped in Figure 12.15. The comparison is between the crop yields for 2050 with climate change and the yields with an unchanged 2000 climate. In the figure we see that the climate models diverge in their predictions. This is not entirely surprising given how precipitation and temperature predictions diverged among the models. We see that in the MIROC 3.2 model the maize yield will generally rise, and in many places that rise will be greater than 25 percent. Given that the MIROC 3.2 model shows very modest temperature increases and significant rainfall increases, this makes sense. Both the CSIRO Mark 3 and the ECHAM 5 models predict mostly yield reductions from climate change. It is much more difficult to generalize from these results, because even in locations where annual rainfall is to increase, we can see yield reductions. One way to understand this is that although the figures show annual rainfall changes and temperature changes for the warmest month, the crop models respond to the

FIGURE 12.15   Yield change under climate change: Rainfed maize in Uganda, 2000–2050, A1B scenario

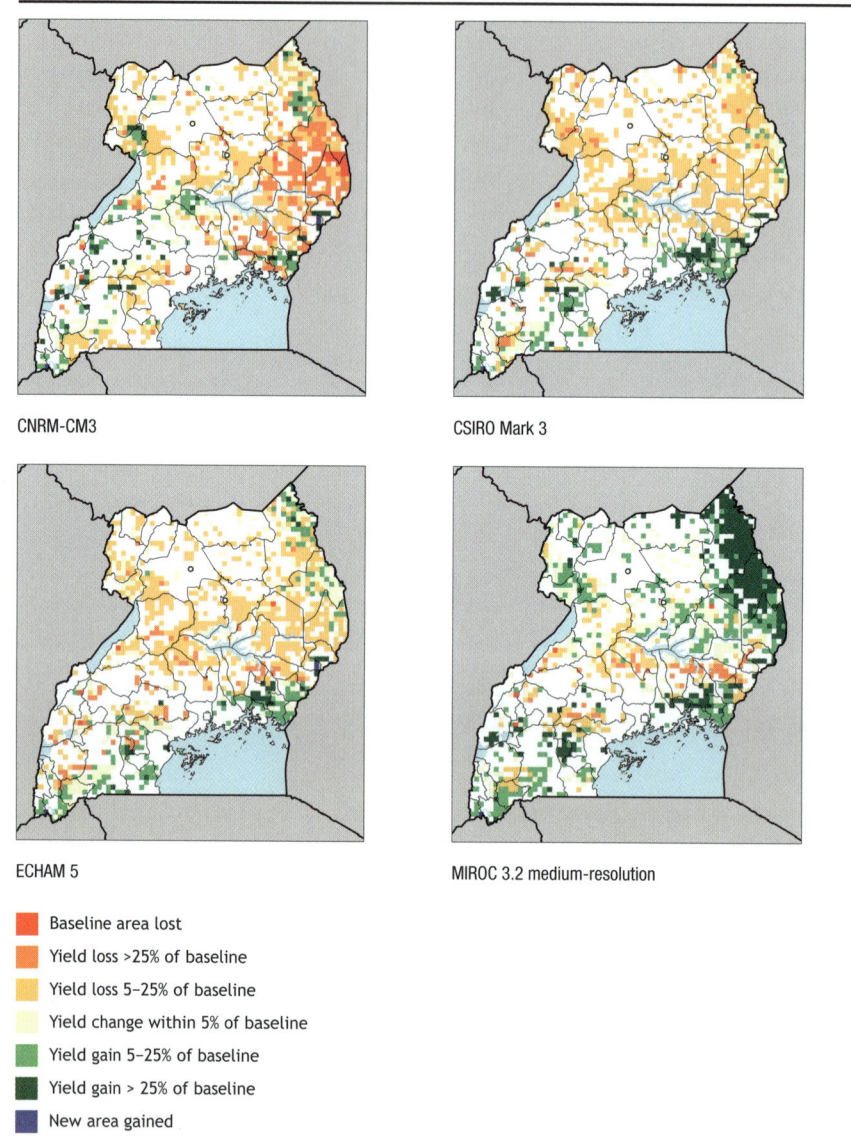

CNRM-CM3

CSIRO Mark 3

ECHAM 5

MIROC 3.2 medium-resolution

- ■ Baseline area lost
- ■ Yield loss >25% of baseline
- ■ Yield loss 5–25% of baseline
- ■ Yield change within 5% of baseline
- ■ Yield gain 5–25% of baseline
- ■ Yield gain > 25% of baseline
- ■ New area gained

Source: Authors' calculations.

Notes: A1B = greenhouse gas emissions scenario that assumes fast economic growth, a population that peaks midcentury, and the development of new and efficient technologies, along with a balanced use of energy sources; CNRM-CM3 = National Meteorological Research Center–Climate Model 3; CSIRO = climate model developed at the Australia Commonwealth Scientific and Industrial Research Organisation; ECHAM 5 = fifth-generation climate model developed at the Max Planck Institute for Meteorology (Hamburg); GCM = general circulation model; MIROC = Model for Interdisciplinary Research on Climate, developed by the University of Tokyo Center for Climate System Research.

rainfall and temperature of the months during which a crop is growing. As an example, annual rainfall can be reduced, but the rainfall during the growing season might increase.

The crop mode results for the CNRM-CM3 GCM predict severe losses in the east, to the extent that not only do some of the losses exceed 25 percent but also area is lost for maize because the climate is expected to become too inhospitable. Checking previous figures, we note that these areas are predicted to have both a large temperature increase and a large precipitation decrease.

Vulnerability

Figure 12.16 shows the impact of future GDP and population scenarios on the number of malnourished children under age five in Uganda. The figure shows the number of malnourished children under age five increasing initially but beginning to decline between about 2025 and 2040—in other words, getting worse before it gets better. At least in the optimistic and baseline

FIGURE 12.16   Number of malnourished children under five years of age in Uganda in multiple income and climate scenarios, 2010–2050

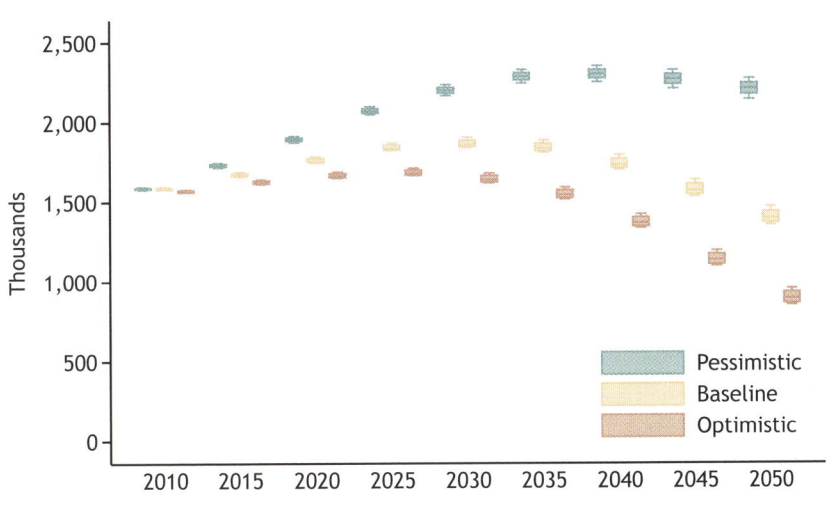

Source: Based on analysis conducted for Nelson et al. (2010).

Note: The box and whiskers plot for each socioeconomic scenario shows the range of effects from the four future climate scenarios.

FIGURE 12.17 Share of malnourished children under five years of age in Uganda in multiple income and climate scenarios, 2010–2050

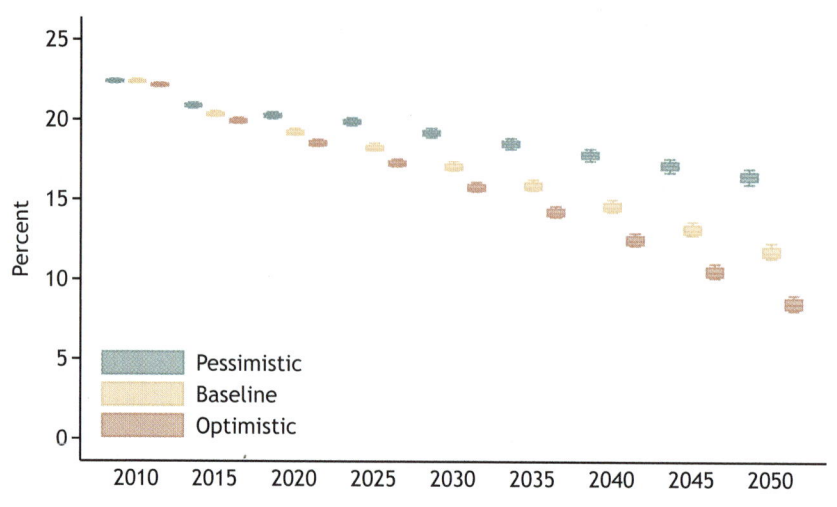

Source: Based on analysis conducted for Nelson et al. (2010).

Note: The box and whiskers plot for each socioeconomic scenario shows the range of effects from the four future climate scenarios.

scenarios, the number of malnourished children will be lower in 2050 than in 2010. With such high population rates, the increasing numbers of malnourished children through the years will result in declining malnutrition rates. Figure 12.17 shows the share of children who have been and will be malnourished from 2010 to 2050.

Figure 12.18 shows the kilocalories per capita available, showing a generally unchanged per capita calorie intake in the pessimistic scenario but greatly improving availability in the baseline scenario and especially in the optimistic scenario. Generally speaking, Uganda is food self-sufficient and a net food exporter, although localized hunger may exist in some remote areas of the country due to security problems, poor road networks, and natural disasters. The main nutrition concern relates not to calorie availability but rather to "hidden hunger," driven by micronutrient deficiency.

Agricultural Outcomes

Figures 12.19 and 12.20 show simulation results from IMPACT associated with key agricultural crops in Uganda. Each featured crop has five graphs for production, yield, area harvested, net exports, and world price.

FIGURE 12.18   Kilocalories per capita in Uganda in multiple income and climate scenarios, 2010–2050

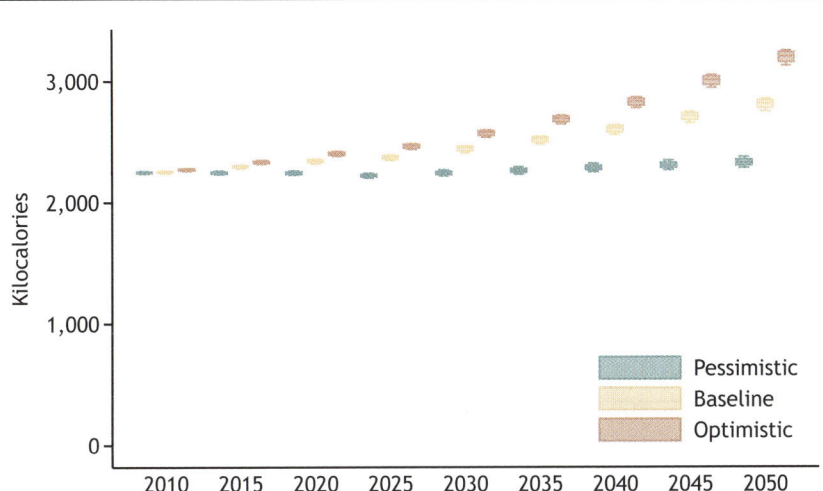

Source: Based on analysis conducted for Nelson et al. (2010).

Note: The box and whiskers plot for each socioeconomic scenario shows the range of effects from the four future climate scenarios.

Figure 12.19 shows yields of maize tripling by 2050, though the area cultivated is projected not to change very much. Net maize exports are projected to grow, with production staying ahead of the consumption demands of a rapidly growing population. The world price of maize is shown to double during that period.

The yield of cassava in Uganda, as depicted in Figure 12.20, is projected to increase by around 80 percent between 2010 and 2050. But with a projected decline in area of around 15 percent, production will expand by only 40 percent. The increase in production will be too little to keep pace with domestic demand, and imports of cassava are projected to rise.

Uganda has recently been selected by the World Bank as the center of excellence in cassava production, and this will spur technology development, dissemination, and adoption, further contributing to yield increases for cassava. Furthermore, as cassava is known to do well in various environments, tolerating drought and high levels of rainfall, its production might be least affected by climate change.

In practice, it is doubtful that Uganda will import substantial amounts of cassava and other roots and tubers owing to the perishability and bulkiness of

FIGURE 12.19 Impact of changes in GDP and population on maize in Uganda, 2010–2050

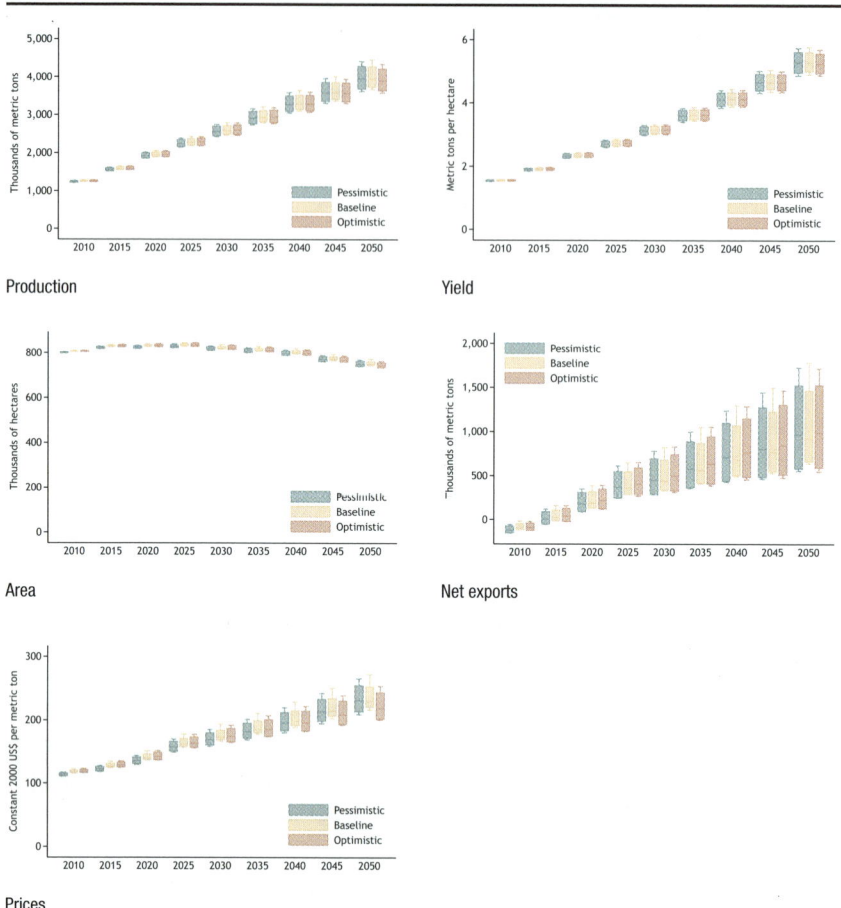

Production

Yield

Area

Net exports

Prices

Source: Based on analysis conducted for Nelson et al. (2010).

Notes: The box and whiskers plot for each socioeconomic scenario shows the range of effects from the four future climate scenarios. GDP = gross domestic product; US$ = US dollars.

these commodities and the current low levels of processing (value addition) technology. A report of an informal cross-border trade survey (UBOS 2006b) reports the following numbers: cassava imports worth $388,067, or 2.1 percent of the total informal imports; sweet potato imports worth $212,822, or 1.1 percent of total informal imports; and potato imports worth $76,066, or 0.4 percent of all informal imports. World prices are likely to remain unchanged over time, and Uganda's production will not affect world root and tuber prices.

FIGURE 12.20   Impact of changes in GDP and population on cassava in Uganda, 2010–2050

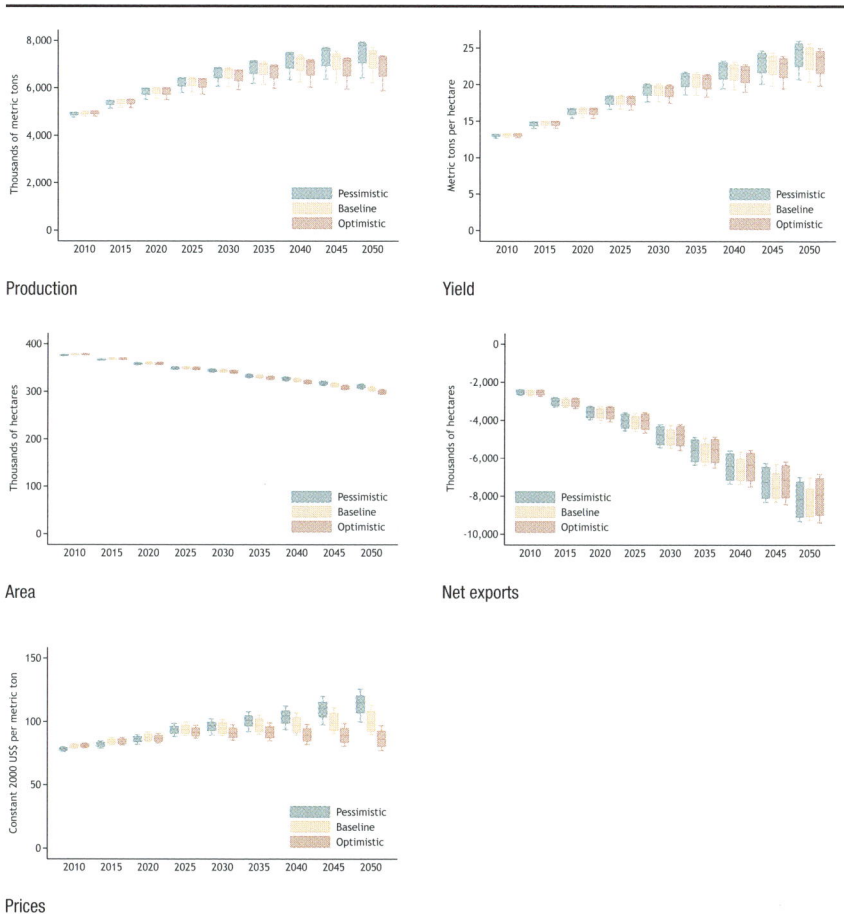

Production

Yield

Area

Net exports

Prices

Source: Based on analysis conducted for Nelson et al. (2010).

Notes: The box and whiskers plot for each socioeconomic scenario shows the range of effects from the four future climate scenarios. GDP = gross domestic product; US$ = US dollars.

Conclusions and Policy Recommendations

Recent data show that the population of Uganda continues to grow at a high rate, remains predominantly rural, and is still dependent on agriculture as its main livelihood source. Although per capita GDP steadily increased over the past decade, the rate has slowed, possibly due to drought and the global economic crisis.

Uganda's universal primary and secondary education policies have increased school enrollment and literacy rates. This trend augurs well for the population's capacity to adapt to and mitigate the impacts of climate change. Life expectancy remains low, reflecting a high disease burden and inadequate sanitation and medical services, factors that would compound Uganda's vulnerability to climate change and constrain its response.

The poverty rate has continued to decline in the face of such challenges as the recent international financial crisis, slower economic growth, and a persistently high rate of population growth. However, poverty remains higher in the rural areas and the northern region, implying that these areas have fewer resources to deal with any adverse impacts of climate change.

Physical infrastructure, critical for agriculture modernization and resilience to climate change, is largely adequate. Except for a few remote areas, most Ugandan communities have easy access to feeder roads, although access to tarmac roads is still limited.

Although regional differences exist, plantains, beans, and cassava remain the most widely grown and consumed commodities. Overreliance on this commodity basket may be compounding the country's vulnerability to climate change through the avenue of crop disease. The traditional crops are increasingly under pressure from banana bacterial wilt and the African cassava mosaic virus (Republic of Uganda, MWE 2007; Kangire et al. 2011).

Uganda has a fairly large and fast-growing population—currently adding about 1.2 million persons per year—that creates challenges for agricultural systems, the environment, employment, markets, and health and limits the country's resilience to climate change. Population projections to 2050 imply that there will be increased population pressure on the environment.

The climate change scenarios used in this chapter are highly diverse in their projections for rainfall and temperature rise. Other studies suggest that rainfall will be more erratic and violent, further disrupting the predominantly rainfed agricultural production system, with coffee predicted to be the most affected. The predicted future climate will also affect the productive infrastructure and will exacerbate the constraints on the other livelihood systems of the majority of Ugandans (Republic of Uganda, Department of Disaster Preparedness et al. 2007).

Despite the likely challenges of climate change, too little has been done so far to reduce the potential impacts on food security and increase the resilience of poor rural communities and households. Policies have been made, but few actions have been taken.

The research community is striving to generate the information needed to increase community awareness and to guide policy formulation designed to enhance the resilience of those most vulnerable to climate change and to mitigate its impacts. The likely agricultural scenarios for climate change in Uganda show production increases for the majority of foodcrops due to acreage expansion for some and technological advancement for others. Disease pressure will increase for the major staples (MWE 2007). Maize exports and cassava imports are predicted to rise; the export picture is less clear for other commodities. The livelihood and food security implications are still unclear, underscoring the need for regular analyses to document the emerging trends.

References

Bartholome, E., and A. S. Belward. 2005. "GLC2000: A New Approach to Global Land Cover Mapping from Earth Observation Data." *International Journal of Remote Sensing* 26 (9–10): 1959–1977.

CIESIN (Center for International Earth Science Information Network), Columbia University, IFPRI (International Food Policy Research Institute), World Bank, and CIAT (Centro Internacional de Agricultura Tropical). 2004. *Global Rural–Urban Mapping Project (GRUMP), Alpha Version: Population Density Grids.* Palisades, NY, US: Socioeconomic Data and Applications Center (SEDAC), Columbia University. http://sedac.ciesin.columbia.edu/gpw.

Deressa, T., and R. M. Hassan. 2009. "Economic Impact of Climate Change on Crop Production in Ethiopia: Evidence from Cross-Section Measures." *Journal of African Economics* 18 (4): 529–554.

FAO (Food and Agriculture Organization of the United Nations). 2010. FAOSTAT. Rome. http://faostat.fao.org.

IPCC (Intergovernmental Panel on Climate Change). 2001. *Climate Change 2001: Impacts, Adaptations, Vulnerability.* Contribution of Working Group II to the third assessment report of the IPCC. Geneva: United Nations Environment Programme / World Meteorological Organization.

Jones, P. G., P. K. Thornton, and J. Heinke. 2009. *Generating Characteristic Daily Weather Data Using Downscaled Climate Model Data from the IPCC's Fourth Assessment.* Project report for the International Livestock Research Institute. Geneva: International Panel on Climate Change.

Kangire, A., P. V. Asten, J. Verhagen, and I. Koomen. 2011. "Towards Climate Smart Agriculture: Lessons from a Coffee x Banana Case; Experiences from Research for Policy Support in Uganda. Accessed April 17, 2012. www.naro.go.ug/ or www.iita.org/ or http://portals.wi.wur.nl/climatechange/.

Lehner, B., and P. Döll. 2004. "Development and Validation of a Global Database of Lakes, Reservoirs, and Wetlands." *Journal of Hydrology* 296 (1–4): 1–22.

Millennium Ecosystem Assessment. 2005. *Ecosystems and Human Well-being: Synthesis.* Washington, DC: Island Press. www.maweb.org/en/Global.aspx.

Nelson, G. C., M. W. Rosegrant, A. Palazzo, I. Gray, C. Ingersoll, R. Robertson, S. Tokgoz, et al. 2010. *Food Security, Farming, and Climate Change to 2050: Scenarios, Results, Policy Options.* Washington, DC: International Food Policy Research Institute.

NEMA (National Environment Management Authority). 2006–07. *State of the Environment Report for Uganda.* Kampala: NEMA House.

——. 2008. *State of the Environment Report for Uganda.* Kampala: NEMA House.

——. 2009. *State of the Environment Report for Uganda.* Kampala: NEMA House.

Oxfam. 2008. *Survival of the Fittest: Pastoralism and Climate Change in East Africa.* Oxfam Briefing Paper 116. Oxford, UK: Oxfam International.

Republic of Uganda. 2010. *National Development Plan (2010/11–2014/15).* Kampala.

Republic of Uganda, Department of Disaster Preparedness and Refugees, Office of the Prime Minister, and Ministry of Water and the Environment. 2007. "Impacts of El Niño in Selected Districts of Uganda." Accessed September 25, 2010. www.meteo-uganda.net.

Republic of Uganda, MAAIF (Ministry of Agriculture, Animal Industry and Fisheries) and MFPED (Ministry of Finance, Planning and Economic Development). 2000. *Plan for Modernization of Agriculture (PMA): Eradicating Poverty in Uganda.* Kampala.

Republic of Uganda, MFPED (Ministry of Finance, Planning and Economic Development). 2010. *The Background to the Budget 2010/11 Fiscal Year: Strategic Priorities to Accelerate Growth, Employment and Socio-Economic Transformation for Prosperity.* Kampala.

Republic of Uganda, MWE (Ministry of Water and Environment). 2007. *Climate Change: Uganda National Adaptation Programmes of Action.* Kampala.

Republic of Uganda, MWLE (Ministry of Water, Lands, and the Environment). 2006. *National Biomass Study.* Forestry Department. Kampala.

Republic of Uganda, NPA (National Planning Authority). 2010. *National Development Plan (2010/11–2014/15).* Kampala.

UBOS (Uganda Bureau of Statistics). 2006a. *Statistical Abstract.* Kampala: Statistics House.

——. 2006b. *The Informal Cross Border Trade Survey Report.* Kampala: Statistics House.

——. 2009. *Statistical Abstract.* Kampala: Statistics House.

——. 2010. *Uganda National Household Survey 2009/2010.* Abridged report (November). Kampala: Statistics House.

UBOS (Uganda Bureau of Statistics) and ILRI (International Livestock Research Institute). 2003–04. *Where Are the Poor: Mapping Patterns of Well-Being in Uganda 1992 and 1999.* Entebbe.

UNDP (United Nations Development Programme). 2005. *Uganda Human Development Report 2005: Linking Environment to Development; A Deliberate Choice.* New York.

UNEP (United Nations Environment Programme) and IUCN (International Union for Conservation of Nature). 2009. World Database on Protected Areas (WDPA) Annual Release. No longer available online.

UNPOP (United Nations Secretariat, Department of Economic and Social Affairs, Population Division). 2009. *World Population Prospects: The 2008 Revision.* Accessed April 6, 2010. http://esa.un.org/unpp.

Wood, S., G. Hyman, U. Deichmann, E. Barona, R. Tenorio, Z. Guo, S. Castano, O. Rivera, E. Diaz, and J. Marin, 2010. "Sub-national Poverty Maps for the Developing World Using International Poverty Lines: Preliminary Data Release." Accessed May 6. http://povertymap.info.

World Bank. 2009. *World Development Indicators.* Accessed May 2011. http://data.worldbank.org/data-catalog/world-development-indicators.

———. 2010. *Economics of Adaptation to Climate Change: Synthesis Report.* Washington, DC. Accessed July 17, 2012. http://climatechange.worldbank.org/content/economics-adaptation-climate-change-study-homepage.

You, L., and S. Wood. 2006. "An Entropy Approach to Spatial Disaggregation of Agricultural Production." *Agricultural Systems* 90 (1–3): 329–347.

You, L., S. Wood, and U. Wood-Sichra. 2006. "Generating Global Crop Distribution Maps: From Census to Grid." Paper presented at the International Association of Agricultural Economists Conference, Brisbane, Australia, August 11–18.

———. 2009. "Generating Plausible Crop Distribution and Performance Maps for Sub-Saharan Africa Using a Spatially Disaggregated Data Fusion and Optimization Approach." *Agricultural Systems* 99 (2–3): 126–140.

Chapter 13

SUMMARY AND CONCLUSIONS

Michael Waithaka, Miriam Kyotalimye, Timothy S. Thomas,
and Gerald C. Nelson

Agriculture drives the economies and accounts for 43 percent of the annual gross domestic product (GDP) in ten countries of Eastern and Central Africa (ECA): Burundi, Democratic Republic of Congo (DRC), Eritrea, Ethiopia, Kenya, Madagascar, Rwanda, Sudan, Tanzania, and Uganda. In Burundi, DRC, Ethiopia, Sudan, and Tanzania, agriculture accounts for more than 50 percent of GDP, while in Kenya, Eritrea, and Madagascar it accounts for less than 30 percent.

Events demonstrative of extreme weather patterns have become common phenomena, suggesting that climate change is already affecting both ECA and the world. Examples of these include prolonged droughts in parts of Ethiopia, Kenya, and Tanzania in 2011 and devastating floods in Brazil, Pakistan, and New Zealand in 2010 (*Economist* 2010), as well as in parts of Kenya, Tanzania, and Uganda. Climate change will have far-reaching consequences for the poor and marginalized groups, among which the majority depend on agriculture for their livelihoods and have a lower capacity to adapt. Weather-related crop failures and livestock deaths in addition to loss of property are already causing economic losses and undermining food security in ECA. This situation is likely to become more desperate and to threaten the very survival of the most vulnerable farmers as global warming continues. Feeding the increasing populations in the subregion, with one of the highest population growth rates in the world (averaging 3.7 percent in nine counties), requires a radical transformation of agriculture over the next four decades. A major challenge is increasing agricultural production among resource-poor farmers without exacerbating environmental problems and simultaneously coping with climate change.

Challenges commonly faced that limit adaptation to climate change include the following: fragile ecosystems; weak infrastructure and economies; poor performance of agriculture; the dependence of food security on rainfall; the reliance of more than 70 percent of the population in the ten countries on climate-sensitive resources for their livelihood; severe poverty and deteriorating livelihoods; limited reliable, accurate, and updated statistical information; inadequate capacities at state and local levels; and lack of policy coordination.

Model Predictions

The climate models (general circulation models, or GCMs) we have used differ on the amount of annual precipitation change that can be expected and the location of that change.[1] The wettest model for ECA is the MIROC 3.2 model, which predicts an increase of more than 100 millimeters per year for over half of the region and only small areas of rainfall decline, notably in western DRC and southern Madagascar. The CSIRO Mark 3 model is probably the driest of the GCMs, predicting a significant rainfall decrease for a large part of DRC, Ethiopia, and Madagascar but no significant change for most areas.

The CSIRO Mark 3 GCM projects the coolest future of the four models. For 90 percent of the region, temperature increases will be in the range of 1°–1.5°C. The MIROC 3.2 model predicts a median value similar to that of the CSIRO Mark 3 model, but the range in the MIROC 3.2 model is much greater. Some areas are predicted to have a change of less than 0.5°C, while other areas are predicted to have a change of more than 2.5°C. Both the CNRM-CM3 and the ECHAM 5 GCMs predict a warmer future than do the other two models; their median value is more than 0.5°C higher than the medians of the other two models, with temperatures generally ranging between 1.5° and 2.5°C and some places, such as northern Sudan, predicted to have changes of up to 3°C. All four GCMs predict that the ten countries will become warmer with different levels of increase in temperature. The rise in temperature will increase evaporation and reduce the soil's moisture, and these may increase the plants' water requirements. Therefore, an increase in temperature would be unfavorable, particularly if associated with lower levels of precipitation and irrigation water.

UN population projections suggest a significant increase in the populations of the ten ECA countries by 2050. In the pessimistic scenario, the populations of all countries in the region will more than double. A similar outcome

1 CNRM-CM3 is National Meteorological Research Center–Climate Model 3. MIROC 3.2 is the Model for Interdisciplinary Research on Climate, developed at the University of Tokyo Center for Climate System Research. CSIRO Mark 3 is a climate model developed at the Australia Commonwealth Scientific and Industrial Research Organisation. ECHAM 5 is a fifth-generation climate model developed at the Max Planck Institute for Meteorology in Hamburg.

is seen in the baseline scenario except in Burundi and Sudan. In the optimistic scenario, the population doubles only in DRC, Ethiopia, Tanzania, and Uganda. Income per capita improves significantly in the optimistic scenario, with increases of up to ninefold in Burundi and down to fourfold in Sudan. This situation is further reflected in the baseline scenario. In the pessimistic scenario, the greatest increase is a meager fourfold in Burundi and Tanzania. In the optimistic scenario, the number of malnourished children decreases for all the countries in ECA. In the pessimistic scenario, the number increases in all countries.

Based on the crop models used, which restrict the analysis by not allowing for adaptation, both the CSIRO Mark 3 and the MIROC 3.2 climate scenarios result in a general increase in maize yields of 5–25 percent of baseline in most parts of the countries of ECA and a yield loss of 5–25 percent in large parts of DRC, Ethiopia, Tanzania, and northern Uganda. Based on both the CSIRO Mark 3 and the MIROC 3.2 climate outcomes using the A1B scenario from the Special Report on Emissions Scenarios of the Intergovernmental Panel on Climate Change (IPCC), sorghum yields will decline by 5–25 percent across ECA.[2] Both models also show a gain in baseline area of 5–25 percent in western DRC, the highlands of Ethiopia, Kenya, Sudan, and Tanzania. What is important to note from the crop models is that in some countries climate change will present opportunities to grow crops in areas where they could not previously have grown. Generally these areas are ones that have been too cold for the crop, but the warming brought about by climate change will make them more suitable for those crops. Sometimes the reason is an increase in precipitation. In any case, these areas may not have very strong legal protection because they have not been threatened with conversion to agriculture under the current climate. Policymakers will need to decide whether to strengthen protection so as to preserve these areas or to strengthen the farmers' abilities to settle in them.

From the global partial equilibrium food and agriculture model, the International Model for Policy Analysis of Agricultural Commodities and Trade (IMPACT), we find that world market prices for maize, rice, sorghum, and wheat increase in all scenarios, while the price of millet is less in 2050 than in 2010. In 2050 the prices of millet, rice, sorghum, and wheat are higher in the pessimistic scenario than in the optimistic scenario. In terms of the impacts

---

2 The A1B scenario is a greenhouse gas emissions scenario that assumes fast economic growth, a population that peaks midcentury, and the development of new and efficient technologies, along with a balanced use of energy sources.

of climate change on prices, when we looked at the median values, millet was least affected, with only a 2.5 percent increase in price due to climate change impacts alone; of the five grains, maize had the highest price rise due to climate change, with a median value of 28.2 percent. Sorghum, rice, and wheat were in between, with increases of 14.5 percent, 19.0 percent, and 21.7 percent, respectively.

The production of maize, millet, and sorghum will increase in all countries of ECA by 2050 except in Ethiopia, Madagascar, and Uganda. The area under cultivation of both millet and sorghum will increase except in Burundi for millet. The productivity of all three crops will increase, mainly due to improved management practices, because the yields projected for 2050 are already attainable with existing local and improved varieties of these crops.

There will also be variation with various model outputs even if they are used under the assumption of similar scenarios. The projected changes in demographic and economic indicators coupled with the projected changes in rainfall and temperature suggest the complexity of the problems facing future agriculture. The projected changes in yield and area of major crops including maize, wheat, and sorghum, coupled with the increasing demand of kilocalories for the growing population, demonstrate the level of vulnerability of the agricultural sector and the challenge of producing enough food. This points to the urgent need to make available alternative adaptation options that will fit into the various plausible scenarios.

The effect of climate change varies across countries depending on geographic, social, economic, cultural, and cultural factors. Countries like DRC, Ethiopia, Kenya, and Tanzania, with diverse agroecologies, relatively large areas, and rapidly growing populations, need to make available and successfully implement robust adaption programs to reach all who are vulnerable. Such programs will likely need to develop a menu of options from which farmers can choose appropriate strategies for their locations, assisted by extension agents who are supported by agricultural researchers. Indeed, not only is food security an explicit concern under climate change; successful adaptation responses in agriculture can be achieved only within the ecological, economic, and social sustainability goals of each country. Owing to their existing vulnerability and potential future changes, the countries of ECA need mechanisms necessary to support adaptation, mitigation, and technology transfer.

The results from the models and scenarios can be used to define future policy directions; therefore, policymakers need to take account of the importance of climate change adaptation in any policy development. It is also important that future development policies favor the institutionalization and mainstreaming of climate change adaptation within the research and development arena.

There is a growing awareness of the potential adverse effects of climate variability and change in both regional institutions and national governments. This has resulted in various initiatives aimed at addressing climate change issues. At the regional level, the East African Community recently published a climate change policy (EAC 2011). The policy is aimed, among other things, at guiding the region on climate change actions, establishing a climate change fund to specifically support adaptation and mitigation activities, developing research institutions of excellence in technology development for climate change adaptation, and mainstreaming climate change in the national development processes. At the national level, the ten countries have already undertaken several activities. They have ratified the United Nations Framework Convention on Climate Change and submitted their initial national communications. They also have National Adaptation Programmes of Action (NAPAs) and a host of other strategies focusing on climate change.

National Adaptation Programmes of Action

NAPAs from the ten countries mention 26 strategies to adapt climate change to agriculture (Nzuma et al. 2010). The strategies common to all member countries include the development and promotion of drought-tolerant and early-maturing crop species and the exploitation of new and renewable energy sources. Most countries have areas that are classified as arid or semiarid and hence need to develop drought-tolerant and early-maturing crops. Strangely, only one country, Ethiopia, recognizes the conservation of genetic resources as an important strategy, although this is potentially important for dealing with drought. Biomass energy resources account for more than 70 percent of the total energy consumption in the ten ECA countries. To mitigate the potential adverse effects of biomass energy depletion, the ten countries plan to harness new and renewable energy sources, including solar power, wind power, hydro and geothermal sources, and biofuels.

Eight of the ten countries cite the promotion of rainwater harvesting as an important adaptation strategy, either on a small scale with small check dams or on a large scale with large dam projects.

The five measures that are common to more than five countries are (1) conservation and restoration of vegetative cover in degraded and mountainous areas; (2) reduction of overall livestock numbers through sale or slaughter; (3) cross-breeding, zero-grazing, and acquisition of smaller livestock (for example, sheep or goats); (4) adoption of traditional methods of natural forest conservation and food use; and (5) establishment of community-based management programs for forests, rangelands, and national parks.

The promotion of environmentally friendly investments and clean development mechanism projects that can be funded through carbon trading is a feature of only one country, Ethiopia.

Three examples of strategies that warrant greater regionwide collaboration are conservation of genetic materials, cited by Burundi; development and promotion of drought-tolerant species, cited by all countries; and soil conservation, cited by Burundi, Rwanda, and Tanzania. To date, the NAPAs of only three countries (Burundi, Rwanda, and Tanzania) have indicated that they carry out these strategies.

Adaptation includes investments in improved land management, adjustment of planting dates, and introduction of new crop varieties; mitigation includes improving energy efficiency and crop yields and increasing carbon storage through new land management techniques. Agricultural adaptation investments remain limited, likely because climate change issues are perceived as long term, whereas political horizons are short term.

Climate-Sensitive Development Policies

Apart from development of technologies as a means of adapting to climate change, action is required at the policy level. Policies are needed to build specific adaptive capacity in some of the most affected areas, to integrate climate change concerns into existing policies, and to ensure that program do not further undermine the resilience of the poor when they are faced with climate change. Some of the commonly cited areas for policy reform mentioned in the chapters of this monograph include the following:

- **Research and extension.** There are already a number of research programs focused on climate change. For example, a project on crops and livestock in Sudan started in 2011. Different types of development projects are also being conducted in DRC, Kenya, Madagascar, Rwanda, Tanzania, and Uganda that consider the issue of climate change. However, more research programs and development projects in different disciplines are required.

- **Rehabilitation of degraded agricultural lands.** This has been highlighted in Eritrea, Ethiopia, Kenya, Sudan, and Uganda, which are identifying adaptable land management practices and tree species and promoting the planting of trees as helpful in preventing soil erosion and controlling and improving the microclimate. In some places, steep slopes with shallow soils are cultivated. Reclamation of degraded lands (sand dunes, gullies, and marginal lands) through the construction of mechanical structures or terraces and the planting of trees could stabilize these structures. In the coastal areas, for instance, plant species tolerant of drought and salinity could be planted both in the major urban centers and in the rural communities.

- **Agro-forestry practices.** The introduction of multipurpose trees integrated with crop production alleviates the problem of animal feed and helps to restore soil fertility, conserve moisture, and increase crop productivity. Tree planting combined with crop production should be promoted at both community and household levels. In the drier areas, livestock feeding and watering points are critical: enriching the rangelands by oversowing ground cover, including species that are drought-tolerant legumes and grasses, can solve the problem of a shortage of animal feed. Protection of wetlands is a concern in Uganda.

- **Coordination and implementation of climate change polices.** This includes harmonization of policies and institutional frameworks affecting climate change adaptation across different approaches and strategies, as is highlighted in Eritrea, Ethiopia, Kenya, Sudan, and Uganda.

- **Capacity building.** There is a need to build human, institutional, infrastructure, and financial capacity for implementing policies directed at climate change, and this need is reflected in the ten countries. It has been raised as a major concern in Burundi, Eritrea, Ethiopia, and Madagascar. Specific needs are training in modeling for climate change, remote sensing, and the development and building of capacity for early warning systems.

- **Irrigation.** This is considered critical in countries with large arid and semi-arid zones, namely Eritrea, Ethiopia, and Kenya.

- **Stemming high rates of population growth.** This is particularly recognized in countries with the fastest growth rates, such as Ethiopia and Madagascar, and in small countries with low growth rates, such as Burundi and Rwanda.

- **Access to and use of land.** Fragmentation of land as well as scattered parcels of land have contributed to land degradation in Burundi and Eritrea. It is important to ensure equitable land tenure by offering tenure for long periods to allow for proper management of land. A number of environment-related policies have been drafted but have yet to be supported with relevant and coherent laws and regulations.

- **Alternative energy sources and risk management.** The use of energy-saving stoves reduces the amount of firewood used and the number of trees that are cut for firewood and charcoal. Other sources of energy such as wind and solar should also be considered. Surprisingly, this was raised as a major concern by Eritrea only. Risk management in the form of insurance is mentioned only in Ethiopia.

In a Nutshell

Climate change predictions point to a rise in minimum temperatures but conflicting projections regarding rainfall. Rainfall is also predicted to be more erratic and violent, further disrupting predominantly rainfed agricultural production systems, with coffee predicted to be the most affected. The predicted future climate will also affect the productive infrastructure of ECA and exacerbate the constraints on other livelihood systems. Options for response to climate change that are noted in the NAPAs are costly and are yet to be implemented.

Crop simulation models and recent observations point to a possible expansion of the crop production zones for staple crops and livestock. The merits of this change include enhancement of the food security of communities in the new production areas, although there will also likely be adverse impacts in the sense that farmgate prices might collapse, undermining household incomes and resilience to climate change.

IMPACT, which includes both climate effects as well as demographic and income effects, predicts that the output of the majority of the foodcrops will increase on account of acreage expansion for some and technological advancement for others. The disease pressure will increase, especially for coffee, cassava, and plantains. Policies and investments are needed to promote agricultural growth with a focus on smallholder productivity in the face of climate change.

The occurrence of the global food crisis has renewed attention to agriculture and spurred increased investment in the sector (Fan, Torero, and Headey 2011; Karugia et al. 2011). Although our modeling in IMPACT suggests that, on average, food production and availability should increase in the future, it is not able to tell us about the year-to-year variation in global food production and availability. In its Fourth Assessment Report the IPCC suggested that droughts may increase with climate change, and if the droughts cover large areas when they occur, shocks to the food system may occur, causing more frequent food crises. Furthermore, the models predicts changes in national production, but apart from a special focus on the small farmers, the gains could be realized mostly in the commercial sector. Therefore, recognizing these distinct possibilities in the future, it is understandable why the ten countries of the EAC call for public policies ensuring that small farmers have opportunities to increase their productivity and income. Some of the investments they propose include investments in crop breeding and livestock research, extension services, improved smallholder access to inputs such as seeds and fertilizer through lower transport and marketing costs, and rural infrastructure. The importance of regional trade as a means of offsetting food shortages is highlighted in Madagascar, Rwanda, and Tanzania.

The agricultural adaptation strategies presented in this monograph outline national priorities for actions to achieve enhanced agricultural and overall development. They are usually put forward by individual countries based on assessments of national needs. These strategies show weaknesses manifested by poor coordination or implementation of the stated objectives. Improved coordination and implementation of strategies and policies will go a long way toward minimizing risks associated with climate change.

Regional Action

Some strategies require regional actions. These include strategies on issues of transboundary resources such as lakes, rivers, and forests and issues that need collective action such as disease control, forecasting and early warning systems, and capacity building. However, attempts to identify strategic priorities for agricultural adaptation at a regional level are lacking. Organizations mandated to develop and implement regional agricultural development programs are just beginning to put together strategies that need further fine-tuning and implementation.

Forces such as globalization, market liberalization, privatization, urbanization, population growth, and climate change are redefining many of the problems facing agricultural policymakers and thus the kinds of policy solutions required. Most of these forces have roots and expressions that extend beyond national boundaries, implying the need for broad perspectives and regional responses (Omamo et al. 2006). Neighboring countries might gain from cooperating with the rest of the world on key problems. For example, when several countries join efforts as a regional bloc they stand to achieve greater negotiating power and leverage than would those countries acting individually in dealing with the World Trade Organization or with other regional groupings. Some countries in a region might be able to act as regional growth centers and pull neighboring countries along with them as they grow. In the spirit of regional integration, countries may buy imports from their neighbors, attract migrant workers, and act as sources of investment capital. These regional trade dynamics can be more powerful if key development policies are synchronized across countries. Finally, some national investments might generate benefits for their neighbors, leading to efficiency gains from regional rather than national investment strategies. For example, agricultural research and development (R&D)—say, in breeding for drought resistance—in one country might lead to spillover benefits for neighboring countries that have similar agroecological conditions. It might be inefficient for each country to undertake wholly independent R&D; significant gains might be achieved from regionally conceived and implemented R&D programs. This approach is already bearing fruit in the regional centers of excellence being developed in ECA for cassava, dairy rice, and wheat (ASARECA 2008).

Climate change also requires networking at local, regional, and global levels. This will promote the development and exchange of information for database development and training. Suitable arrangement of institutions is needed to harmonize and coordinate efforts to adapt to climate change.

It is tempting to imagine that adaptation decisions might wait for models that can provide greater certainty about what might happen where. This is a forlorn hope. Faster computers and new modeling techniques might well provide more details and finer distinctions. But they will not necessarily be more accurate or capable of being shown to be so. Decisions about adaptation will be made in conditions of pervasive uncertainty. So the trick will be to find ways of adapting to many possible future climates, not to tailor expectations to one future in particular. Even then, adaptation can help only up to a point.

Many of the millions of farming households in the ten ECA countries, which make up the bulk of the agricultural labor force, already face more variable weather than do farmers in developed countries. That and a lack of social safety nets makes most of them highly risk averse, which further limits their ability to undertake adaptation strategies such as changing crop varieties and planting patterns. They often prefer strategies with less risk but lower yields. Worse, in bad weather a whole region's crops suffer together. Here as elsewhere, there is a role for insurance to transfer and spread the risks. But getting farmers to invest in such schemes, even with small premiums, is hard. It also requires finding reinsurance for the local insurer because there is a high chance that many claims will be made at once.

Farmers may be cheered by the thought that food prices are likely to rise. For poor farmers, who spend much of their income on food, this will be a mixed blessing, especially if higher frequencies of drought make prices more volatile too. For poor people more generally, it is even worse news.

More than half the world's people already live in cities. Three-fourths or more may do so by midcentury. Encouraging this trend further, at least in some places, may be a useful way of reducing the economy's exposure to climate change. How countries cope with increasing population growth is a matter of concern.

References

ASARECA (Association for Strengthening Agricultural Research in Eastern and Central Africa). 2008. "Establishment of Regional Centers of Excellence in Agricultural Research in Eastern Africa: East African Agricultural Productivity Programme (EAAPP)." Proceedings of a workshop held at Imperial Resort Beach Hotel, Entebbe, Uganda, November 18–21.

EAC (East African Community). 2011. *East African Community Gazette,* vol. AT 1, no. 11, Arusha, August 12.

Economist. 2010. "Briefing: Adapting to Climate Change." *The Economist,* November 27.

Fan, S., M. Torero, and D. Headey. 2011. *Urgent Actions Needed to Prevent Recurring Food Crises.* IFPRI Policy Brief 16. Washington, DC: International Food Policy Research Institute.

Karugia, J., J. Wanjiku, M. Waithaka, and S. Babu. 2011. "Persistence of High Food Prices in Eastern Africa: What Role for Policy?" Unpublished paper. International Food Policy Research Institute, Washington, DC, and Association for Strengthening Agricultural Research in Eastern and Central Africa, Kampala, Uganda.

Nzuma, J., M. Waithaka, R. Mulwa, M. Kyotalimye, and G. Nelson. 2010. *Strategies of Adapting to Climate Change in Rural Sub-Saharan Africa: A Review of Data Sources, Poverty Reduction Strategy Programmes (PRSPs) and National Adaptation Plans for Agriculture (NAPAs) in ASARECA Member Countries.* IFPRI Discussion Paper 01013. Environment and Production Technology Division. Washington, DC: International Food Policy Research Institute.

Omamo, S. W., X. Diao, S. Wood, J. Chamberlain, L. You, S. Benin, U. Wood-Sichra, and A. Tatwangire, A. 2006. *Strategic Priorities for Agricultural Development in Eastern and Central Africa.* IFPRI Report 150. Washington, DC: International Food Policy Research Institute. www.ifpri.org/pubs/ABSTRACT/rr150.asp#dl.

Contributors

Habtamu Admassu (habtamu.admassu@gmail.com), Associate Researcher, Melkassa Agricultural Research Center, Nazareth, Ethiopia

Mutabazi Alphonse (drep@rema.gov.rw), Climate Change Expert, Rwanda Environmental Management Authority

Woldeamlak Araia (woldearaia@yahoo.com), Assistant Professor and Head of Department of Agronomy, Hamelmalo Agricultural College, Eritrea

Juvent Baramburiye (juventbaramburiye@yahoo.fr), Head of Seed Production/Control and Agrobiodiversity Programme, Institut des Sciences Agronomiques du Burundi

Bernard Bashaasha (bashaasha@agric.mak.ac.ug), Associate Professor, Makerere University, Uganda

Menghisteab Gebreselassie (mengish200850@yahoo.com), Lecturer in Agricultural Economics and Registrar, Hamelmalo Agricultural College, Eritrea

Mezgebu Getinet (mezgebug@yahoo.com), Agrometeorologist, Ethiopian Institute of Agricultural Research

Bissrat Ghebru (bissgk@gmail.com), Assistant Professor and Director of the Bureau of Academic Standards and Evaluation of the National Board for Higher Education, Eritrea

Caroline Kilembe (carolkilembe@yahoo.com), Principal Agriculture Officer I, Food Security Department of the Ministry of Agriculture, Food Security and Cooperatives, Tanzania

Miriam Kyotalimye (m.kyotalimye@asareca.org), Programme Assistant for the Policy Analysis and Advocacy Programme, Association for Strengthening Agricultural Research in Eastern and Central Africa, Entebbe, Uganda

Daniel Mason-d'Croz (d.mason-dcroz@cgiar.org), Research Analyst, International Food Policy Research Institute, Washington, DC

Gerald C. Nelson (nelson.gerald.c@gmail.com), Former Senior Research Fellow, International Food Policy Research Institute, Washington, DC, and Professor Emeritus, University of Illinois, Urbana-Champaign

Blandine M. Nsombo (blandinensombo@yahoo.fr), Senior Professor, Faculty of Agronomics Sciences of the University of Kinshasa, Democratic Republic of Congo

Michael Makokha Odera (mikemako67@gmail.com), Ph.D. candidate, Kenyatta University, Nairobi

Woldeselassie Ogbazghi (wogbazghi@yahoo.co.uk), Assistant Professor and Associate Dean for Academic Affairs, Hamelmalo Agricultural College, Eritrea

Amanda Palazzo (palazzo@iiasa.ac.at), Research Scholar, International Institute for Applied Systems Analysis. While working on this monograph, she was a Research Analyst at the International Food Policy Research Institute, Washington, DC

Richard Robertson (r.robertson@cgiar.org), Research Fellow, International Food Policy Research Institute, Washington, DC

Abdelmoneim Taha (ataha54@yahoo.com), Research Economist, Agricultural Economics and Policy Research Center of the Agricultural Research Corporation, Sudan

Ngoga G. Tenge (ngogatenge@gmail.com), Director of Natural Resource Management and Head of the Agrometeorology Research and Extension Program, Rwanda Agricultural Board

Timothy S. Thomas (t.s.thomas@cgiar.org), Research Fellow, International Food Policy Research Institute, Washington, DC

Siza Tumbo (siza.tumbo@gmail.com), Associate Professor in Agricultural Engineering, Sokoine University of Agriculture, Tanzania

Mireille Rahaingo Vololona (rahaingo.mir@gmail.com), Head of the Rural Development Policies Department, Ministry of Agriculture, Madagascar

Michael Waithaka (m.waithaka@asareca.org), Manager for the Policy Analysis and Advocacy Programme, Association for Strengthening Agricultural Research in Eastern and Central Africa, Entebbe, Uganda

Index

Page numbers for entries occurring in figures are followed by an *f;* those for entries in notes, by an *n;* and those for entries in tables, by a *t.*